Surveys and Tutorials in the Applied Mathematical Sciences

Volume 6

Editor-in-Chief
L. Sirovich

Series Editors:
S. Antman, M.P. Brenner, P. Holmes, L. Mahadevan, A. Singer, K.R. Sreenivasan, J. Victor

For further volumes:
http://www.springer.com/series/7219

Surveys and Tutorials in the Applied Mathematical Sciences

Volume 6

The history of science, from its very beginnings, is replete with great achievements facilitated by innovative mathematical thinking. Unfortunately, it is increasingly common for mathematics to be only loosely linked to the science, by an online device called Supplementary Material. This practice is especially common in biology, the heroic science our age, but it also is present in other fields.

The result is usually poorly coupled mathematics that is presented in a manner that fails to convey the central role that analytical thinking plays in achieving the full scientific understanding of a subject. With this in mind, the STAMS series of Springer-Verlag will take on the new goal of countering this tendency, while retaining its existing goal of providing short, up-to-date, readable tutorials and surveys, written in a style accessible to researchers, professionals and graduate students, which can serve as an introduction to recent and emerging subject areas.

In future we therefore will strive to publish manuscripts that emphasize the dual roles of science and mathematics and which achieve the goal of scientific exposition that invites mathematical interest and vice versa.

Larry Sirovich
Editor-in-Chief

Toshikazu Sunada

Topological Crystallography

With a View Towards Discrete Geometric Analysis

Toshikazu Sunada
Professor
School of Science and Technology
Meiji University
1-1-1 Higashi-Mita
Tama-ku, Kawasaki-shi
Kanagawa, Japan

Editor-in-Chief:
L. Sirovich
Laboratory of Applied Mathematics
Department of Bio-Mathematical Sciences
Mount Sinai School of Medicine
New York, NY, USA

ISBN 978-4-431-54176-9   ISBN 978-4-431-54177-6 (eBook)
DOI 10.1007/978-4-431-54177-6
Springer Tokyo Heidelberg New York Dordrecht London

Library of Congress Control Number: 2012953120

© Springer Japan 2013
This work is subject to copyright. All rights are reserved by the Publisher, whether the whole or part of the material is concerned, specifically the rights of translation, reprinting, reuse of illustrations, recitation, broadcasting, reproduction on microfilms or in any other physical way, and transmission or information storage and retrieval, electronic adaptation, computer software, or by similar or dissimilar methodology now known or hereafter developed. Exempted from this legal reservation are brief excerpts in connection with reviews or scholarly analysis or material supplied specifically for the purpose of being entered and executed on a computer system, for exclusive use by the purchaser of the work. Duplication of this publication or parts thereof is permitted only under the provisions of the Copyright Law of the Publisher's location, in its current version, and permission for use must always be obtained from Springer. Permissions for use may be obtained through RightsLink at the Copyright Clearance Center. Violations are liable to prosecution under the respective Copyright Law.
The use of general descriptive names, registered names, trademarks, service marks, etc. in this publication does not imply, even in the absence of a specific statement, that such names are exempt from the relevant protective laws and regulations and therefore free for general use.
While the advice and information in this book are believed to be true and accurate at the date of publication, neither the authors nor the editors nor the publisher can accept any legal responsibility for any errors or omissions that may be made. The publisher makes no warranty, express or implied, with respect to the material contained herein.

Printed on acid-free paper

Springer is part of Springer Science+Business Media (www.springer.com)

Preface

The Russian mathematician P.L. Chebyshev (1815–1897) once said in a general context that the collaboration of theory and practice brings out the most beneficial results in the sciences. N.I. Lobachevskii, one of the discoverers of non-Euclidean geometry, also said, "There is no area of mathematics, however abstract, which may not someday be applied to phenomena of the real world". Their statements pertain to what this work intends to convey to the reader; that is, the author wishes primarily to provide the reader with mathematical insight into modern crystallography, a typical practical science that originated in classifying the observed shapes of crystals. However, the tools we shall employ are not adopted from the traditional theory of crystallographic groups, but are rather from *algebraic topology*, a field in pure mathematics cultivated during the first half of the last century. More specifically, the elementary theory of covering spaces and homology is effectively used in the study of 3D networks associated with crystals. This explains the reason why this book is entitled *Topological Crystallography*.

Further, we formulate a minimum principle for crystals in the framework of *discrete geometric analysis*, which provides us with the concept of *standard realizations*, a canonical way proposed by the author and his collaborator Motoko Kotani in 2000 to place a given crystal structure in space so as to produce the most symmetric microscopic shape. In spite of its purely mathematical nature (thus having nothing to do with physical and chemical aspects of crystals), this concept, combined with homology theory, turns out to fit with a systematic design and enumeration of crystal structures, an area of considerable scientific interest for many years. Incidentally, crystallographers proposed a similar concept in their recent studies to determine the ideal symmetry of a crystal net and to analyze its topological structure.

The objects in topological crystallography are not necessarily restricted to structures of atomic scale, visible only through special devices. Ornamental patterns having crystallographic symmetry in art, nature, and architectures also fall within the scope of this book. Indeed, many interesting *forms* (*katachi* in Japanese) that are potentially useful for artistic designs in various areas can be generated from standard realizations.

Meanwhile, standard realizations show up in the asymptotic behaviors of random walks on *topological crystals*, the abstraction of crystal structures, and are closely related to a discrete analogue of *Abel–Jacobi maps* in algebraic geometry. These remarkable aspects of standard realizations, which are discussed in the final part of this book, indicate that topological crystallography is neither an outdated nor an isolated field in mathematics; it vigorously interacts with other areas in pure mathematics which have been intensively developed in the last decade. Thus this book, though devoted to a single application of mathematics, takes the reader to various mathematical fields.

The main target of this book is, naturally enough, both mathematicians (including graduate and even undergraduate students) and a wide circle of practical scientists (especially crystallographers and design scientists in art and architecture as well) who want to know how ideas and theories developed in pure mathematics are applied to a practical problem. This broad spectrum of readership will justify the style of our exposition in which basic material in mathematics occupies the first half of the book.

This work has grown out of the lecture notes that I prepared for my lectures at Meiji University during the academic year 2011–2012. I am grateful to Davide M. Proserpio, Jean-Guillaume Eon, and Michael O'Keeffe for fruitful discussions and for providing me with relevant references in chemical crystallography. I also thank Hisashi Naito and my daughter Kayo for producing the beautiful CG images of several hypothetical crystals. This work could not have been done without the friendly help and advice of several people, especially Polly Wee Sy and Tadao Oda. I take great pleasure in thanking them.

Kawasaki, Japan   Toshikazu Sunada

Contents

1 **Introduction** 1

Part I Prerequisites for Modern Crystallography

2 **Quotient Objects** 11
   2.1   Equivalence Relations 12
   2.2   Group Actions 13
   2.3   Notes 18

3 **Generalities on Graphs** 21
   3.1   Graphs 22
   3.2   Morphisms and Automorphisms 24
   3.3   Quotient Graphs 26
   3.4   Paths 30
   3.5   Homotopy 32
   3.6   Bipartite Graphs 33
   3.7   Notes 34

4 **Homology Groups of Graphs** 37
   4.1   Chain Groups 38
   4.2   Homology Groups 39
   4.3   The Structure of Homology Groups 43
   4.4   Enumeration of Finite Graphs 47
   4.5   Automorphisms and Homology 48
   4.6   Notes 50

5 **Covering Graphs** 53
   5.1   Definition 54
   5.2   Covering Transformation Groups 56
   5.3   Fundamental Groups 60
   5.4   Universal Covering Graphs 63

5.5	Construction of Universal Covering Graphs	66
5.6	Notes	68

Part II Geometry of Crystal Structures

6 Topological Crystals 73
   6.1 Generalities on Abelian Covering Graphs 76
   6.2 Topological Crystals 81
   6.3 Symmetry of Topological Crystals 86
   6.4 Notes 89

7 Standard Realizations 93
   7.1 Periodic Realizations 94
   7.2 Projection and Reduction 100
   7.3 Idea 102
   7.4 Harmonic Realizations 104
   7.5 Standard Realizations 108
   7.6 Various Properties of Standard Realizations 112
   7.7 Notes 115

8 Explicit Construction 125
   8.1 General Construction 125
   8.2 Computations 129
   8.3 Examples 131
   8.4 Notes 146

Part III Advanced Topics

9 Random Walks on Topological Crystals 155
   9.1 Simple Random Walks 155
   9.2 Integral Formula 158
   9.3 How to Get the Asymptotic Expansion 161
   9.4 Proof of Theorem 9.2 166
   9.5 Bipartiteness and the Eigenvalue μ_N 169
   9.6 Symmetry 171
   9.7 Notes 175

10 Discrete Abel–Jacobi Maps 181
   10.1 Classical Algebraic Geometry 182
   10.2 Discrete Set-up 183
   10.3 Discrete Abel's Theorem 186
   10.4 Intersection Matrix and Tree Number 191
   10.5 Notes 198

Finale 201

Appendix		205
1	Sets and Maps	205
2	Groups and Homorphisms	207
3	Vector Spaces and Linear Operators	212
4	Free Groups	216
5	Crystallographic Groups	217

References 221

Index 225

List of Symbols

$\mathbb{Z}$	Additive group of integers
$\mathbb{Q}$	Set of rational numbers
$\mathbb{R}$	Set of real numbers
$\mathbb{Z}^n$	Additive group of n-ple integers
$\mathbb{R}^n$	n-Dimensional Euclidean space
$\mathbb{Z}_n (= \mathbb{Z}/n\mathbb{Z})$	Additive group of integers modulo n
$|A|$	Number of elements in a finite set A
$A \times B$	Cartesian product of sets A and B
$A \backslash B$ or $A - B$	Difference of A and B
$f : A \longrightarrow B$	A map of A into B
$f(C)$	Image of C ($\subset A$) by f
Image f	Image of a map f
$f^{-1}(D)$	Inverse image of D ($\subset B$) by f
$f|C$	Restriction of a map $f : A \longrightarrow B$ to a subset C of A
$f \circ g$	Composition of $g : A \longrightarrow B$ and $f : B \longrightarrow C$
I	Identity map
$\dim W$	Dimension of a vector space W
$\operatorname{rank} A$	Rank of an abelian group A
$W_1 \oplus W_2$	Direct sum of abelian groups (or vector spaces) W_1 and W_2
$W_1 + W_2$	$\{w_1 + w_2 | \ w_1 \in W_1, \ w_2 \in W_2\}$
$\langle \mathbf{a}, \mathbf{b} \rangle$	Inner product of vectors $\mathbf{a}, \mathbf{b}$
$\|\mathbf{a}\| = \langle \mathbf{a}, \mathbf{a} \rangle^{1/2}$	Norm of a vector $\mathbf{a}$
$\operatorname{Ker} T$	Kernel of a homomorphism T
$W^{\perp}$	Orthogonal complement of a subspace W
$M(m, n)$	Space of matrices of size (m, n)

I_n (or I)	Identity matrix of size n
$\det A$	Determinant of a square matrix A
$\operatorname{tr} A$	Trace of a square matrix A
${}^t A$	Transpose of a matrix A
B_n	n-Bouquet graph
K_n	Complete graph with n vertices
D_d	Graph with two vertices joined by $d+1$ parallel edges
$\deg x$	Degree of a vertex x
$\operatorname{Aut}(X)$	Automorphism group of a graph X
$C_1(X,\mathbb{Z})$	Group of 1-chains on a graph X with coefficients in $\mathbb{Z}$
$H_1(X,\mathbb{Z})$	1-st Homology group of a graph X with coefficients in $\mathbb{Z}$
$C_1(X,\mathbb{R})$	Group of 1-chains on a graph X with coefficients in $\mathbb{R}$
$H_1(X,\mathbb{R})$	1-st Homology group of a graph X with coefficients in $\mathbb{R}$
$b_1(X)$	Betti number of X
$\chi(X)$	Euler number of X
$\langle c \rangle$	1-Chain represented by a path c
$[c]$	Homotopy class of a closed path c
$\pi_1(X,x_0)$	Fundamental group of X with base point x_0
X_0^{uni}	Universal covering graph over a graph X_0
X_0^{ab}	Maximal abelian covering graph over a graph X_0
$E(\Phi,\rho)$	Normalized energy of a periodic realization (Φ,ρ)
Φ^{ab}	Normalized standard realization of X_0^{ab}
$p(n,x,y)$	n-Step transition probability
$\kappa(X_0)$	Tree number of a graph X_0
$[G,G]$	Commutator subgroup of a group G
$G_1 \times G_2$	Group product of two groups G_1 and G_2
$\mathfrak{S}_n$	Symmetry group (group of permutations) of $\{1,2,\ldots,n\}$
$GL_n(\mathbb{R})$	General linear group
$GL_n(\mathbb{Z})$	Group consisting of integral square matrices A with $\det A = \pm 1$
$O(n)$	Orthogonal group
$SO(n)$	Special orthogonal group (rotation group)

Chapter 1
Introduction

> Mathematics allows us to understand the true nature of things by liberating us from the spell of the real world.

Applications of mathematics to crystallography have a long history. The theory of *crystallographic groups* (*space groups* in jargon) is a traditional field dating back to the first half of the nineteenth century, which, needless to say, has been playing a significant role in the classification of crystals in view of the symmetry.[1] Graph theory is another powerful area for the obvious reason that it is used to study the microscopic structure of a crystal[2] (and any molecule) as a 3D (three-dimensional) *network*,[3] in which each atom (or each cluster of atoms) is represented by a vertex of the net, and each edge of the net represents a *bond* (or a polymeric ligand) in the crystal structure.

A mathematical discipline that looks unconventional at first sight, but turns out to be an effective tool when applied to crystallography is *algebraic topology*, a field in geometry which studies structures of topological spaces by assigning algebraic data to topological spaces in order to translate topological problems into algebraic ones that hopefully will be more accessible [45]. This is by no means surprising because a graph is identified with a *cell complex* of one-dimension, and thus it is expected to be able to employ basic tools such as *covering spaces* and *homology*. Indeed, if a topologist is asked "What's the mathematical nature of crystal structures?", his

---

[1] To be precise, the notion of crystallographic groups has its origin in the study of *morphology* (the observed shapes) of crystals. In this sense, the link between mathematics and crystallography is ancient. See the beginning of Chap. 6.

[2] In this book, crystals mean solids composed of atoms arranged in an orderly repetitive array. In crystallography, more general materials such as quasicrystals are counted as crystals.

[3] Wells [109] initiated a systematic study of crystal structures as 3D networks. Applications of graph theory to chemistry could be traced back to 1864 when the Edinburgh chemist Crum Brown proposed representing chemical compounds by graphs.

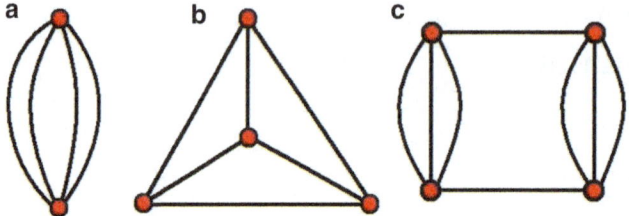

Fig. 1.1 "Seeds" of diamond, K_4 crystal, Lonsdaleite

immediate answer would be "topologically they are infinite-fold abelian covering graphs over finite graphs. The crystal nets (the networks in space associated with crystals) are their periodic realizations".

This bold answer is satisfactory enough, from the theoretical view at least, since among all abelian covering graphs over a given finite graph X_0, there is a *maximal* one from which we may construct, in a unified way, every abelian covering graphs over X_0. In fact, if one starts with a finite graph X_0 to obtain (hypothetical) 3D crystal structures,[4] then the main recipe in this construction is the selection of a subgroup H of $H_1(X_0, \mathbb{Z})$, the first integral homology group of X_0, such that the factor group $H_1(X_0, \mathbb{Z})/H$ is a free abelian group of rank-3, which is eventually going to be the *period lattice* when we realize the covering graph in space. A period lattice[5] here means a lattice group in $\mathbb{R}^3$ leaving a given crystal net invariant when it acts on the space $\mathbb{R}^3$ by translations (a more general definition is given in Sect. 7.1). In particular, the enumeration of topological structures of crystal nets reduces to that of finite graphs and subgroups of homology groups.

In a few cases, maximal abelian covering graphs themselves give 3D crystal structures. For instance, maximal abelian covering graphs over the graphs (A) and (B) in Fig. 1.1 yield the crystal structures of *Diamond*[6] and the hypothetical K_4 *crystal (diamond twin)*,[7] respectively. On the other hand, *Lonsdaleite*[8] (Fig. 1.2[9]) is, as an abstract graph, a *non-maximal* abelian covering graph over the finite graph (C);

---

[4]By a *crystal structure* we mean an abstract graph associated with a crystal net.

[5]The notion of period lattices is a generalization of *Bravais lattices* in crystallography. Precisely speaking, a Bravais lattice is a representative of period lattices when classified by symmetry.

[6]Diamond is an allotrope of carbon which is formed and synthesized at high-pressure and high-temperature conditions, and is known to be less stable than graphite though the conversion rate from diamond to graphite is negligible at ambient conditions. Silicon and germanium adopt similar types of crystal structure.

[7]See Sect. 8.3 and Notes (IV) in Chap. 8 for the detailed account where we explain the reason why the K_4 crystal deserves to be called "diamond twin". The picture of the K_4 crystal is given in Fig. 1.4.

[8]This carbon allotrope, formed when meteorites containing *graphite* strike the earth, is named in honour of crystallographer Kathleen Lonsdale, also referred to as the *hexagonal diamond*.

[9]Source of the figure: WebElements (http://www.webelements.com/).

1 Introduction

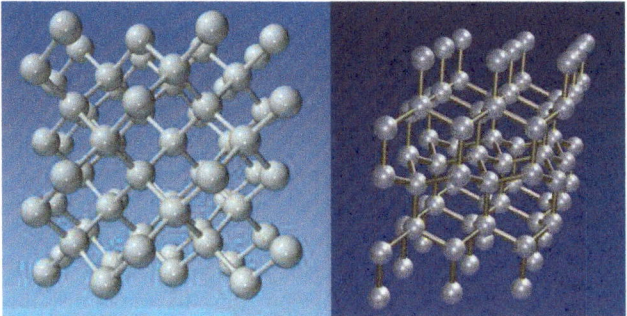

Fig. 1.2 Diamond and Lonsdaleite

indeed, the maximal abelian covering graph over (C) is five-dimensional. Therefore crystal structures of general dimension turn up even when we are handling three-dimensional crystals.

The simple answer above to the question about the mathematical nature of crystal nets is a sort of folklore in the community of mathematicians, and hence it is not attributed to anyone. It is no wonder, however, that Henri Poincaré (1854–1912), founder of algebraic topology, could have easily conceived this answer if he would have witnessed Max von Laue's discovery of crystal structures by the diffraction of X-rays, which was coincidentally accomplished in the last year of Poincaré's life.

Chemical crystallographers adopt the term "periodic graphs" for the underlying topology of crystal structures, based on the same reasoning as mathematicians that a crystal structure is a graph with a translational action which becomes a finite graph when factored out.[10] This difference of nomenclature[11] for the same objects makes crystallographers say "the rich world of periodic graphs has been largely neglected by mathematicians" [50]. The fact is that mathematicians have been interested in general covering spaces over cell complexes, not only of one-dimension but also of higher dimensions, and that our machinery in algebraic topology to tackle this important subject had been well developed long before. For all this, mathematicians' predisposition for liking general theories has hindered them from turning their eyes to crystal structures as illuminating examples of covering spaces.

The primary purpose of this book is to provide the reader with some fundamentals about "topological crystallography"[12] (or "topological methods in crystallography") in order to bridge the gap of knowledge between mathematicians and

---

[10] They also use the term "minimal nets" for maximal abelian covering graphs [9] and the term "cycle spaces" for homology groups [42]. See Notes (V) in Chap. 8.

[11] Even for the description of crystallographic groups there are three main systems of notations; one used in mathematics, and other two (the Schoenflies system and International system) used by chemists and crystallographers.

[12] In chemistry, the term "topological crystal(lography)" is used sometimes in a different context.

crystallographers.[13] Thus no knowledge of algebraic topology and crystallography is presupposed. What the reader needs to be familiar with is basic material in undergraduate mathematics such as *sets, maps, matrices, vector spaces (with inner products)* and an elementary part of group theory *(abelian groups, homomorphism, factor groups,* etc.). Some of these prerequisites are briefly explained in the appendix for the convenience of the reader who is not knowledgeable about modern mathematics.

Moreover, this book is designed, for the most part, to be as self-contained as possible, taking into consideration as the reader practical scientists and undergraduate students with a modest background in mathematics. To accomplish this, we divide the book into three parts. Based on the fundamental material mentioned above, Part I starts with a quick review of the notions of quotient sets (Chap. 2) and graphs (Chap. 3). These notions are fairly standard in pure mathematics, and indispensable in modern crystallography. Subsequently, we provide a comprehensive account of homology (Chap. 4) and covering maps of graphs (Chap. 5) from the combinatorial viewpoint. Since they are usually not treated as independent topics in the literature of algebraic topology and combinatorics, we find it worthwhile to include these chapters here. Furthermore, the notions introduced in Chap. 5 will be useful in a set-up for the analysis and geometry of "non-commutative crystals", a generalization of crystal structures [see [95] and Notes (IV) in Chap. 9].

The contents up to this point are preliminaries to topological crystallography. The heart of the matter begins with Part II. After skimming through the preceding chapters, and if necessary going through the terminology employed as well as our own usage of notations, the reader having mathematics at his/her fingertips may then start from Chap. 6 containing some details about abelian covering graphs.

In this book, the term "topological crystal" is adopted for an infinite-fold abelian covering graph over a finite graph.[14] The reason is to emphasize its abstract nature and at the same time to keep the word "crystal" in order to make it clear that we are addressing the problem of crystals, not the problem of general graphs. In any case, topological crystals are purely mathematical objects "living in the logical world and not in real space", in the sense that they are constructed on the basis of pure reflection.[15]

The issue on how to place (realize) them "canonically" in space is discussed at length in Chap. 7, where we give a down-to-earth account, together with an algorithm (in a loose sense[16]) and many examples, of the mathematical construction

---

[13] See the book [44] by J-G. Eon, W.E. Klee, B. Souvignier and J.S. Rutherford as a reference in crystallography.

[14] In [58], we have used the term "crystal lattice" instead.

[15] This is the so-called *Platonic view*; that is, we mathematicians insist that mathematical entities are abstract in not being spatiotemporally located, and hence lie outside of the real world.

[16] Algorithm means a step-by-step procedure involving a precise set of instructions for what to do next. With access to a computer and with some work in computer graphics based on our algorithm, 3D crystal structures may be displayed on a screen.

1 Introduction

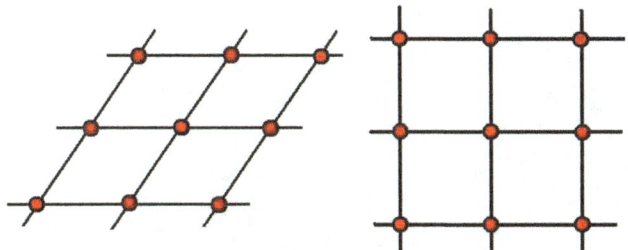

Fig. 1.3 A harmonic realization and the standard realization

stated in Kotani and Sunada [58]. The *standard realization*[17] spelled out here is most symmetric among all realizations (hence deserves to be described as "standard"), and is characterized uniquely by a *minimal principle* for the potential energy (per unit cell) when we look at a realization as a simple system of harmonic oscillators. This may remind the reader of the classical *isoperimetric inequality* characterizing the *round circle*, the most symmetric closed curve.[18]

One remark is in order here. Crystallographers also sought independently canonical ways to place periodic graphs and thus proposed several notions:

1. *Archetype embedding* [42]. This turns out to be the same as the standard realization of the maximal abelian covering graph.
2. *Equilibrium placement* [30]. This notion, inspired by Tutte [102, 103] and suggested by Klein [55], coincides with the *harmonic realization* introduced in [58] as a special case of *harmonic maps* [39,69] [see Notes (IV) in Chap. 7]. Roughly speaking, harmonic realizations (equilibrium placements) are characterized as minimizers[19] of the energy functional *when a period lattice is fixed*.[20] On the other hand, standard realizations are minimizers *without any constraint on period lattices* (except for the one on the volume of unit cells). Therefore standard realizations are special harmonic realizations (Fig. 1.3 illustrates the difference between a general harmonic realization and the standard realization in the case of the quadrangle lattice). Based on the notion of equilibrium placement, Delgado-Friedrichs constructed a powerful algorithm (*SYSTRE*[21]) for the *barycentric drawing*, a special equilibrium placement (see [32]), which seems to coincide with the standard realization as far as several examples are examined (at least in the two-dimensional case).

---

[17] In [96], I used the term "canonical placement".

[18] For a closed curve in the plane, if its perimeter is L and the area that it encloses is A, then $4\pi A \leq L^2$, where equality holds if and only if the curve is a circle.

[19] A *minimizer* of a given function (or functional) is a point (or function) at which the minimum value is attained.

[20] Fixing a period lattice is equivalent to imposing a periodic boundary condition.

[21] The program is available at http://www.gavrog.org/.

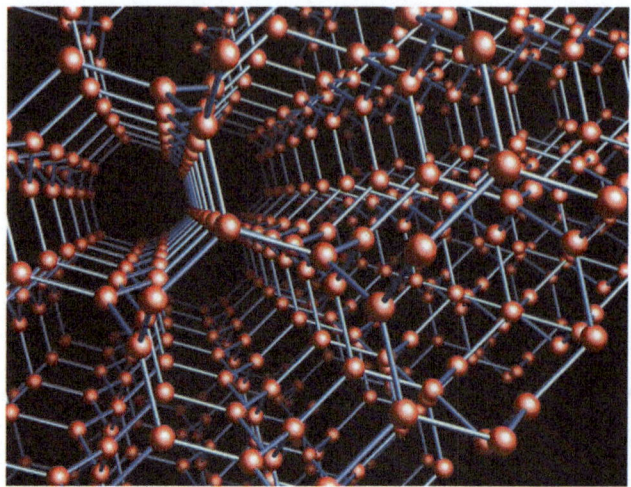

Fig. 1.4 CG image of the K_4 crystal (created by Kayo Sunada)

3. *Archetypical representation* [43]. This is an appropriate orthogonal projection of the archetype embeddings, and turns out to be identical to the standard realization (see Chap. 8).

Admittedly the nature of standard realizations is purely mathematical, and has almost nothing to do with the physical and chemical aspects of real crystals. Thus one can say that the nets obtained by standard realizations are merely toy models of hypothetical crystal,[22] i.e., not necessarily in actual existence. Nevertheless, as will be observed in Chap. 8, the nets corresponding to several typical real crystals turn out to be canonically placed in our sense. Diamond and Lonsdaleite are such examples. The nets obtained from the *face-centered cubic lattice* and the *body-centered cubic lattice* are also standard realizations. Incidentally, the hypothetical K_4 crystal mentioned above is constructed by means of a standard realization (Fig. 1.4).

The nets of real crystals are usually different from the standard realization because the mechanism of crystallization is much more complicated than the one explained by the harmonic-oscillator model for a crystal. Even in this case, however, the standard realization offers, in the light of its uniqueness, a basis for comparison with which one may talk quantitatively about how much the net is distorted (Sect. 7.5).

Apart from crystallography handling forms of Angstrom scale, the concept of standard realization is of some interest from the view of beautiful "ornamental patterns" that are visible to the naked eye, both in nature and art, such as the

---

[22]Once we find a hypothetical crystal, a systematic prediction of its physical properties for appropriate atoms can be carried out by *first principles calculations* used in chemistry. The prediction appealing to the computer power encourages (or discourages) material scientists to synthesize the hypothetical crystals [26].

1 Introduction

Fig. 1.5 Patterns in nature and architecture

honeycomb pattern built by beehives which is used in countless artistic structures. The honeycomb is actually the standard realization of the *hexagonal lattice*,[23] a two-dimensional topological crystal described as the maximal abelian covering graph over the graph with two vertices joined by three parallel edges; thus the net of diamond crystal is regarded as the three-dimensional analogue of the honeycomb; see Sect. 8.3.

What is more, the beauty of shapes is somehow bound up with the mechanical functions of architectural structures e.g., as beehives instinctively know that their honeycomb is the best structure to reach minimal weight and minimal material cost. The picture on the right side in Fig. 1.5 is a light weight rigid structure in architecture (due to Alexander Graham Bell and Buckminster Fuller), which turns out to tie up with the net associated with the face-centered cubic lattice, an example of standard realizations as mentioned above; see Sect. 8.3 (III).

Let us return to the structure of the book. Part III, mainly targeting mathematicians and graduate students, is concerned with two idiosyncratic topics closely related to standard realizations; more specifically, we deal with *random walks* on topological crystals, and a discrete analogue of classical algebraic geometry. I would like to emphasize that a remarkable relation between random walks and standard realizations, the subject in Chap. 9, is my starting point to develop topological crystallography. Chapter 10 introduces the notion of *discrete Abel–Jacobi map* which is relevant to the standard realizations of maximal abelian covering graphs. What we discuss in this final chapter is considered a ramification of algebraic graph theory and also of *tropical geometry*, a relatively new and thriving area.

Each chapter ends with a section entitled "Notes" wherein, except for historical remarks added for the pedagogical reason, we make liberal use of concepts in

---

[23]The usage of the term "lattice" for crystal structures may give rise to confusion because customarily a lattice means a discrete subgroup of $\mathbb{R}^d$ (or more generally a discrete subgroup of a Lie group). But we will follow the convention to use "lattice" for some crystal structures (see Remark in Sect. 2.2).

advanced mathematics. The overall purpose is to inform the reader, somewhat in a rambling style, how the subject discussed in the chapter is related to other fields of mathematics, say, the Ihara zeta function, a graph-theoretic analogue of class field theory, discrete Laplacians, and harmonic maps. Although not necessarily needed for what follows in Parts I and II, it offers the reader with a mathematical bent some additional insights into our subject and a underlying motive. Part III also has a similar nature, and may give a surprise to the reader. He or she will discover that a problem in one area leads us into a quite different area of mathematics.

In writing this book, I followed the conventional style of mathematical books. That is, in accordance with mathematics culture, I will give proofs to almost all claims that the reader might find difficult to verify by himself/herself (except for those in Notes). The reason for doing so is that I want to remind the reader of the significant function of proofs. Proofs in mathematics play the same role as verifications by experiments in other sciences.[24] In view of this, I strongly recommend the reader (even a practical scientist) to try to understand proofs.

---

[24] A "proof" is a logical procedure to derive what we anticipate to be true from what we have already known to be true.

Part I
Prerequisites for Modern Crystallography

Chapter 2
Quotient Objects

Crystallographers employ the concept of "quotient graph" in a systematic enumeration of crystal structures (see for instance [20, 41, 76, 86]). Behind this concept is the periodic nature of crystal structures with respect to parallel translations; that is, a quotient graph in crystallography is nothing but the quotient graph of a 3D network by the translational action of a lattice group; see the next chapter for details.

As a matter of fact the idea of "quotient" shows up in various fields of modern mathematics; say, *quotient vector spaces* in linear algebra, the *cut-and-paste technique* in topology, *coset spaces* and *factor groups* in group theory, to name several. If we identify graphs with one-dimensional cell complexes,[1] then a quotient graph is a special case of *quotient spaces* introduced in topology. Even if our discussion is confined to crystallography, the reader will come across many quotient objects.

Related to the idea of "quotient" is the word "well defined". This rather strange locution is used when we want to define a new object such as a map or an operation among quotient objects and we need to ensure the consistency of the definition. The issue of well-definedness is hardly avoided in modern mathematics. Beginners are expected to get used to it by verifying many examples appearing in our discussion.[2]

Hereafter the reader is assumed to know set-theoretic language and some basic notions in group theory. I recommend the reader to consult Appendices 1 and 2 for the meaning of the terminology used in this chapter.

---

[1] A cell complex is a topological space obtained by gluing together certain basic building blocks called cells. Here an n-dimensional cell is a topological space that is homeomorphic to an n-dimensional closed ball.

[2] According to my experience, the issue of well-definedness is often a hindrance to students who are struggling to learn modern mathematics.

2.1 Equivalence Relations

The meaning of "quotient" is roughly interpreted as follows. Suppose we are given an aggregation of objects (animals or plants for instance). We classify these objects according to certain properties, and form groups by gathering classified objects. Following this procedure, we give a name to each group (mammals, birds, reptiles, etc., for animals). Then the set of names is what we call the *quotient set*.

One can make this plain-language explanation rigorous by introducing the notion of equivalence relations as follows. Let A be a set with a *relation* $\sim$ among elements in A. More precisely, we are given a subset R of the cartesian product $A \times A$ with which the statement "$(x,y) \in R$" is read "x is related to y" and is written $x \sim y$. A relation $\sim$ satisfying the following three conditions is said to be an *equivalence relation*.

1. $x \sim x$ for every x.
2. If $x \sim y$, then $y \sim x$.
3. If $x \sim y$ and $y \sim z$, then $x \sim z$.

When $x \sim y$ for an equivalence relation $\sim$, we say that x and y are *equivalent* (what we have in mind is a statement like "x and y have the same property"). In essence, an equivalence relation is a generalization of equality "$=$" which, of course, satisfies the above conditions.

The equivalence relation yields a partition of A as follows. For $x \in A$, we denote by $E(x)$ the *equivalence class* of x; that is, the subset consisting of all elements equivalent to x. Then $x \in E(x)$, and either $E(x) \cap E(y) = \emptyset$ (empty set) or $E(x) = E(y)$, which implies that different equivalence classes do not overlap, and they cover the whole set A, thereby inducing a partition of A. Conversely, a partition of A yields an equivalence relation in a natural manner.

Now consider the set of equivalence classes for which we use the notation $A/\sim$, and call it the *quotient set* with respect to the relation $\sim$. Therefore each equivalence class, originally a subset of A, becomes a single element in the quotient set $A/\sim$. The *canonical projection* is the map $p : A \longrightarrow A/\sim$ that carries $x \in A$ into the equivalence class of x.

Useful later on in our discussion is the following observation. Let A and B be two sets with equivalence relations $\sim_A$ and $\sim_B$, respectively, and let $f : A \longrightarrow B$ be a map having the property that if $x \sim_A y$, then $f(x) \sim_B f(y)$. For the canonical projections $p_A : A \longrightarrow A/\sim_A$ and $p_B : B \longrightarrow B/\sim_B$, there is a map $F : A/\sim_A \longrightarrow B/\sim_B$ such that $F \circ p_A = p_B \circ f$. The following diagram, called a *commutative diagram*, will help the reader understand visually this equality of maps.

$$
\begin{array}{ccc}
A & \xrightarrow{f} & B \\
{\scriptstyle p_A}\downarrow & & \downarrow{\scriptstyle p_B} \\
A/{\sim_A} & \xrightarrow{F} & B/{\sim_B}
\end{array}
$$

The question is how to define F. The equality $F \circ p_A = p_B \circ f$ to be satisfied by F provides us with a hint on how to do it. Namely, for each $\alpha \in A/{\sim_A}$, taking $x \in A$ with $p_A(x) = \alpha$, we put $F(\alpha) = p_B(f(x))$. Caution! To define $F(\alpha)$, we choose x from the equivalence α. What we must ascertain is that F is "well-defined", namely $p_B(f(x))$ does not depend on the choice of x. But this is a consequence of the property $x \sim_A y \implies f(x) \sim_B f(y)$. The map F defined in this way is said to be *induced* from f.

"Well-definedness" (or consistency of definitions) is always an issue that we must pay attention to when we want to construct induced maps of quotient sets out of maps of original sets. The reader having learned group theory will have experienced the well-definedness of induced homomorphisms of factor groups (see the next section).

The word "canonical" (or "natural") is also frequently used in mathematics. Broadly speaking, an object defined without bringing in any additional data is said to be canonical. For instance, for a finite-dimensional vector space W, we need a basis or an inner product of W to define a linear isomorphism of W onto its dual space[3] W^*, while a linear isomorphism between W and $(W^*)^*$ can be constructed without such things. Thus one can say that W is *canonically* isomorphic to $(W^*)^*$. The canonical projection $p : A \longrightarrow A/\sim$ is a very example of canonical objects.

A subset D of A such that $p|D : D \longrightarrow A/\sim$, the restriction of the map p to D, is a bijection[4] (one-to-one correspondence), is said to be a *fundamental set*. In other words, D is a subset of A obtained by selecting an element from each equivalence class and gathering them.[5] An element in D is literally a *representative* of the equivalence class containing it. The choice of a fundamental set is, of course, not unique.

2.2 Group Actions

A typical equivalence relation, occupying a key role in crystallography, comes from a *group action* on a set. From now on, we assume familiarity with the concept of *group* (see Appendix 2).

---

[3] See Appendix 3 for the terminology in this paragraph.

[4] See Appendix 1 for the definition.

[5] We need *Axiom of Choice*, one of the premises in the axiomatic set theory, to make sure that a fundamental set *does exist*.

A group G is said to *act on* a set A (or A is said to be a *G-set*) when we are given a map $\varphi : G \times A \longrightarrow A$ satisfying the followings.

1. If 1 is the identity element of G, then $\varphi(1,x) = x$ for all $x \in A$.
2. $\varphi(g_1, \varphi(g_2,x)) = \varphi(g_1 g_2, x)$ for all $g_1, g_2 \in G$ and $x \in A$.

Writing $\varphi(g,x)$ in the form gx, we have easy-to-see expressions of (1) and (2); $1x = x$, $g_1(g_2 x) = (g_1 g_2)x$.

Example 2.1. We let S_A be the group consisting of all bijections of A onto itself. The map $\varphi : S_A \times A \longrightarrow A$ defined by $\varphi(\sigma,x) = \sigma(x)$ yields an action of S_A on A. The group S_A is called the *symmetry group* (or *permutation group*) of A, and elements of S_A are called *permutations* of A. When $A = \{1,2,\ldots,n\}$, we write S_n for S_A. Giving an action of G on A is equivalent to giving a homomorphism of G into S_A. □

Given an action of G on A, we define the relation $\sim$ on A by setting

$$x \sim y \iff \text{there exists } g \in G \text{ such that } y = gx.$$

The relation $\sim$ defined in this way is an equivalence relation. In fact,

$$x = 1x,$$
$$y = gx \implies x = g^{-1}y,$$
$$y = gx, \ z = hy \implies z = hgx.$$

The equivalence class of x is $Gx = \{gx|\ g \in G\}$, which we call the *orbit*[6] through x. The quotient space $A/\sim$, customarily expressed as A/G, is called the *orbit space*.

A subset B of a G-set A is said to be *G-invariant* (or *G-stable*) if

$$gB(= \{gb|\ b \in B\}) = B$$

for every $g \in G$. In a natural manner, B itself is a G-set.

Let A and B be two G-sets, and let $p_A : A \longrightarrow A/G$ and $p_B : B \longrightarrow B/G$ be the canonical projections. A map $f : A \longrightarrow B$ satisfying $f(gx) = gf(x)$ for every $x \in A$ and $g \in G$ is said to be *G-equivariant*. For such a map f, there is a canonical map $F : A/G \longrightarrow B/G$ such that $F \circ p_A = p_B \circ f$.

---

[6]The classical meaning of "orbit" is the journey that a planet makes around the sun or another planet. This is generalized as the time evolution of a pointmass in a dynamical system, which is described in the abstract way by an action of the additive group $\mathbb{R}$ of real numbers on a configuration space.

2.2 Group Actions

$$\begin{array}{ccc} A & \xrightarrow{f} & B \\ p_A \downarrow & & \downarrow p_B \\ A/G & \xrightarrow{F} & B/G. \end{array}$$

Indeed, this is a special case of what we have seen in the previous section. In particular, for a G-invariant subset B of A, the inclusion map $i : B \longrightarrow A$ induces a map $i' : B/G \longrightarrow A/G$, which turns out to be injective.[7] Using i', we identify B/G with a subset of A/G.

More generally, let A and B be two sets on which groups G and H act respectively, and let $\rho : G \longrightarrow H$ be a homomorphism. If $f : A \longrightarrow B$ is a map satisfying $f(gx) = \rho(g)f(x)$ (such f is also said to be G-equivariant), then there is a map $F : A/G \longrightarrow B/H$ such that $F \circ p_A = p_B \circ f$. The importance of the notion of equivariant maps will become clear when we discuss realizations of topological crystals (Sect. 7.1).

We shall give a few examples of group actions and their orbit spaces. The first two examples are fundamental in group theory.

Example 2.2. The *adjoint action* of G on itself is defined by $\varphi(g,h) = ghg^{-1}$. Two elements $g_1, g_2 \in G$ are said to be *conjugate* if they belong to the same orbit. The orbit through h is called the *conjugacy class* of h, and denoted by $[h]$, namely $[h] = \{ghg^{-1}| \, g \in G\}$.

A subgroup H is *normal* if it is invariant by the adjoint action, i.e., $gHg^{-1} \subset H$ for every $g \in G$. □

Example 2.3. The *left action* on G of a subgroup H of G is given by $\varphi(h,g) = hg$. An orbit of g is called a (right) *coset*, and denoted by Hg. The orbit space, symbolically G/H, is referred to as the *coset space*. When the set G/H is finite, we call $|G/H|$ the *index* of the subgroup H.

For a normal subgroup N, the coset space G/N has a natural group structure induced from that of G. Indeed, we have $NaNb = Nab$ which says that the multiplication $(Na, Nb) \mapsto Nab$ in G/N is well defined. G/N is called the *factor group* (or quotient group) of G modulo N. The canonical projection $p : G \longrightarrow G/N$ is a surjective homomorphism[8] whose kernel coincides with N. We call p the *canonical homomorphism*. For a homomorphism $f : G \longrightarrow G_1$, the kernel of f (denoted by $\text{Ker} f$) is a normal subgroup of G, and f induces an injective homomorphism $F : G/N \longrightarrow G_1$ such that $F \circ p = f$, where $N = \text{Ker} f$ and $F(Ng) = f(g)$ (*the first isomorphism theorem*; see Appendix 2).

---

[7] See Appendix 1 for the definition of "injective map".
[8] See Appendix 1 for the definition of "surjective map".

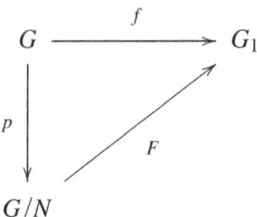

For two normal subgroups N_1, N_2 of G such that $N_1 \subset N_2$, the correspondence $N_1 g \mapsto N_2 g$ yields a surjective homomorphism of G/N_1 onto G/N_2 whose kernel is N_2/N_1. □

Example 2.4. Let A be an additive group (See Appendix 2). For a subgroup B of A, we write $a_1 \equiv a_2 \pmod{B}$ when $a_1 - a_2 \in B$. This is an equivalence relation, and its quotient set is nothing but the factor group A/B. When $A = \mathbb{Z}$ and $B = n\mathbb{Z} = \{kn|\ k \in \mathbb{Z}\}$, we write $a_1 \equiv a_2 \pmod{n}$ instead of $a_1 \equiv a_2 \pmod{n\mathbb{Z}}$, and say that a_1 and a_2 are congruent modulo n. We also employ the notation $\mathbb{Z}_n$ for $\mathbb{Z}/n\mathbb{Z}$. □

Let us give an example of an orbit space appearing in crystallography.

Example 2.5. Consider a *lattice group* L in the Euclidean space[9] $\mathbb{R}^d$; that is, an additive group expressed as

$$L = \{k_1 \mathbf{a}_1 + \cdots + k_d \mathbf{a}_d|\ k_1, \ldots, k_d \text{ are integers}\},$$

by using a basis $\mathbf{a}_1, \ldots, \mathbf{a}_d$ of the vector space $\mathbb{R}^d$ (which is said to be a $\mathbb{Z}$-*basis* of L). The group L acts on $\mathbb{R}^d$ by translations: $\mathbf{x} \mapsto \mathbf{x} + \sigma$ ($\sigma \in L$, $\mathbf{x} \in \mathbb{R}^d$). As a fundamental set (a *unit cell* in crystallography), we may take

$$D_L = \{t_1 \mathbf{a}_1 + \cdots + t_d \mathbf{a}_d|\ 0 \leq t_1 < 1, \ldots, 0 \leq t_d < 1\},$$

which we call a *fundamental parallelotope* for L (a generalization of parallelogram in plane and parallelepiped in space).

The orbit space $\mathbb{R}^d/L$ is not merely a set, but it also inherits various structures of $\mathbb{R}^d$. Firstly, it possesses a topology (a qualitative structure of "far and near").[10] As a topological space, it is the d-dimensional torus.[11] Figure 2.1 demonstrates how we get the two-dimensional torus from a fundamental set (parallelogram) in $\mathbb{R}^2$. In this construction, both pairs of opposite edges of the parallelogram are identified in $\mathbb{R}^2/L$.

---

[9] *Euclidean space* means a finite-dimensional vector space with an inner product. In this sense, $\mathbb{R}^d$ with the standard inner product is an Euclidean space.

[10] More generally, the quotient set $A/\sim$ of a topological space A has a natural topology such that the canonical projection is continuous.

[11] The d-dimensional torus is the product space $S^1 \times \cdots \times S^1$ of d circles.

2.2 Group Actions

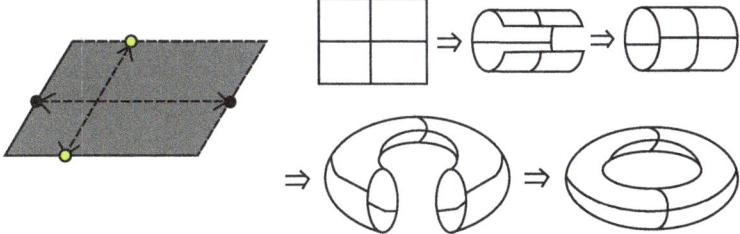

Fig. 2.1 How to get the two-dimensional torus

Secondly, $\mathbb{R}^d/L$ inherits the Euclidean structure so that it is a surface provided with a flat metric[12] on which Euclidean geometry can be locally argued (however, even for the two-dimensional case, we cannot visualize it as a surface in the space $\mathbb{R}^3$ with the metric induced from the Euclidean metric). Thirdly, $\mathbb{R}^d/L$ is an abelian group with the addition induced from the vector sum in $\mathbb{R}^d$ (actually $\mathbb{R}^d/L$ is a factor group of $\mathbb{R}^d$ modulo L).

One might think that since the fundamental set D_L is identified with the flat torus $\mathbb{R}^d/L$ via the canonical projection $\pi : \mathbb{R}^d \longrightarrow \mathbb{R}^d/L$, it is not necessary to make use of $\mathbb{R}^d/L$ in the discussion. Indeed, D_L is visible for $d = 2, 3$ (and one may somehow imagine D_L even for the case $d \geq 4$), but the flat torus $\mathbb{R}^d/L$ is not visible even for $d = 2, 3$. From the mathematical view, however, the flat torus $\mathbb{R}^d/L$ is a much more natural object than D_L because D_L depends on the choice of a $\mathbb{Z}$-basis $\mathbf{a}_1, \ldots, \mathbf{a}_d$ of L, while $\mathbb{R}^d/L$ is uniquely determined by L itself. An advantage of $\mathbb{R}^d/L$ will be seen when we relate problems in crystallography to ones in differential geometry, as explained in the later discussion.

If L is a group generated by linearly independent vectors $\mathbf{a}_1, \ldots, \mathbf{a}_s$ with $s < d$, we get the *generalized flat cylinder* $\mathbb{R}^d/L$, which is topologically the product of the s-dimensional torus and the $(d-s)$-dimensional Euclidean space. In the case $d = 2$ and $s = 1$, it is realized in $\mathbb{R}^3$ as a circular cylinder.[13] More explicitly, $\mathbb{R}^2/\mathbb{Z}\mathbf{a}$ is realized as

$$C = \left\{ (x,y,z) \in \mathbb{R}^3 \mid x^2 + y^2 = \frac{\|\mathbf{a}\|^2}{4\pi^2} \right\},$$

where $\|\mathbf{a}\|$ stands for the length (norm) of the vector $\mathbf{a}$. A circular cylinder appears in crystallography as a model surface on which *carbon nanotubes* are represented (see Example 6.4 and the reamrk at the end of Sect. 7.1). □

Remark. As mentioned in the introduction, a lattice group is referred to as a *period lattice* if it is associated with a crystal net as a group of translations leaving the net invariant. The reader of this book should be aware of the distinction between lattice

---

[12] Precisely speaking, by a metric, we mean a Riemannian metric. A metric is *flat* if its curvature vanishes.

[13] The metric on a circular cylinder induced from the Euclidean metric is flat. This is because the circular cylinder is obtained by deforming the flat plane without expansion and contraction.

groups and crystals. I was told that there is a lot of confusion about this distinction in many textbooks and among practicing scientists. Nevertheless, in this book, we follow the convention to use the term "lattice" for some crystal structures and for crystals themselves, like "hexagonal lattice" and "honeycomb lattice". □

We go back to general group actions. Let A and B be sets with G-actions. The product group $G \times G$ acts on $A \times B$ by $(g_1, g_2)(a, b) = (g_1 a, g_2 b)$. The orbit space $(A \times B)/(G \times G)$ is identified with $(A/G) \times (B/G)$.

We may think of $A \times B$ as a G-set with the *diagonal* G-action defined by $g(a,b) = (ga, gb)$. Let C be another set with a G-action, and let $f_A : C \longrightarrow A$ and $f_B : C \longrightarrow B$ be G-equivariant maps. Define $f : C \longrightarrow A \times B$ by $f(c) = \bigl(f_A(c), f_B(c)\bigr)$. The homomorphism $\rho : G \longrightarrow G \times G$ defined by $\rho(g) = (g, g)$ satisfies $f(gc) = \rho(g) f(c)$, whence f induces a map

$$F : C/G \longrightarrow (A/G) \times (B/G).$$

We say that a group G acts *transitively* on A if the orbit space A/G consists of a single element; in other words, an action of G on A is transitive if, for any $x, y \in A$, there exists $g \in G$ with $gx = y$. We also say that G acts *freely* on A if $g \in G$ with $gx = x$ for some $x \in A$ must be the identity element 1. A free and transitive action is called *simply transitive*. For example, the left action of G on itself is simply transitive.

These terminologies will be employed in the theory of covering graphs (Chap. 5).

2.3 Notes

(I) The set of natural numbers $1, 2, 3, \ldots$ is a quotient set that human beings had already acquired in the remote past.

A simple, but the most convincing story about how they invented natural numbers is this: Even when they did not yet have the slightest idea of numbers, they had already a way to "count" things nonetheless. For instance, ancient people were supposed to be able to check whether their cattle put out to pasture returned safely to their shed, without any having been lost. The idea is to use *one-to-one correspondence*. Preparing, in advance, some identical things such as small stones or sticks of twigs, in oredr to make a one-to-one correspondence between the cattle and the things, they could make sure that no cattle were missing by making a one-to-one correspondence again when the cattle returned from pasture.

They extended the manner of one-to-one correspondence to counting things in various aggregations. If the size of an aggregation was small, they counted things on their fingers. Such an experience had been inherited from generation to generation for centuries, and presumably made ancient people aware that there is "something in common" behind, which does not depend on individual aggregation.

2.3 Notes

In order to capture "something", they needed to develop the ability to think in abstractions, and finally got a way of identifing as an entity all aggregations among which there are one-to-one correspondences. This entity is nothing but a natural number. Once reached this stage, it did not take much time for them to start to assign names (as one, two, three,... in English) and to develop primitive arithmetic.

The philosopher A.N. Whitehead said "the first man who noticed the analogy between a group of seven fish and a group of 7 days made a notable advance in the history of thought".

As with natural numbers, integers, fractions (rational numbers), and real numbers are also quotient objects in our sense. So is the notion of vectors in plane and space that we learn at school. It should be emphasized that the *transfinite set theory* established by G. Cantor for giving mathematical content to the actual infinite is based on the idea of one-to-one correspondence, just the same idea that ancient people used to count things.

(II) Topology is, in a word, the branch of geometry that studies the properties preserved through deformations, twistings, and stretchings of objects, thereby without regrad for the measurement of lengths or angles. The term "topology" was first used by J.B. Listing, a disciple of C.F. Gauss (1777–1855), in his treatise "Vorstudien zur Topologie" (1847). He used it as a synonym for the "geometry of positions".[14] It was G.W. Leibniz (1646–1716), if traced back to the origin of topology, who for the first time suggested establishing the geometry of positions as a field without any measurements or calculations (1674). The first work that embodied Leibniz's suggestion is Euler's solution to the problem of the seven Königsberg bridges (1735). Euler's idea eventually evolved into graph theory (see the next chapter and [10]).

The successor of Listing is Poincaré, a prominent French mathematician, who is also famous for his pioneer work on the qualitative theory of dynamical systems and special relativity. He investigated higher dimensional figures by using algebraic techniques. The voluminous paper *Analysis situs* published in 1895 contains the ground-breaking work which set up a starting point of the new field called today algebraic topology. In this paper, the concept of homology was introduced by unifying and generalizing the fragmented ideas of Euler, Riemann, Jordan, and Betti.

(III) As Example 2.5 suggests, the idea of quotient objects is of vital importance in topology. It is used for giving a logical articulation of things derived by such topological operations as "glueing" figures and "contraction" of some part of a figure to a point. As for two-dimensional figures, we are able to explain intuitively how we are manipulating them by handwaving. The reader who has some idea

---

[14]The reader might be curious about why the word "topology" can be a synonym. Although the word cannot be found in ordinary dictionaries, the prefix "topo" is used to express "position"; for instance, "topography" means the physical features of an area of land, especially the position of its rivers, mountains, etc. The Greek word "topos" meaning an idyllic place is the origin of "topo". Accordingly topology means literally "theory of position".

Fig. 2.2 From a cylinder to a circular cone

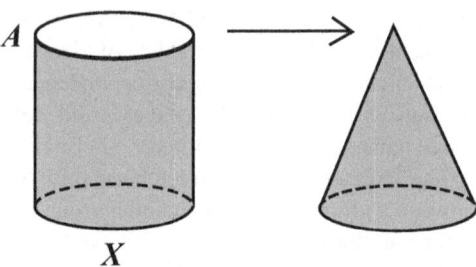

of elementary topology should recall, for instance, how we construct the Möbius band and the Klein bottle. But in the case of higher dimensional figures or general topological spaces, we need a logical construction to make the argument rigorous. This is the place where the idea of quotient comes in.

For example, the space obtained by the contraction of a subset A in a topolological space X is the quotient space for the equivalence relation defined by

$$\begin{cases} \text{if } x \notin A, \text{ then } x \sim y \iff x = y \\ x \sim y \text{ for all } x, y \in A. \end{cases}$$

Figure 2.2 illustrates how a circular cone is obtained from a cylinder by a contraction.

An example of glueing figures by means of equivalence relations is seen in the construction of a of 1-dimensional cell complex associated with an abstract graph (see Sect. 3.1). The idea of contractions will be used in Sect. 7.2.

Chapter 3
Generalities on Graphs

A graph is an abstract concept describing a set of objects where some pairs of objects are connected by links.[1] Practically, we represent a graph by a diagram in a plane or space consisting of points (vertices) and lines (edges). Lines are allowed to intersect each other. In some applications, say communication networks and electric circuits, a diagram may be associated in a direct way with the objects considered. A crystal structure also provides us with a visible diagram in space when we represent atoms and chemical bonds by points and lines,[2] respectively. In many other cases, however, a graph structure is not explicitly visible at first hand.

Remember that, when David Hilbert (1862–1943) reconstructed Euclidean geometry by reexamining the system of axioms (postulates) in Euclid's *Elements*,[3] he decided to leave points and straight lines undefined because if we would try to define them individually by following our intuition, then we would encounter an infinite regress.[4] Instead, he extracted a few relations between points and straight lines to make up a new system of axioms. What this signifies is that visibility does not play any essential roles in mathematics (even in geometry) when a theory is to be logically built.[5] We shall follow his point of view for an abstract definition of graphs to avoid confusion caused by intuition.

This chapter only scratches the surface of graph theory as a minimum requirement for modern crystallography. The reader should refer to Bollobas [15] or Diestel [36] for a full account.

---

[1] As mentioned in Notes in the previous chapter, the notion of graphs was implicitly used by Leonhard Euler (1707–1783).

[2] Strictly speaking, the line representing a bond is virtual. See the beginning part of Chap. 7.

[3] *Uber die Grundlagen der Geometrie*, 1899.

[4] In Euclid's *Elements*, we find the following definitions. "A point is that which has no part". "A line is breadthless length". "A straight line is a line which lies evenly with the point on itself".

[5] I am, of course, not denying the significance of intuition in ongoing studies. Without intuitive thinking, theorems in geometry could not have been formulated. Also we must emphasize the importance of pictures in understanding any "formal" proof.

3.1 Graphs

Formally a *graph* is an ordered pair $X = (V, E)$ of disjoint sets V and E with two maps $i : E \longrightarrow V \times V$, $\iota : E \longrightarrow E$ such that

$$\iota^2 = I_E \quad \text{(the identity map of } E\text{)}$$
$$\iota(e) \neq e, \quad i(\iota(e)) = \tau(i(e)), \quad (e \in E),$$

where $\tau : V \times V \longrightarrow V \times V$ is the map defined by $\tau(x, y) = (y, x)$. Putting $i(e) = (o(e), t(e))$, we obtain the maps $o : E \longrightarrow V$ and $t : E \longrightarrow V$ with the relation $t(\iota(e)) = o(e)$ $(e \in E)$. When both V and E are finite, X is called a *finite* graph.

A brief amplification of this definition of Bourbaki's style is helpful in establishing terminology and notation. In what follows, elements of V are called *vertices* or *points* of X, and elements of E are called *(directed) edges* of X. The map i is called the *incidence map*, while the map ι is called the *inversion map*. Note that the inversion map gives rise to an action of $\mathbb{Z}_2 = \mathbb{Z}/2\mathbb{Z}$ on E. An *undirected edge* is an element of the orbit space $E/\mathbb{Z}_2$. In other words, an undirected edge is obtained by identifying e and $\iota(e)$. We say that $o(e)$ is the *origin*, $t(e)$ is the *terminus*, and $\iota(e)$ is the *inversion* of e (or *inverse edge*). We usually write $\iota(e)$ in the form $\overline{e}$. With these notations, the equality $i(\iota(e)) = \tau(i(e))$ is expressed as $o(\overline{e}) = t(e), t(\overline{e}) = o(e)$.

In a diagram representing a graph, we express each directed edge by an arrow sitting on an arc, as shown in Fig. 3.1. Thus e and $\overline{e}$ are expressed by the same arc having arrows with opposite direction. However we usually omit arrows when we draw the diagram of a graph.

A vertex x is said to be *adjacent* to a vertex y provided that there exists an edge e with $o(e) = x$ and $t(e) = y$. We also say that *e joins x and y*.

As indicated in the diagram of a graph, every graph can be realized as a *cell complex* of dimension ≤ 1. More precisely, given a graph $X = (V, E)$, take the disjoint union $V \cup (E \times [0, 1])$ and introduce the equivalence relation $\sim$ defined by

$$o(e) \sim (e, 0), \quad t(e) \sim (e, 1), \quad (e, t) \sim (\overline{e}, 1 - t) \text{ for } 0 \leq t \leq 1.$$

Then the *realization* of X as a cell complex is the quotient space $V \cup (E \times [0, 1])/\sim$ so that undirected edges (respectively vertices) are identified with 1-cells (respectively 0-cells).

Our graphs are allowed to have *loop edges* (e with $o(e) = t(e)$) and *parallel edges* ($e_1 \neq e_2$ with $o(e_1) = o(e_2), t(e_1) = t(e_2)$) (Fig. 3.2). A *combinatorial graph* is a graph without loop edges and parallel edges.

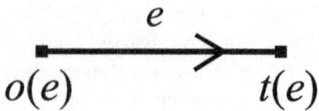

Fig. 3.1 Directed edge

3.1 Graphs

Fig. 3.2 Loop edge and parallel edges

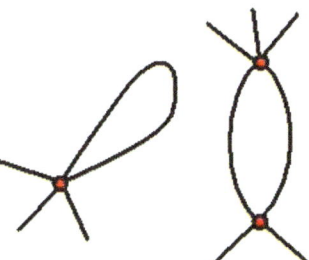

Let V_1 and E_1 be subsets of V and E, respectively. If $i(E_1) \subset V_1 \times V_1$ and $\overline{E_1} = E_1$, then the pair $X_1 = (V_1, E_1)$ forms a graph. We say that the graph X_1 is a *subgraph* of X, and write $X_1 \subset X$.

The set of directed edges with origin $x \in V$ is denoted by E_x; i.e., $E_x = \{e \in E \mid o(e) = x\}$. Throughout this book, we only consider *locally finite graphs*; that is, graphs $X = (V, E)$ such that E_x is finite for every $x \in V$.

The *degree* (or *valence* in chemistry) of x, symbolically $\deg x$, is the number of elements in E_x, i.e., $\deg x = |E_x|$. The degree of X is the maximum of $\{\deg x \mid x \in V\}$. We find that $E = \bigcup_{x \in V} E_x$ and $E_x \cap E_y = \emptyset$ for $x \neq y$. This leads to the useful fact that the single summation $\sum_{e \in E}$ is written as the double summation $\sum_{x \in V} \sum_{e \in E_x}$. In particular, for a finite graph,

$$\sum_{x \in V} \deg x = |E|. \tag{3.1}$$

Notice that the number of undirected edges is equal to $|E|/2$.

When $\deg x$ does not depend on x; say $\deg x \equiv k$ for some constant k, the graph X is said to be a *regular graph* of degree k. An importance of regular graphs becomes apparent when we handle crystals consisting of a single kind of atom. For instance, the network of a carbon crystal of the sp^3 bond type[6] such as diamond and Lonsdaleite is a regular graph of degree 4, while that of a carbon crystal of the sp^2 bond type is a regular graph of degree 3.

In some cases, we are given a diagram of points and directed lines. An abstraction of this diagram is the notion of *directed graph*, a pair $X^o = (V, E^o)$ with an *incidence map* $i^o : E^o \longrightarrow V \times V$. Putting $i^o(e) = (o(e), t(e))$, we get two maps o and t of E^o into V.

Given a graph $X = (V, E)$, we obtain a directed graph $X^o = (V, E^o)$ by selecting a subset E^o of E such that $E^o \cup \overline{E^o} = E$ and $E^o \cap \overline{E^o} = \emptyset$. The subset E^o is said to be an *orientation* of X (Fig. 3.3).

---

[6]When a molecule is formed by atoms, the electron cloud of one atom and the electron cloud of another atom interact to form a new electron cloud. This formation, depending on the orbitals overlapping, is known as *hybridisation*, and is classified as sp, sp^2 and sp^3, etc.

Fig. 3.3 Orientation

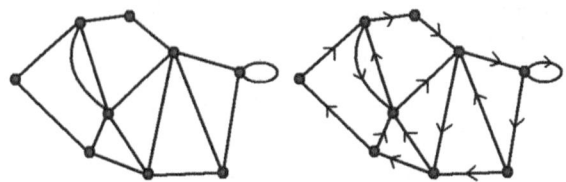

Conversely, given a directed graph $X^o = (V, E^o)$, we may construct an ordinary graph (V, E) with the orientation E^o. For this we define E to be the disjoint union of E^o and its copy $\overline{E^o}$, and extend i^o to E by setting $i(e') = \tau i^o(e)$, where $e' \in \overline{E^o}$ is the *twin* of $e \in E^o$. The inversion map ι is defined by $\iota(e) = e'$ and $\iota(e') = e$. In the diagram of a directed graph, this is simply the procedure to remove the arrow on each edge.

3.2 Morphisms and Automorphisms

In general, a *morphism* is an abstraction of a function or map between two sets. For two graphs $X_1 = (V_1, E_1)$ and $X_2 = (V_2, E_2)$, a morphism $f : X_1 \longrightarrow X_2$ is a correspondence of vertices and edges preserving the adjacency relation among them. More precisely, it is a pair $f = (f_V, f_E)$ of maps

$$f_V : V_1 \longrightarrow V_2, \quad f_E : E_1 \longrightarrow E_2$$

such that $o(f_E(e)) = f_V(o(e))$, $t(f_E(e)) = f_V(t(e))$, and $\overline{f_E(e)} = f_E(\bar{e})$ for every $e \in E_1$. By abuse of notation, we frequently use the same symbol f for both f_V and f_E for simplicity.

Obviously the *image*

$$f(X_1) = (f_V(V), f_E(E))$$

is a subgraph of X_2. If $X_1 = (V_1, E_1)$ is a subgraph of X, the pair of inclusion maps $i = (i_V, i_E) : X_1 \longrightarrow X$ is a morphism.

One can define the *composition* of two morphisms in a natural manner. In the case that f_V and f_E are bijections, f is called an *isomorphism*, and X_1 and X_2 are said to be *isomorphic*. For an isomorphism $f = (f_V, f_E)$, the inverse (morphism) of f is $f^{-1} = (f_V^{-1}, f_E^{-1})$. When $X_1 = X_2 = X$, an isomorphism is called an *automorphism*. The totality of automorphisms of X, symbolically Aut(X), forms a group,[7] which we call the *automorphism group* of X. The identity element of Aut(X) is the identity morphism $I = (I_V, I_E)$.

---

[7]In the case that X is an infinite graph, it is often natural to think of Aut(X) as a topological group with "compact-open topology" (see Sect. 9.6).

3.2 Morphisms and Automorphisms

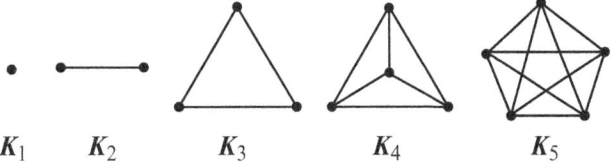

Fig. 3.4 Complete graphs

Fig. 3.5 The bouquet graph

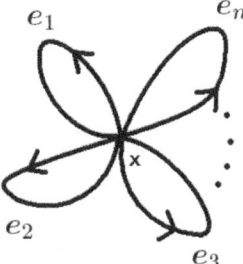

Needless to say, Aut(X) is used to quantify symmetry of the graph X; namely, the more elements Aut(X) has, the more symmetric X is. In topological crystallography, automorphism groups appear as an abstract counterpart to crystallographic groups (see Sects. 6.3 and 9.6).

A homomorphism $\varphi : \mathrm{Aut}(X) \longrightarrow \mathcal{S}_V$ is defined by setting $\varphi\big((g_V, g_E)\big) = g_V$. If X is a combinatorial graph, then φ is injective. For a finite graph X, the automorphism group Aut(X) is a finite group.

Suppose that X is a finite graph with $|V| = n$. Assigning numbers $1, 2, \ldots, n$ to vertices, and identifying V with $\{1, 2, \ldots, n\}$, we obtain a homomorphism $\varphi : \mathrm{Aut}(X) \longrightarrow \mathcal{S}_n$ (see Example 3.1).

Example 3.1. A *complete graph* is a graph in which each pair of distinct vertices is connected by a unique edge (Fig. 3.4). We use the symbol K_n for the complete graph with n vertices. Clearly $\varphi : \mathrm{Aut}(K_n) \longrightarrow \mathcal{S}_n$ is an isomorphism. □

Example 3.2. An *n-bouquet graph* B_n is a graph with a single vertex and n loop edges (Fig. 3.5). Let us describe the structure of the automorphism group of B_n. Select an orientation $\{e_1, \ldots, e_n\}$ of B_n. We put $G = \mathrm{Aut}(B_n)$ and

$$K = \{g \in G|\ \text{there is}\ \sigma \in \mathcal{S}_n\ \text{such that}\ g(e_i) = e_{\sigma(i)}\ (i = 1, \ldots, n)\}.$$

Clearly K is a subgroup of G isomorphic to $\mathcal{S}_n$.

We identify $\mathcal{S}_2$ with the multiplicative abelian group $\{1, -1\}$, and put

$$e_i^{\varepsilon_i} = \begin{cases} e_i & (\varepsilon_i = 1) \\ \overline{e}_i & (\varepsilon_i = -1) \end{cases}$$

Fig. 3.6 Graph with two vertices joined by n parallel edges

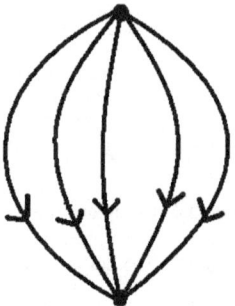

for $\varepsilon = (\varepsilon_1, \ldots, \varepsilon_n) \in \mathcal{S}_2^n = \mathcal{S}_2 \times \cdots \times \mathcal{S}_2$. If we define

$$H = \{g \in G | \text{ there is } \varepsilon \in \mathcal{S}_2^n \text{ such that } g(e_i) = e_i^{\varepsilon_i} \ (i = 1, \ldots, n)\},$$

then H is a subgroup of G isomorphic to $\mathcal{S}_2^n$, and $H \cap K = \{1\}$. Moreover, for $h \in H$ and $g \in G$, we find that $ghg^{-1} \in H$, and hence H is a normal subgroup of G. On the other hand, each $g \in G$ gives rise to a permutation of undirected edges. Hence there exist $\sigma \in \mathcal{S}_n$ and $\varepsilon = (\varepsilon_1, \cdots, \varepsilon_n) \in \mathcal{S}_2^n$ such that $g(e_i) = e_{\sigma(i)}^{\varepsilon_{\sigma(i)}}$. Defining $h \in H$ and $k \in K$ by

$$h(e_i) = e_i^{\varepsilon_i}, \quad k(e_i) = e_{\sigma(i)} \quad (i = 1, \ldots, n),$$

we find that $hk(e_i) = e_{\sigma(i)}^{\varepsilon_{\sigma(i)}} = g(e_i)$, and hence $g = hk$.

In general, a group G is said to be the *semi-direct product* of a normal subgroup H and a subgroup K if the following two conditions are fulfilled.

1. $H \cap K = \{1\}$.
2. $G = HK$, i.e., for any $g \in G$, there exist $h \in H$, $k \in K$ with $g = hk$ (note that such h, k are uniquely determined in view of (1); for, if $g = hk = h_1 k_1$ ($h, h_1 \in H$, $k, k_1 \in K$), then $h_1^{-1} h = k_1 k^{-1}$, $h_1^{-1} h = k_1 k^{-1} \in H \cap K = \{1\}$, whence $h = h_1, k = k_1$).

Thus $G = \mathrm{Aut}(B_n)$ is the semi-direct product of $\mathcal{S}_2^n$ and $\mathcal{S}_n$. □

Example 3.3. Let X be a graph with two vertices joined by n parallel edges (Fig. 3.6). Then $\mathrm{Aut}(X)$ is isomorphic to the product group $\mathcal{S}_2 \times \mathcal{S}_n$. The proof is left as an exercise for the reader. □

3.3 Quotient Graphs

At the beginning of the previous chapter, we mentioned that crystallographers imported the idea of quotient graphs of crystal nets to enumerate crystal structures in a systematic way. The group actions in their set-up are the ones by parallel translations. This section handles quotient graphs by general group actions, in consideration of applications to the theory of covering graphs (Chap. 5).

3.3 Quotient Graphs

We say that a *group G acts on a graph* $X = (V, E)$ when a homomorphism of G into $\text{Aut}(X)$ is given. Thus an action on X gives rise to actions of G on both V and E. For simplicity, we write the actions of $g \in G$ on $x \in V$ and $e \in E$ in the form gx and ge, respectively. Evidently $o(ge) = go(e)$ and $t(ge) = gt(e)$; that is, the maps $o, t : E \longrightarrow V$ are G-equivariant.

We say that G acts on X *without inversion* if, for any $g \in G$, there is no edges e with $ge = \bar{e}$. If, in addition, G acts freely on V, then the action is said to be *free* (this being the case, G also acts freely on X as an action on a cell-complex).

Theorem 3.1. *Let $X = (V, E)$ be a graph on which a group G acts without inversion, and let $V_1 = V/G$ and $E_1 = E/G$, the orbit spaces for the G-actions. Then there exists a unique graph structure on the pair $X_1 = (V_1, E_1)$ such that the pair $\omega = (\omega_V, \omega_E)$ of the canonical projections*

$$\omega_V : V \longrightarrow V_1, \quad \omega_E : E \longrightarrow E_1$$

form a morphism of graphs.

To prove this, let $i_1 : E_1 \longrightarrow V_1 \times V_1$ be the map induced from the incidence map $i : E \longrightarrow V \times V$ (see Sect. 2.2), and let $\iota_1 : E_1 \longrightarrow E_1$ be the map induced from the inversion map $\iota : E \longrightarrow E$. Clearly under the maps ι_1 and i_1, (V_1, E_1) is a graph and ω is a morphism. To see the uniqueness of the graph structure on (V_1, E_1), let $i' : E_1 \longrightarrow V_1 \times V_1$ and let $\iota' : E_1 \longrightarrow E_1$ be an incidence map and an inversion map for which ω is a morphism. Define

$$\omega \times \omega : V \times V \longrightarrow V_1 \times V_1$$

by $(\omega \times \omega)(x, y) = \big(\omega(x), \omega(y)\big)$. Then for $e \in E$, we have $i'(\omega e) = (\omega \times \omega) \circ i(e) = i_1(\omega e)$ and $\iota'(\omega e) = \omega \circ \iota(e) = \iota_1(\omega e)$. Since $\omega_E : E \longrightarrow E_1$ is surjective, we have $i' = i_1$ and $\iota' = \iota_1$, as desired.

The graph X_1 in the theorem above is the *quotient graph* of X by the G-action, and is denoted by X/G. The morphism $\omega : X \longrightarrow X/G$ is called the *canonical morphism*.

In the following theorem, we keep the situation in the above theorem.

Theorem 3.2. *Let $Y = (V_Y, E_Y)$ be a graph on which a group H acts without inversion, and let $F = (F_V, F_E) : X \longrightarrow Y$ be a morphism. Suppose that there is a homomorphism $\rho : G \longrightarrow H$ such that $f \circ g = \rho(g) \circ f$ for every $g \in G$. If we denote by $\omega_Y : Y \longrightarrow Y/H$ the canonical morphism, then there exists a morphism $f : X/G \longrightarrow Y/H$ such that $F \circ \omega = \omega_Y \circ f$.*

$$\begin{array}{ccc} X & \xrightarrow{f} & Y \\ \omega \downarrow & & \downarrow \omega_Y \\ X/G & \xrightarrow{F} & Y/H \end{array}$$

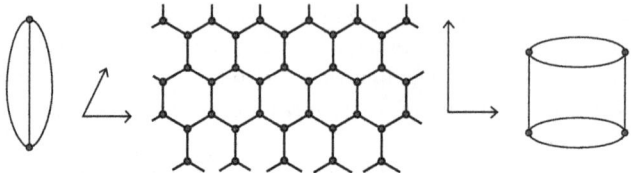

Fig. 3.7 The honeycomb lattice and its quotient graphs

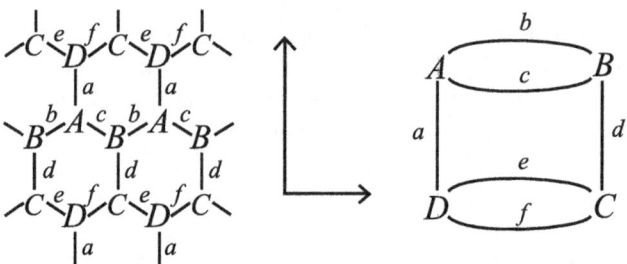

Fig. 3.8 How to get the quotient graph

What we have to prove is that the induced maps $f_V : V/G \longrightarrow V_Y/H$ and $f_E : E/G \longrightarrow E_Y/H$ comprise a morphism $f = (f_V, f_E)$. The proof is routine and omitted.

We shall give three examples of quotient graphs which bear relevance to crystallography. The first two examples are figures in a plane, thereby not spatial figures associated with ordinary crystals, but still objects being treated in crystallography.[8]

Example 3.4. The *honeycomb lattice*[9] depicted in the middle of Fig. 3.7, also called the *hexagonal lattice* when we consider it as an abstract graph, has a free action by a lattice group generated by two vectors.[10]

Two kinds of actions (represented by arrows) are shown in the figure. The quotient graphs are the finite graphs drawn on the left and right. The action and its quotient graph drawn on the left show that the honeycomb lattice is considered a two-dimensional analogue of the net associated with the diamond crystal (see Sect. 8.3).

For the convenience of the reader, let us explain the concrete way, by using Fig. 3.8, as to how to get the quotient graph. Using some symbols, we first label vertices and edges of a given graph with a group action; say, A, B, C, D for vertices

---

[8] For instance, the honeycomb lattice depicted in Fig. 3.7 appears as the crystal structure of *graphene*, an allotrope of carbon. The term graphene was coined by Hanns-Peter Boehm (1962). The Nobel Prize in Physics for 2010 was awarded to Andre Geim and Konstantin Novoselov "for groundbreaking experiments regarding the two-dimensional material graphene".

[9] This is also called a *tortoise shell pattern*.

[10] Such a lattice is called a period lattice of the honeycomb lattice.

3.3 Quotient Graphs

Fig. 3.9 The dice lattice and its quotient graph

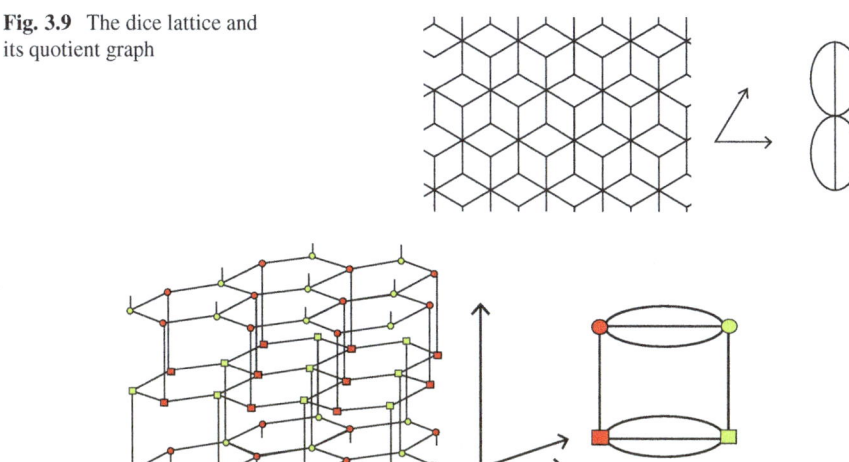

Fig. 3.10 The graphite-like realization of Lonsdaleite and its quotient graph

and a,b,c,d,e,f for edges in the honeycomb lattice, in such a way that the same label is assigned to edges (respectively vertices) in the same orbit. We then gather the labels of vertices and join them by labeled edges, keeping the adjacency relation. What you get is the quotient graph.

Moreover a glance at Fig. 3.8 reveals that one can attach an "accessory" to the quotient graph X_0 of a crystal net. That is, one can assign a (unique) vector to each directed edge of X_0. For example, the vector $\overrightarrow{AbB}$ appearing in the honeycomb lattice is assigned to the directed edge AbB in X_0. The idea of vector assignment, which turns out to be key to the study of crystal structures, is applied to an arbitrary crystal net (Sect. 7.1). □

Example 3.5. Figure 3.9 is the so-called *dice lattice* (induced from the *rhombille tiling*) which has a free action by the lattice group generated by two vectors (represented by arrows). The quotient graph is the one on the right. □

Example 3.6. The *graphite-like* network in space depicted in Fig. 3.10 has a free action by the lattice group generated by three vectors (represented by arrows).[11] The quotient graph is the one on the right. This graphite-like graph is isomorphic to the network structure of Lonsdaleite. □

---

[11] Graphite as a mineral is one of the allotropes of carbon, and the most stable form under standard conditions. It has a layered, planar structure. In each layer, the carbon atoms are arranged in a honeycomb lattice.

The reader should construct some more examples to be convinced the usefulness of the idea of quotient graph in crystallography. Plenty of 3D examples of crystal nets are exhibited in Sect. 8.3.

3.4 Paths

In topological crystallography the notion of paths plays a significant role, along with homology theory and the theory of covering graphs, the subjects in the next two chapters.

A *path* in a graph $X = (V, E)$ is a sequence of directed edges

$$c = (e_1, e_2, \ldots, e_n)$$

such that $t(e_i) = o(e_{i+1})$ (Fig. 3.11). We write $o(c) = c(0) = o(e_1)$ for the origin of c, and $c(k) = t(e_k)$. We put $t(c) = t(e_n)$ for the terminus of c and $|c| = n$ for the *length* of c. It is often said that c *joins* $o(c)$ and $t(c)$. We also write $\bar{c} = (\overline{e_n}, \ldots, \overline{e_1})$ for the *inversion* of c. A path c with $o(c) = t(c)$ is said to be *closed*.

We regard a vertex $x \in V$ as a path of length 0, which is called the *point path* at x and denoted by $\emptyset_x$.

Notice that a path of length n in X is a morphism from the *segment graph* $[0, n]$ with vertices $0, 1, \ldots, n$ into X. Likewise, a closed path of length n is a morphism from the *circuit graph* C_n (the circle with n vertices) into X (Fig. 3.12).

A graph is *connected* if for any two vertices x and y, there is a path c with $o(c) = x$ and $t(c) = y$. Note that X is connected if and only if X as a cell complex is connected (actually, arcwise connected). Our graphs are supposed to be connected unless otherwise stated.

The following fact is sometimes useful.

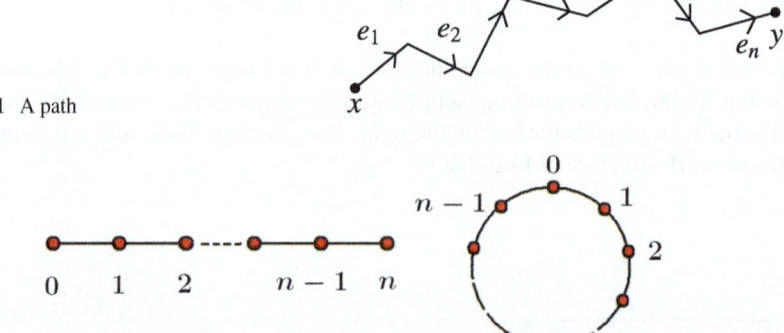

Fig. 3.11 A path

Fig. 3.12 The segment graph and circuit graph

3.4 Paths

Fig. 3.13 Separating edge

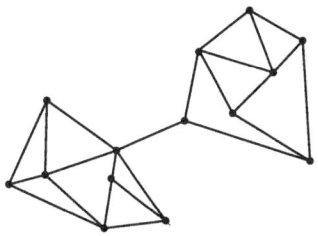

Lemma 3.4.1. *For a non-empty subset $A \subset V$, if*

$$x \in A \Rightarrow t(e) \in A$$

holds for every $e \in E_x$, then $A = V$.

An undirected edge is said to *separate* X if the graph obtained by removing the edge (and by leaving the vertices unchanged) is disconnected. Such an edge is called a *separating edge* (Fig. 3.13). Obviously loop edges and parallel edges are not separating edges. If X has no separating edges, then $\deg x \geq 2$ for every vertex x.

Given two paths $c_1 = (e_1, e_2, \ldots, e_n)$ and $c_2 = (e'_1, e'_2, \ldots, e'_m)$ with $t(c_1) = o(c_2)$, the *product* $c_1 \cdot c_2$ is the path $(e_1, e_2, \ldots, e_n, e'_1, e'_2, \ldots, e'_m)$. We set $\emptyset_x \cdot c = c$ ($x = o(c)$) and $c \cdot \emptyset_x = c$ ($x = t(c)$). If c_1, c_2, c_3 are paths with $t(c_1) = o(c_2)$, and $t(c_2) = o(c_3)$, then $c_1 \cdot (c_2 \cdot c_3) = (c_1 \cdot c_2) \cdot c_3$.

For a morphism $f : X \longrightarrow X_1$ and a path $c = (e_1, \ldots, e_n)$, we put $f(c) = (f(e_1), \ldots, f(e_n))$, which is a path in X_1. Evidently

$$f(c_1 \cdot c_2) = f(c_1) \cdot f(c_2).$$

A path $c = (e_1, e_2, \ldots, e_n)$ is a *geodesic path* (or simply a *geodesic*) if $e_i \neq \overline{e_{i+1}}$ for $i = 1, \ldots, n-1$, i.e., if c has no *back-tracking parts* anywhere.

If a geodesic $c = (e_1, e_2, \ldots, e_n)$ is closed and $e_n \neq \overline{e_0}$, then c is said to be a *closed geodesic*. A *circuit* in X, referred to as a *ring* in chemistry, is a closed geodesic without self-intersection. The *girth* is the minimal length of circuits.

A *tree* is a graph having no circuits. A subgraph X_1 of a graph X is called a *subtree* if X_1 is a tree. Among all subtrees in a graph X, we can find a maximal one with respect to inclusion,[12] which we call a *spanning tree* of X. As Fig. 3.14 shows, a graph can have many spanning trees.

It is easy to see that if a subtree T contains all vertices of X, then T is a spanning tree. Conversely, a spanning tree $T = (V_T, E_T)$ contains all vertices of X.

---

[12] When X is infinite, we need Zorn's lemma to deduce the existence. Zorn's lemma is known to be equivalent to Axiom of Choice.

Fig. 3.14 Spanning trees

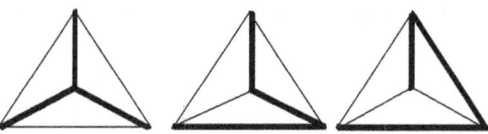

The following three statements for a graph X are equivalent:

1. X is a tree.
2. For any two vertices x and y, there is one and only one geodesic joining x and y.
3. Every edge separates X.

As for separating edges of a graph X, we have:

(a) Every edge separates X into at most two connected components.
(b) An edge e does not separate X if and only if there exists a circuit in X containing e.
(c) Let E_1 be the set of edges that separate X into two connected components. The graph Y obtained by gathering all edges in E_1 is a *forest* (a disjoint union of trees).
(d) By contracting each connected component of the complement of Y to one point, we get a tree.

The proof goes as follows.

(a) Suppose that an edge e separates X into more than two components, say X_1, X_2, X_3. Let x_i be a vertex of X_i and let c be a path joining x_1 and x_2. The edge e is contained in c. A path joining x_1 and x_3 must also go through the edge e so that the vertex x_3 is joined with x_1 or x_2 by a path in the complement of e. This is a contradiction.
(b) Suppose that e does not separate X. Take a minimal geodesic path c with $o(c) = t(e), t(c) = o(e)$ in the graph obtained by removing e. Then $e \cdot c$ is a circuit. The converse is obvious.
(c) If there is a circuit in Y, then an edge in Y does not separate X.
(d) The argument is straightforward.

3.5 Homotopy

In algebraic topology, the term "homotopy" is used to express "continuous deformations", by which geometric objects such as curves and surfaces are identified in a loose sense. Here we explain homotopy of paths in a graph from a combinatorial point of view.

Let c_1, c_2 be two paths in X. We write $c_2 \mapsto c_1$ when c_2 has two successive edges e, f with $f = \bar{e}$, and c_1 is obtained from c_2 by removing e and f (namely, c_1 is obtained by removing a *back-tracking* part of c_2). When $c = (e, \bar{e})$, we write $c \mapsto \emptyset_{o(c)}$. If $c_2 \mapsto c_1$, then $o(c_1) = o(c_2)$ and $t(c_1) = t(c_2)$.

Let $x, y \in V$. We write $C(x,y)$ for the set of all paths c with $o(c) = x$ and $t(c) = y$. A *loop* with *base point* x means a path in $C(x,x)$. We say that c_1 and c_2 in $C(x,y)$ are *homotopic*, and write $c_1 \sim c_2$ if there is a finite sequence of paths $\{c^{(1)}, c^{(2)}, \ldots, c^{(k)}\}$ with $c^{(i)} \in C(x,y)$ such that $c_1 = c^{(1)}$, $c_2 = c^{(k)}$ and $c^{(i+1)} \mapsto c^{(i)}$ or $c^{(i)} \mapsto c^{(i+1)}$.

It is obvious that $\sim$ is an equivalence relation on the set $C(x,y)$. Equivalence classes of this relation are called *homotopy classes*.

Below are five fundamental properties of homotopy that follow directly from the definition:

1. Suppose $c_1, c_2 \in C(x,y)$, $c'_1, c'_2 \in C(y,z)$. If $c_1 \sim c_2$ and $c'_1 \sim c'_2$, then $c_1 \cdot c'_1 \sim c_2 \cdot c'_2$.
2. $c \cdot \bar{c} \sim \emptyset_x$ ($x = o(c)$).
3. If $c_1 \sim c_2$, the parity[13] of $|c_1|$ and $|c_2|$ is the same.
4. Each homotopy class contains a unique geodesic.
5. For a geodesic path c_0, if $c \sim c_0$, then either $c = c_0$ or there exists a sequence of paths $c_1, c_2, \ldots, c_n$ such that $c_n = c$ and $c_i \mapsto c_{i-1}$ ($i = 1, 2, \ldots, n$).

The last property is shown by induction on the length $|c|$.

3.6 Bipartite Graphs

A graph is said to be *bipartite* if one can paint the vertices using two colors in such a way that any adjacent vertices have different colors (Fig. 3.15). More precisely, a graph $X = (V, E)$ is bipartite when there exists a *bipartition* $V = A \cup B$ such that the end points of each edge never belong to the same set.

Bipartite graphs appear in crystal structures consisting of two kinds of atoms which are alternately bound together, such as a system of Boron and Nitride.

A graph X is bipartite if and only if every closed path in X has even length. Indeed, a bipartite graph has no closed paths with odd length. For the converse, we define the equivalence relation $\sim$ on V by setting

$$x \sim y \Leftrightarrow \text{there exists a path joining } x \text{ and } y \text{ with even length.}$$

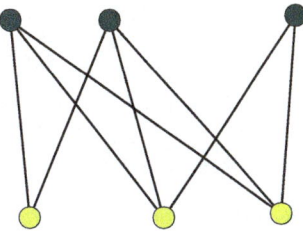

Fig. 3.15 A bipartite graph

---

[13] The even or odd quality of an integer.

We see straight away that there are at most two equivalence classes. The result derived from the assumption is that there exist just two equivalence classes, say A and B, yielding a bipartition of X. In view of this proof, we observe that the bipartition $V = A \cup B$ of a bipartite graph is unique.

A tree is bipartite since every closed path in a tree has even length.

Let G be a group of automorphisms acting on a bipartite graph $X = (V, E)$ with the bipartition $V = A \cup B$. Each $g \in G$ yields a partition $\{gA, gB\}$. By uniqueness of the bipartition, g yields a permutation of $\{A, B\}$. Therefore we get a homomorphism of G into the symmetry group S_2. The kernel of this homomorphism is $G_0 = \{g \in G|\ gA = A\}$, which coincides with G or is a subgroup of G of index-two.

Suppose that X is a bipartite graph with bipartition $\{A, B\}$ on which a group G acts without inversion. If A and B are G-invariant, the quotient graph X/G is also a bipartite graph with bipartition $\{A/G, B/G\}$. Indeed, if $o(\omega(e)) \in A/G$, then there exist $g \in G$ and $a \in A$ with $go(e) = a$, and hence $o(e) = g^{-1}a \in A$. Here $\omega : X \longrightarrow X/G$ is the canonical morphism. By hypothesis, $t(e) \in B$, from which it follows that $t(\omega(e)) \in B/G$.

3.7 Notes

(I) Various two-dimensional crystal structures are typically seen in ornamental patterns printed on Japanese garments, especially on *yukata* (unlined cotton kimono). Figure 3.16 exhibits some of them, in which the reader may identify the honeycomb lattice, the square lattice, the kagome lattice[14] and the triangular lattice.

These kinds of ornamental patterns appear also on the ceilings and walls of tombs in Egypt, in Persian manuscripts, in wallpapers tiling in the Alhambra, and in many

Fig. 3.16 Patterns

---

[14] See Example 6.5 in Sect. 6.2 and Fig. 8.1 in Sect. 8.3.

3.7 Notes

other places and at different times. These patterns have been studied by virtue of symmetries and are very helpful in better understanding their archaeological art and artifacts.

(II) The notion of geodesics is a discrete analogue of geodesics on Riemannian manifolds (see [68]). A *geodesic* on a Riemannian manifold M is a curve which locally minimizes the length. If M is complete, then any two points are joined by a geodesic. Furthermore, each homotopy class of continuous curves joining x and y contains a geodesic. If, in addition, M is non-positively curved, such a geodesic is uniquely determined. Thus the property (4) in Sect. 3.5 tells us that graphs are regarded as a discrete analogue of non-positively curved Riemannian manifolds. This observation is useful when we develop *discrete geometric analysis*, a hybrid field of several traditional disciplines: graph theory, geometry, theory of discrete groups, and probability [95].

Graph theory forms a small but beautiful realm in the "universe" of mathematics which somehow reminds us of *Bonsai*, the famous Japanese traditional art of botanical miniature. In this art, real plants such as pine and *ume* (apricot) trees are beautifully and skillfully miniaturized in a terse style, and their conciseness stirs up our imagination about the real beauty of their original appearances. A graph, as such, is a simple geometric object consisting only of vertices and edges. Nevertheless it holds numerous analogues of concepts in other mathematical fields, Riemannian geometry (as mentioned above), number theory, harmonic analysis and algebraic geometry, which can be thought of, like bonsai, as miniaturized objects of the original ones (see Chap. 10 for an analogue of algebraic geometry).

In the mean time, graph theory has been applied to many practical sciences: electric circuits, communication networks, web search engines, inter-industry relations analysis, computer science, etc. Moreover, graph-theoretic ideas are thought to be useful when we treat the vast amounts of data, for instance in *systems biology* which is a biology-based interdisciplinary field that focuses on complex interactions in biological systems. Another potential application is the *self-assembly problem* which deals with basic biological structures.

Chapter 4
Homology Groups of Graphs

Ancient Greek mathematicians tried to establish their theory of area and volume by means of "geometric algebra".[1] Namely, in order to compare the area (or volume) of two figures, they made up an algebraic system with addition and subtraction performed among a class of figures, e.g., polygons or polyhedra. Figure 4.1 illustrates a way to prove using geometric algebra that the area of a triangle is equal to one-half of the area of a rectangle with the same base and height.

In a sense, the concept of homology group (and fundamental group discussed in the next chapter) is considered a modern version of geometric algebra because an algebraic system is constructed from figures (cells) in a cell complex. Historically, the germination of homology theory is seen in the work by Riemann and Poincaré on algebraic functions. Following the same path as the development of abstract algebra, homology theory came to the fore in algebraic topology in the first half of the twentieth century.

For a general topological space, the definition of homology groups is considerably sophisticated, and could be a stumbling block for the reader to figure out what is going on. To make matters worse, computing homology groups is not easy. Fortunately, confining ourselves to the case of graphs, we can easily explain the homology group, and compute it without any technical difficulties. We also would like to emphasize that the homology theory of graphs has a special feature which is not shared by the general theory.

What should be noted is that the notion of homology groups not only underlies topological crystallography, but also can be used to interpret many graph-theoretical concepts, in spite of the simplicity of the definition. Thus it is worth to set up a chapter for homology theory of graphs.

---

[1] This is of course a modern interpretation of what they did. Greek mathematicians had no idea of *algebra* in today's sense. The word "algebra" was derived from the title "*ilm-al-jabra wa'l muqabalah*" (the science of reduction and cancellation) of the book written by Mohammed ibn Mûsâ al-Khowârizmi (780?–850?).

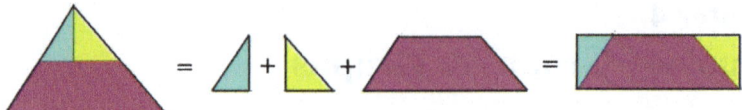

Fig. 4.1 Geometric algebra

The reader is assumed to have knowledge of some basic facts of abelian groups. If not, he/she should consult Appendix 2. Throughout this chapter, graphs are assumed to be connected unless otherwise stated.

4.1 Chain Groups

Let $X = (V, E)$ be a graph. In the first place, V is just a set, and has no algebraic structure. Nevertheless we may build an algebraic system based on V.

Consider a sum over vertices of the form:

$$\sum_{x \in V} a_x x \quad (a_x \in \mathbb{Z}),$$

where $a_x = 0$ for all but finitely many x (we use the symbol $\sum^{\text{finite}}$ for such a summation, but if it is clear from the context that we are considering a finite sum, then we omit "finite" in the symbol). We call such a sum a 0-*chain* (with coefficients in $\mathbb{Z}$). This sum and the product $a_x x$ (*integer* × *vertex*) are formal so that one should not worry about their geometric meaning.[2]

The set of 0-chains is denoted by $C_0(X, \mathbb{Z})$ with which we equip the structure of an abelian group obeying the following rule.

1. $\sum_{x \in V}^{\text{finite}} a_x x = 0$ if and only if $a_x = 0$ for every $x \in V$,
2. $\sum_{x \in V}^{\text{finite}} a_x x \pm \sum_{x \in V}^{\text{finite}} b_x x = \sum_{x \in V}^{\text{finite}} (a_x \pm b_x) x.$

In other words, $C_0(X, \mathbb{Z})$ is a *free abelian group* with $\mathbb{Z}$-basis V (see Appendix 2).

Likewise, we define the abelian group $C_1(X, \mathbb{Z})$ consisting of finite formal sums over directed edges

$$\sum_{e \in E} a_e e \quad (a_e \in \mathbb{Z}),$$

---

[2] The reader who is used to down-to-earth objects might feel that the notion of chain is too abstract, and thus might have difficulty to comprehend. The definition is surely far from the classical flavor of mathematics. For this reason, a beginner in modern mathematics is often perplexed, feeling as if he/she is walking on air. However once he/she gets used to modern ways, he/she will come to understand that it makes matters more transparent.

which we call the group of 1-*chains* (with coefficients in $\mathbb{Z}$). Here again $a_e = 0$ for all but finitely many e. This time, however, we impose the relation $\bar{e} = -e$ so that a way to express a 1-chain, such as the one given above, is not unique.[3] If we fix an orientation E^o of X; that is, a subset $E^o \subset E$ satisfying $E^o \cap \overline{E^o} = \emptyset$, $E^o \cup \overline{E^o} = E$, then E^o is a $\mathbb{Z}$-basis of $C_1(X,\mathbb{Z})$, and so any 1-chain is expressed as $\sum_{e \in E^o} a_e e$ in a unique way. We also see

$$\sum_{e \in E} a_e e = \sum_{e \in E^o} (a_e - a_{\bar{e}}) e.$$

4.2 Homology Groups

Now we define the homology group. Somewhat imprecisely stated, the homology group of a graph is introduced to describe the adjacency relation between vertices and edges in an algebraic way. To be exact, we start with the *boundary operator* (homomorphism)

$$\partial : C_1(X,\mathbb{Z}) \longrightarrow C_0(X,\mathbb{Z})$$

defined by putting $\partial(e) = t(e) - o(e)$ for $e \in E$ and extending it to $C_1(X,\mathbb{Z})$ as a homomorphism. We should note that $\partial \bar{e} = \partial(-e)$, and hence the definition of ∂ is compatible with the relation $\bar{e} = -e$. The 1-st homology group $H_1(X,\mathbb{Z})$ is defined to be the kernel of ∂:

$$H_1(X,\mathbb{Z}) = \operatorname{Ker} \partial \ (= \{\alpha \in C_1(X,\mathbb{Z}) | \ \partial \alpha = 0\}).$$

A peculiar feature of the homology group $H_1(X,\mathbb{Z})$ for a graph X, which turns out to be crucial in the later discussion, is that it is a subgroup of $C_1(X,\mathbb{Z})$; this is actually not the case for general topological spaces [see Notes (II) in this chapter].

Let E^o be an orientation of X. Then

$$\partial \left(\sum_{e \in E^o} a_e e \right) = 0$$

if and only if

$$\sum_{\substack{e \in E^o \\ t(e)=x}} a_e = \sum_{\substack{e \in E^o \\ o(e)=x}} a_e \qquad (x \in V). \tag{4.1}$$

Remark. If we regard a finite graph X as an electric circuit and a_e as the intensity of an electric current on the wire corresponding to the edge e, then (4.1) is nothing but Kirchhoff's current law stating that the sum of current into a junction (vertex) equals

---

[3] Precisely speaking we are considering the factor group of the free abelian group with $\mathbb{Z}$-basis E modulo the subgroup generated by $\{e + \bar{e} | \ e \in E\}$.

the sum of current out of the junction. In this sense, Kirchhoff's work on electric circuits is considered a prototype of homology theory. See Notes (II) in Chap. 7. □

A *homology class* or a *1-cycle* is an element of $H_1(X,\mathbb{Z})$. As seen in the next section, for a finite graph X, $H_1(X,\mathbb{Z})$ is a free abelian group of *finite rank* (in general, the *rank* of a free abelian group A is the number of elements in a $\mathbb{Z}$-basis of A). The (first) *Betti number* $b_1(X)$ is defined to be the rank of $H_1(X,\mathbb{Z})$, which is an indicator of the "complexity" of finite graphs (see Sect. 4.4).

A morphism $f = (f_V, f_E) : X_1 = (V_1, E_1) \longrightarrow X = (V, E)$ induces a homomorphism $f_* : H_1(X_1, \mathbb{Z}) \longrightarrow H_1(X, \mathbb{Z})$. Indeed, if we define $f_{0*} : C_0(X_1, \mathbb{Z}) \longrightarrow C_0(X, \mathbb{Z})$ and $f_{1*} : C_1(X_1, \mathbb{Z}) \longrightarrow C_1(X, \mathbb{Z})$ by

$$f_{0*}\left(\sum_x a_x x\right) = \sum_x a_x f_V(x),$$

$$f_{1*}\left(\sum_e a_e e\right) = \sum_e a_e f_E(e),$$

then $\partial \circ f_{1*} = f_{0*} \circ \partial_1$, where ∂_1 is the boundary operator for X_1.

$$\begin{array}{ccc} C_1(X_1, \mathbb{Z}) & \xrightarrow{\partial_1} & C_0(X_1, \mathbb{Z}) \\ f_{1*} \downarrow & & \downarrow f_{0*} \\ C_1(X, \mathbb{Z}) & \xrightarrow{\partial} & C_0(X, \mathbb{Z}). \end{array}$$

Thus for $\alpha \in \operatorname{Ker} \partial_1$, we have $\partial(f_{1*}(\alpha)) = f_{0*}(\partial_1 \alpha) = 0$ and $f_{1*}(\alpha) \in \operatorname{Ker} \partial$ so that $f_*(\operatorname{Ker} \partial_1) \subset \operatorname{Ker} \partial$. Restricting f_{1*} to $\operatorname{Ker} \partial_1$, we get the homomorphism $f_* : H_1(X_1, \mathbb{Z}) \longrightarrow H_1(X, \mathbb{Z})$.

It is a simple matter to observe that if $X_1 = (V_1, E_1)$ is a subgraph of X, then the pair of inclusion maps $i = (i_V, i_E) : (V_1, E_1) \longrightarrow (V, E)$ induces an injective homomorphism $i_* : H_1(X_1, \mathbb{Z}) \longrightarrow H_1(X, \mathbb{Z})$.

There is a close relation between homology classes and closed paths. For a path $c = (e_1, \ldots, e_n)$ in X, the symbol $\langle c \rangle$ expresses the 1-chain $e_1 + \cdots + e_n$. If c is closed, then

$$\partial \langle c \rangle = \big(t(e_1) - o(e_1)\big) + \big(t(e_2) - t(e_2)\big) + \cdots + \big(t(e_n) - o(e_n)\big)$$
$$= \big(t(e_n) - o(e_1)\big) + \big(t(e_1) - o(e_2)\big) + \cdots + \big(t(e_{n-1}) - o(e_n)\big) = 0,$$

and hence $\langle c \rangle$ is a homology class.

The homology class of a point path $\emptyset_x$ is understood to be zero.

An important fact we frequently use is that *every homology class α is represented by a closed path*. To show this, we first argue that there exist closed paths $c_1, \ldots, c_k$ such that $\alpha = \langle c_1 \rangle + \cdots + \langle c_k \rangle$. Express α as

$$\alpha = e_1 + \cdots + e_N \quad (e_i \in E)$$

4.2 Homology Groups

Fig. 4.2 Outgoing and incoming edges

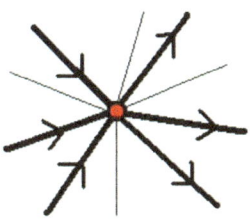

with $e_i \neq \bar{e}_j$ ($i \neq j$). Since $\partial \alpha = 0$, we have

$$o(e_1) + \cdots + o(e_N) = t(e_1) + \cdots + t(e_N).$$

This implies that for $e_1, \ldots, e_N$, the number of edges coming in x coincides with the number of edges going out from x for every x (Fig. 4.2). Therefore one can find a permutation σ of $\{1, 2, \ldots, N\}$ such that

$$t(e_i) = o(e_{\sigma(i)}) \quad (i = 1, \ldots, N).$$

Denote by $\langle \sigma \rangle_{\mathbb{Z}}$ the cyclic subgroup of the symmetry group $\mathcal{S}_N$ generated by σ, and let $\{1, \ldots, N\} = \langle \sigma \rangle_{\mathbb{Z}} i_1 \cup \cdots \cup \langle \sigma \rangle_{\mathbb{Z}} i_k$ be the decomposition into orbits for the $\langle \sigma \rangle_{\mathbb{Z}}$-action on $\{1, \ldots, N\}$. Writing

$$\langle \sigma \rangle_{\mathbb{Z}} i_h = \{i_h, \sigma(i_h), \ldots, \sigma^{n_h - 1}(i_h)\},$$

($\sigma^{n_h}(i_h) = i_h$), we put

$$c_h = (e_{i_h}, e_{\sigma(i_h)}, \ldots, e_{\sigma^{n_h - 1}(i_h)}) \quad (h = 1, \ldots, k).$$

Then c_h is a closed path, and

$$\alpha = \langle c_1 \rangle + \cdots + \langle c_k \rangle.$$

Now pick up a vertex x_0, and let c'_i be a path with $o(c'_i) = x_0$ and $t(c'_i) = o(c_i)$. Then we obtain a closed path

$$c = c'_1 \cdot c_1 \cdot \overline{c'_1} \cdot c'_2 \cdot c_2 \cdot \overline{c'_2} \cdots c'_k \cdot c_k \cdot \overline{c'_k}. \tag{4.2}$$

The closed path c represents the homology class α since the c'_i ($i = 1, \ldots, k$) are cancelled out in $C_1(X, \mathbb{Z})$ (Fig. 4.3).

The 0-th homology group $H_0(X, \mathbb{Z})$ is the factor group $C_0(X, \mathbb{Z})/\text{Image } \partial$. It turns out that $H_0(X, \mathbb{Z})$ is always isomorphic to $\mathbb{Z}$. To verify this, define the homomorphism

$$\varepsilon_0 : C_0(X, \mathbb{Z}) \longrightarrow \mathbb{Z}$$

Fig. 4.3 Representing a homology class by a closed path

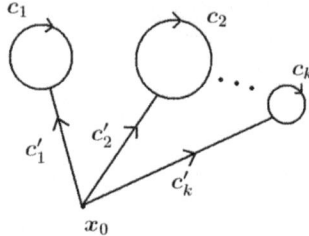

by setting

$$\varepsilon_0\left(\sum_{x\in V} a_x x\right) = \sum_{x\in V} a_x.$$

Evidently ε_0 is surjective. If $\sum_{x\in V} a_x x \in \mathrm{Ker}\,\varepsilon_0$, then

$$\sum_{x\in V} a_x x = \sum_{x\in V} a_x(x - x_0),$$

and hence $\{x - x_0|\ x \in V,\ x \neq x_0\}$ comprises a $\mathbb{Z}$-basis of $\mathrm{Ker}\,\varepsilon_0$. On the other hand, if c is a path joining x_0 and x, then $x - x_0 = \partial c$ so that $\mathrm{Image}\,\partial = \mathrm{Ker}\,\varepsilon_0$. By virtue of the first isomorphism theorem, $H_0(X,\mathbb{Z})$ is isomorphic to $\mathbb{Z}$.

In relation to the homomorphism ε_0, we present a technical theorem which will be used later.

Theorem 4.1. *$H_1(X,\mathbb{Z})$ is a direct summand of $C_1(X,\mathbb{Z})$; that is, there exists a subgroup W of $C_1(X,\mathbb{Z})$ such that $C_1(X,\mathbb{Z})$ is isomorphic to the direct sum $H_1(X,\mathbb{Z}) \oplus W$. In particular, any homomorphism $f : H_1(X,\mathbb{Z}) \longrightarrow K$ into an additive group K extends to a homomorphism $F : C_1(X,\mathbb{Z}) \longrightarrow K$.*

Proof. Using the fact that $\mathrm{Ker}\,\varepsilon_0$ is a free abelian group, and that the factor group $C_1(X,\mathbb{Z})/H_1(X,\mathbb{Z})$ is isomorphic to $\mathrm{Image}\,\partial = \mathrm{Ker}\,\varepsilon_0$ (the first isomorphism theorem), we conclude that $H_1(X,\mathbb{Z})$ is a direct summand of $C_1(X,\mathbb{Z})$ (see Appendix 2). Since $\alpha \in C_1(X,\mathbb{Z})$ can be expressed as a sum $\beta + \beta'$ ($\beta \in H_1(X,\mathbb{Z}), \beta' \in W$) in a unique way, putting $F(\alpha) = f(\beta)$, we obtain an extension F of f. □

So far we have handled the homology group with coefficients in $\mathbb{Z}$. In a similar fashion, $H_1(X,\mathbb{R})$, the 1-st homology group with coefficients in $\mathbb{R}$, is defined by replacing $\mathbb{Z}$ by $\mathbb{R}$ in the definition. The group $H_1(X,\mathbb{R})$ is actually a vector space over $\mathbb{R}$. For a finite graph X, $H_1(X,\mathbb{Z})$ is a lattice group in $H_1(X,\mathbb{R})$, and hence $b_1(X)$ coincides with the dimension of $H_1(X,\mathbb{R})$. As will be explained in Chap. 8, the homology group $H_1(X,\mathbb{R})$ plays a crucial role in the construction of *standard realizations* of topological crystals.

4.3 The Structure of Homology Groups

The structure of the homology group $H_1(X, \mathbb{Z})$ is quite simple. In fact, it is a free abelian group as we will see. The starting point is the following fact.

Theorem 4.2. *If X is a tree, then $H_1(X, \mathbb{Z}) = \{0\}$.*

Proof. It is enough to prove that $\langle c \rangle = 0$ in $H_1(X, \mathbb{Z})$ for every closed path c. Let $x = o(c) = t(c)$. Since there exists no geodesic loops in X except for point geodesics $(\emptyset_x)$, we can find a sequence of loops $c_0, c_1, \ldots, c_n$ with $c = c_0 \mapsto c_1 \mapsto \cdots \mapsto c_n \mapsto \emptyset_x$ (see Sect. 3.5). Hence $\langle c \rangle = \langle c_1 \rangle = \cdots = \langle c_n \rangle = \langle \emptyset_x \rangle = 0$. □

For a general graph $X = (V, E)$, we shall construct a $\mathbb{Z}$-basis of $H_1(X, \mathbb{Z})$ consisting of homology classes represented by circuits. Let T be a spanning tree of X, and fix an orientation E^o of X. Let $E^o(T)$ be the set of edges $e \in E^o$ belonging to T (Fig. 4.4).

Consider the free abelian subgroup W of $C_1(X, \mathbb{Z})$ with $\mathbb{Z}$-basis $E^o \backslash E^o(T)$:

$$W = \left\{ \sum_{e \in E^o \backslash E^o(T)}^{\text{finite}} a_e e \,\middle|\, a_e \in \mathbb{Z} \right\}.$$

We shall demonstrate that the homomorphism J of $H_1(X, \mathbb{Z})$ into W defined by

$$J\left(\sum_{e \in E^o} a_e e \right) = \sum_{e \in E^o \backslash E^o(T)} a_e e$$

is an isomorphism. For each $e \in E^o \backslash E^o(T)$, we take a geodesic c'_e in T such that $o(c'_e) = t(e)$ and $t(c'_e) = o(e_e)$. The path $c_e = c'_e \cdot e$ is a circuit in X, and $J(\langle c_e \rangle) = e$, which implies that J is surjective. Next suppose $J(\alpha) = 0$ for $\alpha = \sum_{e \in E^o} a_e e \in \text{Ker } \partial$. Then $\alpha = \sum_{e \in E^o(T)} a_e e$, so $\alpha \in H_1(T, \mathbb{Z}) = \{0\}$. Thus J is an injective homomorphism.

The argument above ensures that $\{\langle c_e \rangle \mid e \in E^o \backslash E^o(T)\}$ comprises a $\mathbb{Z}$-basis of $H_1(X, \mathbb{Z})$. In particular, if X is finite, then $|E^o \backslash E^o(T)| = b_1(X)$, namely the first Betti number is equal to the number of undirected edges not belonging to a spanning tree.

From now on, $X = (V, E)$ is assumed to be *finite*.

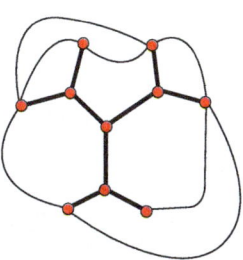

Fig. 4.4 Edges not in a spanning tree

Example 4.1. Let X' be the graph obtained by removing all loop edges $e_1, \overline{e_1}$, $\ldots, e_k, \overline{e_k}$ from a finite graph X. Then T is also a spanning tree of X', and so $b_1(X) = b_1(X') + k$. Moreover one can verify that if $\{\alpha_1, \ldots, \alpha_h\}$ is a $\mathbb{Z}$-basis of $H_1(X', \mathbb{Z})$, then $\{\alpha_1, \ldots, \alpha_h, e_1, \ldots, e_k\}$ is a $\mathbb{Z}$-basis of $H_1(X, \mathbb{Z})$ (note that $e_i \in H_1(X, \mathbb{Z})$). □

Example 4.2. Let $E^o = \{e_1, \ldots, e_n\}$ be an orientation of the n-bouquet graph B_n. Then, $e_1, \ldots, e_n$ constitute a $\mathbb{Z}$-basis of $H_1(B_n, \mathbb{Z})$. □

Example 4.3. Let K_n be the complete graph with n vertices $x_0, x_1, \ldots, x_{n-1}$. Denoting by e_i the edge such that $o(e_i) = x_0$, $t(e_i) = x_i$, we have $E_{x_0} = \{e_1, \ldots, e_{n-1}\}$. Let e_{ij} be a directed edge with $o(e_{ij}) = x_i$ and $t(e_{ij}) = x_j$ $(1 \leq i < j \leq n-1)$. Apparently the subgraph $T = (V_T, E_T)$ with $V_T = V$ and $E_T^o = E_{x_0}$ is a spanning tree of K_n. Therefore putting $c_{ij} = (e_i, e_{ij}, \overline{e_j})$, we have a $\mathbb{Z}$-basis $\{\langle c_{ij}\rangle\}_{1 \leq i < j \leq n-1}$ of $H_1(K_n, \mathbb{Z})$. The number of undirected edges not contained in this tree is $\binom{n-1}{2}$, whence

$$b_1(K_n) = \binom{n-1}{2} = \frac{(n-1)(n-2)}{2}.$$

□

When X is a *plane graph*[4] without vertices of degree-one, we have an alternative way to construct a $\mathbb{Z}$-basis of the homology group $H_1(X, \mathbb{Z})$. Notice that X divides the plane into a finitely many bounded connected small regions; say, $D_1, \ldots, D_k$. The union

$$D_0 = \bigcup_{i=1}^{k} D_i$$

is also a bounded connected region. The boundary of D_i $(i = 0, 1, \ldots, k)$ yields a circuit c_i in X. We will give c_i the counterclockwise parametrization as a simple closed curve in the plane (see Fig. 4.5). Then $\{\langle c_1 \rangle, \ldots, \langle c_k \rangle\}$ is a $\mathbb{Z}$-basis of $H_1(X, \mathbb{Z})$ (hence $k = b_1(X)$). In verifying this, it is enough to prove that the homomorphism f of $\mathbb{Z}^k$ into $H_1(X, \mathbb{Z})$ defined by

$$f(n_1, \ldots, n_k) = \sum_{i=1}^{k} n_i \langle c_i \rangle$$

is an isomorphism.

Let $e \in E$ be an edge in the inside of D_0. There exist just two small regions, D_i, D_j $(1 \leq i, j \leq k)$, whose boundaries contain e. Taking a look at the orientation on c_i and c_j, we conclude that the coefficient of e in $\sum_{i=1}^{k} n_i \langle c_i \rangle$ is $\pm(n_i - n_j)$. On the other

---

[4] A plane graph is a graph drawn on the plane in such a way that its edges intersect only at their endpoints, while a *planar graph* is a graph that *can be* embedded in the plane.

4.3 The Structure of Homology Groups

Fig. 4.5 A plane graph

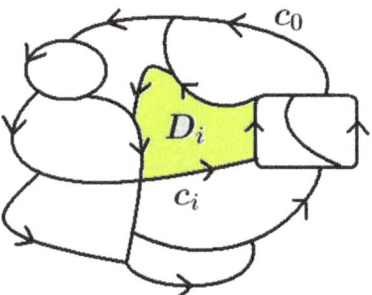

hand, for an edge e on the boundary of D_0, there exists just one small region, D_i ($i = 1, \ldots, k$), whose boundary contains e. This being the case, the coefficient of e in $\sum_{i=1}^{k} n_i \langle c_i \rangle$ is $\pm n_i$. Putting these observations together, we infer that if $\sum_{i=1}^{k} n_i \langle c_i \rangle = 0$, then $n_1 = \cdots = n_k = 0$. Hence f is an injective homomorphism.

Next we prove that f is surjective. Because $H_1(X, \mathbb{Z})$ has a $\mathbb{Z}$-basis consisting of circuits, it suffices to show that for an arbitrary circuit c (with counterclockwise parametrization), there exists $(n_1, \ldots, n_k)$ with $f(n_1, \ldots, n_k) = \langle c \rangle$. The circuit c encloses several small regions, say, $D_{i_1}, \ldots, D_{i_h}$. If we put

$$n_i = \begin{cases} 1 & \text{if } i = i_j \text{ for some } j \\ 0 & \text{otherwise,} \end{cases}$$

then, in $\sum_{i=1}^{k} n_i \langle c_i \rangle$, edges inside c are cancelled out so that

$$f(n_1, \ldots, n_k) = \langle c \rangle,$$

as desired.

We have explained two methods to construct a $\mathbb{Z}$-basis $\{\alpha_1, \ldots, \alpha_b\}$ ($b = b_1(X)$) of $H_1(X, \mathbb{Z})$. In some cases, we need to confirm whether a given family $\{\beta_1, \ldots, \beta_b\}$ of homology classes is a $\mathbb{Z}$-basis. For this, writing

$$\beta_i = \sum_{j=1}^{b} a_{ij} \alpha_j \qquad (a_{ij} \in \mathbb{Z}),$$

we compute the determinant $\det A$ of the square matrix A whose $(ij)^{\text{th}}$ entry is a_{ij}. Then $\{\beta_1, \ldots, \beta_b\}$ is a $\mathbb{Z}$-basis of $H_1(X, \mathbb{Z})$ if and only if $\det A = \pm 1$.

Fig. 4.6 How to find a ℤ-basis

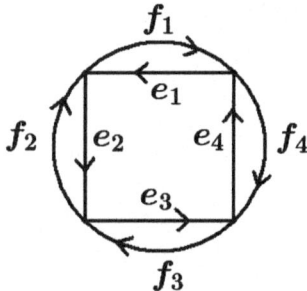

Example 4.4. Consider the plane graph X depicted in Fig. 4.6. From what we have observed above, it follows that

$$\alpha_1 = e_1 + f_1, \quad \alpha_2 = e_2 + f_2, \quad \alpha_3 = e_3 + f_3, \quad \alpha_4 = e_4 + f_4$$
$$\alpha_5 = e_1 + e_2 + e_3 + e_4$$

comprise a ℤ-basis of $H_1(X, \mathbb{Z})$. On the other hand, we consider the family $\beta_1, \ldots, \beta_5$ defined by

$$\beta_1 = e_1 + f_1, \quad \beta_2 = e_2 + f_2, \quad \beta_3 = e_3 + f_3, \quad \beta_4 = e_1 + e_2 + e_3 + e_4$$
$$\beta_5 = f_1 + f_2 + f_3 + f_4.$$

Since

$$\beta_1 = \alpha_1, \quad \beta_2 = \alpha_2, \quad \beta_3 = \alpha_3, \quad \beta_4 = \alpha_5,$$
$$\beta_5 = \alpha_1 + \alpha_2 + \alpha_3 + \alpha_4 + \alpha_5,$$

the matrix A is given by

$$\begin{pmatrix} 1 & 0 & 0 & 0 & 0 \\ 0 & 1 & 0 & 0 & 0 \\ 0 & 0 & 1 & 0 & 0 \\ 0 & 0 & 0 & 0 & 1 \\ 1 & 1 & 1 & 1 & -1 \end{pmatrix}$$

It is easy to see $\det A = -1$ so that $\{\beta_1, \beta_2, \beta_3, \beta_4, \beta_5\}$ is a ℤ-basis. □

Separating edges in X do not play any role in the homology group $H_1(X, \mathbb{Z})$. To be exact, for the connected components $X_1, \ldots, X_n$ of the graph obtained by removing all separating edges, we have

$$H_1(X, \mathbb{Z}) = H_1(X_1, \mathbb{Z}) \oplus \cdots \oplus H_1(X_n, \mathbb{Z}).$$

In fact, separating edges e in c are cancelled out when we consider the cycle $\langle c \rangle$ because e and its inverse edge $\bar{e}$ appear evenly in c.

Finally, we shall give a criterion of bipartiteness in terms of homology groups. Let X be a (possibly infinite) graph. For a 1-chain $\alpha = \sum_{e \in E} a_e e \in C_1(X, \mathbb{Z})$, the parity of $\sum_{e \in E} a_e$ does not depend on the choice of an expression; that is, if

$$\sum_{e \in E} a_e e = \sum_{e \in E} b_e e,$$

then

$$\sum_{e \in E} a_e \equiv \sum_{e \in E} b_e \pmod{2}.$$

Thus we may assign to α the equivalence class $\varepsilon_1(\alpha)$ of $\sum_{e \in E} a_e$ in $\mathbb{Z}_2$. We call $\varepsilon_1(\alpha)$ the *parity* of α. Certainly $\varepsilon_1 : C_1(X, \mathbb{Z}) \longrightarrow \mathbb{Z}_2$ is a homomorphism.[5]

Using this terminology, we have the following theorem (we use (4.2) for the proof).

Theorem 4.3. *A graph X is bipartite if and only if $\varepsilon_1(\alpha) = 0$ for every homology class $\alpha \in H_1(X, \mathbb{Z})$.*

Remark. With regard to the parity of homology classes, it is more natural to deal with it in $C_1(X, \mathbb{Z}_2)$, the group of 1-chains with coefficients in $\mathbb{Z}_2$. In this group, the computation is performed under the rule $e = \bar{e} = -e$. □

4.4 Enumeration of Finite Graphs

The Betti number can be used to enumerate finite graphs (and also hypothetical crystal structures; Sect. 6.2).

To begin with, we shall establish the well-known equality[6]:

$$|V| - |E|/2 = 1 - b_1(X), \tag{4.3}$$

where the left-hand side of (4.3) is the *Euler number* for which we use the notation $\chi(X)$. We first observe that this is true if X is a tree; namely $|V| - |E|/2 = 1$, which is checked by using mathematical induction on the number of vertices. For a general

---

[5] The correspondence $\sum_{e \in E} a_e e \mapsto \sum_{e \in E} a_e$ does *not* yield a "well-defined" homomorphism of $C_1(X, \mathbb{Z})$ into $\mathbb{Z}$.

[6] This is a graph version of the famous formula due to Euler (1750) which asserts that if v, e, f are the number of vertices, edges and faces of a convex polyhedron, respectively, then $v - e + f = 2$. Historically Euler's formula is considered another starting point of "topology".

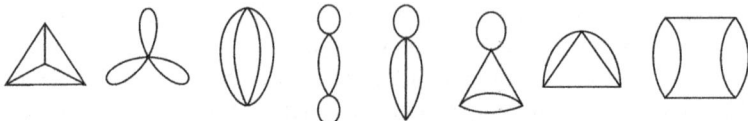

Fig. 4.7 Several graphs X with $b_1(X) = 3$

X, we take a spanning tree T. Recall that $b_1(X)$ is equal to the number of undirected edges not belonging to a spanning tree (Sect. 4.3). Thus

$$|E(T)|/2 + b_1(X) = |E|/2.$$

Substituting $|E(T)|/2 = |V(T)| - 1 = |V| - 1$ in the left-hand side, we get equality (4.3).

The Betti number does not change when we put new vertices on edges, or when we remove a vertex x with $\deg x = 2$. This is so because when a new vertex is put on an edge, the number of undirected edges increases just by one.

Example 4.5. Let X be a graph having no separating edges.[7] If $\deg x \geq 3$ for every $x \in V$ and $b_1(X) = 3$, then X must be one of the graphs in Fig. 4.7. In fact, from $|V| - |E|/2 = 1 - 3 = -2$ and $3|V| \leq |E|$ [use (3.1)], we have $|V| \leq 4$. Enumerating all possible cases, we get the claim. □

Theorem 4.4. *For a fixed integer $b \geq 0$, there are only finitely many finite graphs X (up to isomorphisms) such that $\deg x \geq 3$ for all vertex x and $b_1(X) = b$.*

Indeed, by the same reasoning as above, we have $|V| \leq 2(b-1)$ (use $3|V| \leq |E|$ and $2|V| - |E| = 2(1-b)$) and also $|E|/2 \leq 3b - 3$. Therefore the number of vertices and edges is bounded.

There are plenty of finite graphs X with $b_1(X) = 4$. Figure 4.8 exhibits some of them.

Restricting ourselves to the case of regular graphs with degree three or four, we have the following relations which can be used when we enumerate the crystal structures of hypothetical carbon crystals.

1. $|V| = 2(b_1(X) - 1)$, $|E|/2 = 3(b_1(X) - 1)$ in the case of degree three,
2. $|V| = b_1(X) - 1$, $|E|/2 = 2(b_1(X) - 1)$ in the case of degree four.

4.5 Automorphisms and Homology

We will now give a relationship between the automorphism group and the homology group of a finite graph.

---

[7] Separating edges give rise to a pathological phenomenon when we consider topological crystals associated with a finite graph.

4.5 Automorphisms and Homology

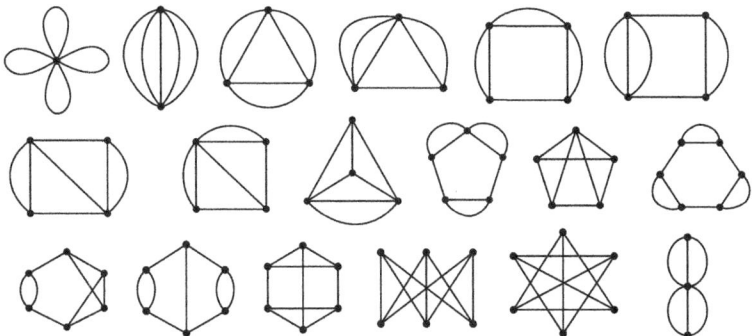

Fig. 4.8 Several graphs X with $b_1(X) = 4$

Let $X = (V, E)$ be a finite graph having no separating edges. We assume that X is not isomorphic to the circuit graph.

Theorem 4.5. *Let* $\mathrm{Aut}(H_1(X, \mathbb{Z}))$ *be the group of automorphisms of the homology group* $H_1(X, \mathbb{Z})$, *and let* $g_* : H_1(X, \mathbb{Z}) \longrightarrow H_1(X, \mathbb{Z})$ *be the automorphism induced from* $g \in \mathrm{Aut}(X)$. *Then the correspondence*

$$g \in \mathrm{Aut}(X) \longmapsto g_* \in \mathrm{Aut}(H_1(X, \mathbb{Z}))$$

is an injective homomorphism.[8]

What we have to prove is that if $g_* = I$, then $g = I$. Its proof is carried out along the following steps.

(a) Since $g_* = I$, we have

$$ge_1 + \cdots + ge_n = e_1 + \cdots + e_n$$

for any circuit $c = (e_1, \ldots, e_n)$. Therefore there is an element $h \in \mathbb{Z}/n\mathbb{Z}$ such that $ge_i = e_{i+h}$ $(i \in \mathbb{Z}/n\mathbb{Z})$. In other words, $(ge_1, \ldots, ge_n)$ is obtained by a cyclic permutation of $(e_1, \ldots, e_n)$. In particular, if e is a loop edge, then $ge = e$.

(b) Suppose $g \neq I$. One finds a non-loop edge $e_1 \in E$ with $ge_1 \neq e_1$. By the assumption that X has no separating edges, there exists a circuit $c = (e_1, \ldots, e_n)$ starting with e_1.

(c) Since X is non-circuit, there is a vertex x on c such that $\deg x \geq 3$. Without loss of generality, we may assume $o(c) = x$. Clearly $gx \neq x$.

(d) Take an edge e_1' not contained in c such that $o(e_1') = x$. If e_1' is a loop edge, then $ge_1' = e_1'$, and $gx = x$, thereby obtaining a contradiction. If e_1' is a non-loop edge,

---

[8] We may equip $H_1(X, \mathbb{R})$ with a natural inner product such that g_* is an orthogonal transformation (see Sect. 8.1).

one can find a circuit $c' = (e'_1, \ldots, e'_m)$, again by the assumption that X has no separating edges. If x is only the vertex that c and c' share, then $gx = x$, thereby again a contradiction. If not, there is a part $(e'_1, \ldots, e'_s)$ ($s < m$) of c' such that $t(e'_s)$ is on c and $t(e_i)$ ($i = 1, \ldots, s-1$) is not on c.

(e) Let $t(e'_s) = t(e_t)$. Then $(e_1, \ldots, e_t, \overline{e'_s}, \ldots, \overline{e'_1})$ is a circuit, and hence $(ge_1, \ldots, ge_t, g\overline{e'_s}, \ldots, g\overline{e'_1})$ is obtained by a cyclic permutation of $(e_1, \ldots, e_t, \overline{e'_s}, \ldots, \overline{e'_1})$. This is a contradiction.

This completes the proof of our assertion.

4.6 Notes

(I) As a "dual" of homology, one can introduce the notion of *cohomology group*. For an additive group A, put

$$C^0(X,A) = \{f : V \longrightarrow A\},$$
$$C^1(X,A) = \{\omega : E \longrightarrow A \mid \omega(\overline{e}) = -\omega(e) \ (e \in E)\}$$

which we call the group of 0-*cochains* and the group of 1-*cochains*, respectively. Define the *coboundary operator* $d : C^0(X,A) \longrightarrow C^1(X,A)$ by $(df)(e) = f(t(e)) - f(o(e))$. Then the cohomology groups $H^i(X,A)$ ($i = 0, 1$) are

$$H^0(X,A) = \text{Ker } d, \quad H^1(X,A) = C^1(X,A)/\text{Image } d.$$

The duality between homology and cohomology is embodied in the fact that there is a canonical isomorphism between $H^i(X,A)$ and $\text{Hom}(H_i(X,\mathbb{Z}), A)$, the group of homomorphisms of $H_i(X,\mathbb{Z})$ into A.

Although homology theory is exclusively employed in this work, we encounter, once in a while, the scene where the idea of cohomology makes our discussion transparent (see Chap. 10 in particular).

(II) In Poincaré's seminal paper *Analysis situs* (1895) the concept of homology class was introduced for a cell complex. The group structure on the set of homology classes, which Poincaré failed to indicate, was studied by Emmy Noether and independently by L. Vietoris and W. Mayer in the period 1925–1928. Since then the homology group and its generalizations have occupied a central position in algebraic topology.

The *singular homology theory* developed by S. Eilenberg in 1944 allows us to introduce the homology groups for arbitrary topological spaces (as a general reference, we refer to Vick [107]). In fact, the k-th homology group $H_k(X,\mathbb{Z})$ of a topological space X is defined for any integer $k \geq 0$. The procedure of its construction starts with abelian groups $C_k(X,\mathbb{Z})$ (the group of singular k-chains) and a series of homomorphisms (called the chain complex)

4.6 Notes

$$\cdots \xrightarrow{\partial_{k+1}} C_k(X,\mathbb{Z}) \xrightarrow{\partial_k} \cdots \xrightarrow{\partial_2} C_1(X,\mathbb{Z}) \xrightarrow{\partial_1} C_0(X,\mathbb{Z}) \xrightarrow{\partial_0} 0$$

such that $\partial_k \circ \partial_{k+1} = 0$. Roughly speaking, $C_k(X,\mathbb{Z})$ is formed by what we call "singular k-simplices", that is, continuous maps from the k-simplex into X. The equality $\partial_k \circ \partial_{k+1} = 0$ implies that Image $\partial_{k+1} \subset \text{Ker } \partial_k$; so one may consider the factor group

$$H_k(X,\mathbb{Z}) = \text{Ker } \partial_k / \text{Image } \partial_{k+1}.$$

This is the k-th homology group.

The most pivotal fact in this theory is the homotopy invariance of homology groups. A continuous map $f : X_1 \longrightarrow X_2$ is said to be an *homotopy equivalence* if there exists a continuous map $g : X_2 \longrightarrow X_1$ such that both $g \circ f$ and $f \circ g$ are continuously deformed to the identity maps of X_1 and X_2, respectively. A homotopy equivalence $f : X_1 \longrightarrow X_2$ induces an isomorphism $f_* : H_k(X_1,\mathbb{Z}) \longrightarrow H_k(X_2,\mathbb{Z})$. One may apply this fact to a graph X to establish that the homology group $H_1(X,\mathbb{Z})$ is isomorphic to $H_1(B_n,\mathbb{Z})$ where n is the number of undirected edges not contained in a spanning tree T, and B_n is the n-bouquet graph. Indeed, the contraction of T to a single point gives rise to a homotopy equivalence $f : X \longrightarrow B_n$.

Chapter 5
Covering Graphs

Another central concept in topological crystallography is the "covering map". Roughly speaking, a covering map in general is a surjective continuous map between topological spaces which preserves the local topological structure (see Notes in this chapter for the exact definition). Figure 5.1 exhibits a covering map from the line $\mathbb{R}$ onto the circle S^1, which somehow suggests an idea why we use the term "covering".[1] The canonical projection $\pi : \mathbb{R}^d \longrightarrow \mathbb{R}^d/L$, introduced in Example 2.5, is another example of a covering map.

The theory of covering maps is regarded as a geometric analogue of Galois theory in algebra that describes a beautiful relationship between field extensions[2] and Galois groups. Indeed, in a similar manner as Galois theory, covering maps are completely governed by what we call *covering transformation groups*. In this chapter, sacrificing full generality, we shall deal with covering maps for graphs from a combinatorial viewpoint. See [107] for the general theory.

Our concern in crystallography is only special covering graphs (more specifically *abelian covering graphs* introduced in Chap. 6). However the notion of general covering graphs turns out to be of particular importance when we consider a "non-commutative" version of crystals, a mathematical object related to discrete groups [see Notes (IV) in Chap. 9]. Some problems in statistical mechanics are also well set up in terms of covering graph.

---

[1] More explicitly, the covering map is given by

$$x \in \mathbb{R} \mapsto e^{2\pi\sqrt{-1}x} \in S^1 = \{z \in \mathbb{C} |\ |z| = 1\}.$$

[2] A *field* is an algebraic system with addition, subtraction, multiplication and division such as rational numbers, real numbers and complex numbers.

Fig. 5.1 Idea of covering maps

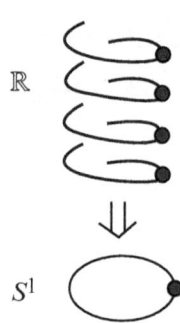

5.1 Definition

Let $X = (V, E)$ and $X_0 = (V_0, E_0)$ be connected graphs. A morphism $\omega : X \longrightarrow X_0$ is said to be a *covering map* if it preserves local adjacency relations between vertices and edges; more precisely,

(i) $\omega : V \longrightarrow V_0$ is surjective.
(ii) For every $x \in V$, the restriction $\omega|E_x : E_x \longrightarrow E_{0,\omega(x)}$ is a bijection.

Given a covering map $\omega : X \longrightarrow X_0$, we call X a *covering graph* over the *base graph* X_0. The *fiber* over $x \in V_0$ is the inverse image:

$$\omega^{-1}(x) = \{y \in V \mid \omega(y) = x\}.$$

If $|\omega^{-1}(x)|$ is finite for a vertex $x \in V_0$, then ω (or X) is said to be *finite-fold*, otherwise *infinite-fold*.

For a finite-fold covering map ω, the number $|\omega^{-1}(x)|$ does not depend on $x \in V_0$. In fact, if we put $n(x) = |\omega^{-1}(x)|$, then, for $e \in E_0$, we find that $|\omega^{-1}(e)| = |\omega^{-1}(o(e))| = n(o(e))$, and also $|\omega^{-1}(e)| = |\omega^{-1}(t(e))| = n(t(e))$, and hence $n(o(e)) = n(t(e))$. Since X_0 is connected, $n(x)$ is constant[3] (Lemma 3.4.1). A covering map ω with $n(x) = k$ for some constant k is said to be *k-fold*.

Example 5.1. In Fig. 5.2, define the map ω by setting $\omega(1) = \omega(7) = a$, $\omega(2) = \omega(8) = b$, $\omega(3) = \omega(5) = c$, $\omega(4) = \omega(6) = d$. Then ω is a twofold covering map.
□

A useful aspect of a covering map $\omega : X \longrightarrow X_0$ is that it has the *unique path-lifting property*, which claims that for a path $c_0 = (e_{0,1}, \ldots, e_{0,n})$ in X_0 and a vertex x in X with $\omega(x) = o(c_0)$, there exists a unique path $c = (e_1, \ldots, e_n)$ in X (called a *lifting* of c_0) such that $\omega(c) = c_0$ and $o(c) = x$. To see this, we first apply condition

---

[3] For an infinite-fold covering map, if we mean by $|\omega^{-1}(x)|$ the *cardinality* of the set $\omega^{-1}(x)$, we have the same conclusion. Here two sets A and B in general are said to have the same cardinality, symbolically $|A| = |B|$, if there exists a one-to-one correspondence between them.

5.1 Definition

Fig. 5.2 Example of covering map

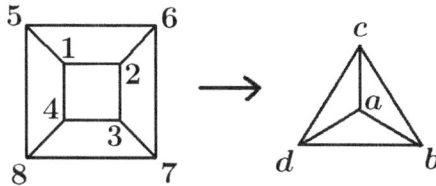

(ii) to get a unique $e_1 \in E_x$ with $\omega(e_1) = e_{0,1}$, and then successively apply (ii) to obtain $e_i \in E_{t(e_{i-1})}$ with $\omega(e_i) = e_{0,i}$ ($i = 2, \ldots, n$). The uniqueness of c is obvious from the discussion.

In general, a *lifting* of a morphism $f_0 : Y = (V_Y, E_Y) \longrightarrow X_0$ is a morphism $f : Y \longrightarrow X$ satisfying $f_0 = \omega \circ f$.

Theorem 5.1 (Unique Lifting Theorem). *If two liftings f_1 and f_2 of f satisfy $f_1(y_0) = f_2(y_0)$ for some $y_0 \in Y$, then $f_1 = f_2$.*

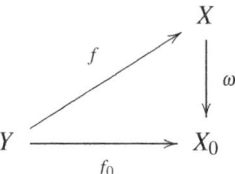

Proof. Let $e \in E_Y$ with $o(e) = y_0$. We have $\omega(f_1(e)) = \omega(f_2(e)) = f_0(e)$ and

$$o(\omega(f_1(e))) = o(\omega(f_2(e))) = o(f_0(e)) = f_0(o(e)) = f_0(y_0).$$

Thus applying condition (ii) in the definition of covering map, we find that $f_1(e) = f_2(e)$ and $f_1(t(e)) = f_2(t(e))$. Repeating this argument, we observe that $f_1(e) = f_2(e)$ for every $e \in E_Y$ and $f_1(y) = f_2(y)$ for every $y \in V_Y$; thereby proving $f_1 = f_2$ (we use the assumption that Y is connected). □

In general, a lifting over all of Y does not have to exist (see Sect. 5.3 for a criterion so that a lifting exists).

The Euler number behaves well under covering maps; that is, for a k-fold covering graph $X = (V, E)$ over a finite base graph X_0, we get

$$\chi(X) = k\chi(X_0)$$

since $|V| = k|V_0|$ and $|E| = k|E_0|$. Equivalently,

$$b_1(X) = k(b_1(X_0) - 1) + 1.$$

5.2 Covering Transformation Groups

The theory of covering maps is combined with group theory via the notion of covering transformations. This fact is vital to develop topological crystallography.

Let $\omega : X \longrightarrow X_0$ be a covering map. An automorphism σ of X is said to be a *covering transformation* (or deck transformation) provided that $\omega \circ \sigma = \omega$. By definition, σ preserves each fiber of ω, i.e., $\sigma(\omega^{-1}(x)) = \omega^{-1}(x)$ for every $x \in V_0$, and σ is a lifting of the covering map ω as the following commutative diagram indicates:

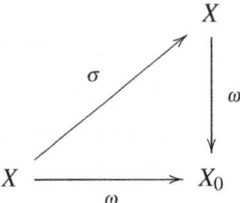

The totality of covering transformations, symbolically $G(\omega)$, forms a subgroup of $\mathrm{Aut}(X)$, which we call the *covering transformation group* of ω.

We show that $G(\omega)$ acts freely on the graph X. Let $\sigma = (\sigma_V, \sigma_E) \in G(\omega)$. Suppose that there exists a vertex $x \in V$ with $\sigma_V x = x$. Since $\sigma_V x = I_V(x)$ (I_V being the identity map of V), from the unique lifting theorem, it results in $\sigma_V = I_V$. To check that $G(\omega)$ acts on X without inversion, suppose $\sigma_E(e) = \bar{e}$. Then

$$\omega(e) = \omega(\sigma_E e) = \omega(\bar{e}) = \overline{\omega(e)},$$

thereby giving rise to a contradiction.

A covering map $\omega : X \longrightarrow X_0$ is said to be *regular* provided that the action of $G(\omega)$ on every fiber $\omega^{-1}(x)$ is transitive (hence simply transitive). Thus, if ω is a finite-fold regular covering map, then the order of the group $G(\omega)$ (that is, $|G(\omega)|$) is equal to $|\omega^{-1}(x)|$.

Example 5.2. A twofold covering map $\omega : X \longrightarrow X_0$ is always regular. To show this, define the automorphism $\sigma = (\sigma_V, \sigma_E)$ of X by

$$\begin{cases} \sigma_V x = y, \ \sigma_V y = x & (\omega(x) = \omega(y), \ x \neq y), \\ \sigma_E e_x = e_y, \ \sigma_E e_y = e_x & (\omega(e_x) = \omega(e_y), \ e_x \in E_x, \ e_y \in E_y, \ x \neq y). \end{cases}$$

We may check that σ is a covering transformation with $\sigma^2 = I$, and that $G(\omega) = \{I, \sigma\}$. Obviously $G(\omega)$ acts transitively on each fiber. □

We shall prove that *for a regular covering map $\omega : X \longrightarrow X_0$, the quotient graph $X/G(\omega)$ is isomorphic to X_0*. From the definition of quotient graphs and regular covering maps, it follows that ω induces a bijection of $V/G(\omega)$ onto V_0. To prove that the induced map of $E/G(\omega)$ into E_0 is a bijection, suppose

5.2 Covering Transformation Groups

$\omega(e) = \omega(e')$ $(e, e' \in E)$. Since $\omega(o(e)) = \omega(o(e'))$, the regularity of ω allows us to find $\sigma \in G(\omega)$ such that $o(e') = \sigma o(e) = o(\sigma e)$. Using $\omega(\sigma e) = \omega(e)$ and condition (ii) for covering maps, we have $e' = \sigma e$. The bijections obtained in this way preserve the adjacency relation of graphs. Thus we have an isomorphism between $X/G(\omega)$ and X_0.

We shall collect several fundamental facts on covering graphs.

Theorem 5.2. *Suppose that a group G acts freely on a graph X. Then the canonical morphism $\omega : X \longrightarrow X_0 = X/G$ is a regular covering map whose covering transformation group is G (namely, every regular covering map is obtained as a canonical morphism associated with a free action on a graph).*

Proof. We first show that ω is a covering map. Both $\omega : V \longrightarrow V_0$ and $\omega : E \longrightarrow E_0$ are surjective. Thus, we only have to verify that for every $x \in V$, the map $\omega|E_x : E_x \longrightarrow E_{0\omega(x)}$ is a bijection. For $e_0 \in E_{0\omega(x)}$, there is $e \in E$ such that $\omega(e) = e_0$. Observe
$$\omega(o(e)) = o(\omega(e)) = o(e_0) = \omega(x),$$
so $o(e)$ and x are in the same orbit. This being so, there exists $g \in G$ with $go(e) = x$. We thus have $ge \in E_x$ and $\omega(ge) = \omega(e) = e_0$. This implies that $\omega|E_x$ is surjective. To show that $\omega|E_x$ is an injection, suppose $\omega(e) = \omega(e')$ for $e, e' \in E_x$. Let $g \in G$ be an element with $e' = ge$. Then $x = o(e') = o(ge) = gx$. Because G acts freely on V, we have $g = 1$ and $e' = e$.

Obviously each element in G acts on X as a covering transformation; therefore, $G \subset G(\omega)$, and ω is a regular covering map since G acts transitively on each $\omega^{-1}(x)$. To prove $G = G(\omega)$, let $\sigma \in G(\omega)$ and $x \in V$. There exists $g \in G$ with $\sigma x = gx$. The transformation $g^{-1} \circ \sigma \in G(\omega)$ has a fixed point, whence $g^{-1} \circ \sigma = I$ and $\sigma = g$. □

Theorem 5.3. *Let $\omega : X \longrightarrow X_0$ be a regular covering map, and let Y be a graph. If two morphisms $f_1, f_2 : Y \longrightarrow X$ are liftings of a morphism $f_0 : Y \longrightarrow X_0$, then there exists a unique element σ in $G(\omega)$ such that $f_2 = \sigma \circ f_1$.*

Proof. Pick $y_0 \in Y$, and let $\sigma \in G(\omega)$ with $f_2(y_0) = \sigma \circ f_1(y_0)$ where, to find such σ, we use the regularity of ω together with $\omega(f_1(y_0)) = \omega(f_2(y_0)) = f_0(y_0)$. Two morphisms $\sigma \circ f_1$ and f_2 are liftings of f_0; thus applying the unique lifting theorem, we have $f_2 = \sigma \circ f_1$. The uniqueness of σ is a consequence of the fact that $G(\omega)$ acts freely on X. □

Theorem 5.4. *Let $\omega : X \longrightarrow X_0$ and $\omega_Y : Y \longrightarrow Y_0$ be regular covering maps, and let $f : X_0 \longrightarrow Y_0$ and $F : X \longrightarrow Y$ be morphisms satisfying $F \circ \omega_Y = f \circ \omega$. Then there exists a homomorphism $\rho : G(\omega) \longrightarrow G(\omega_Y)$ such that $F \circ \sigma = \rho(\sigma) \circ f$ for every $\sigma \in G(\omega)$.*

$$\begin{array}{ccc} X & \xrightarrow{F} & Y \\ \omega \downarrow & & \downarrow \omega_Y \\ X_0 & \xrightarrow{f} & Y_0 \end{array}$$

Proof. Both F and $F \circ \sigma$ are liftings of $f \circ \omega$ for the covering map ω_Y so that according to the above theorem, there exists a unique $\mu \in G(\omega_Y)$ with $F \circ \sigma = \mu \circ F$. If we define ρ by putting $\mu = \rho(\sigma)$, then $\rho(1) = 1$ and

$$\rho(\sigma_1 \sigma_2) F(x) = F(\sigma_1 \sigma_2 x) = \rho(\sigma_1) F(\sigma_2 x) = \rho(\sigma_1) \rho(\sigma_2) F(x).$$

Hence $\rho(\sigma_1 \sigma_2) = \rho(\sigma_1) \rho(\sigma_2)$, and ρ is a homomorphism. □

The following theorem exhibits resemblance between the theory of covering graphs and Galois theory.

Theorem 5.5. *Let $\omega_1 : X \longrightarrow X_1$, $\omega_2 : X_1 \longrightarrow X_0$ be covering maps.*

(a) *The composition $\omega = \omega_2 \circ \omega_1 : X \longrightarrow X_0$ is also a covering map, and $G(\omega_1)$ is a subgroup of $G(\omega)$.*
(b) *If ω is a regular covering map, then so is ω_1.*
(c) *Suppose that ω is regular. The covering map ω_2 is regular if and only if $G(\omega_1)$ is a normal subgroup of $G(\omega)$. In this case the factor group $G(\omega)/G(\omega_1)$ is isomorphic to $G(\omega_2)$.*

Proof. (a) is obvious. We call ω_1 a *subcovering map* of ω.

(b) is verified as follows. Let $x, y \in X$ with $\omega_1(x) = \omega_1(y)$. Then $\omega(x) = \omega(y)$ so that there exists $\sigma \in G(\omega)$ such that $y = \sigma x$. Certainly $\omega_1 \circ \sigma$ and ω_1 are liftings of ω for the covering map ω_2, and $\omega_1 \circ \sigma(x) = \omega_1(y) = \omega_1(x)$. Therefore, in view of the unique lifting theorem, we see that $\omega_1 \circ \sigma = \omega_1$. This implies $\sigma \in G(\omega_1)$, and hence ω_1 is regular.

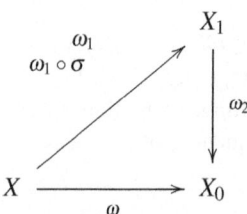

The proof of (c) is a bit complicated. First, suppose that $G(\omega_1)$ is a normal subgroup of $G(\omega)$. Let $\sigma \in G(\omega)$, and let x' be a vertex in X with $\omega_1(x') = x$ (respectively an edge e' in X with $\omega_1(e') = e$). Define an action of $G(\omega)$ on X_1 by

$$\sigma \cdot x = \omega_1(\sigma x'), \quad x \in V_1 \qquad (5.1)$$

$$\sigma \cdot e = \omega_1(\sigma e'), \quad e \in E_1. \qquad (5.2)$$

5.2 Covering Transformation Groups

To see that this actually yields an action, we first check that the right-hand side of (5.1) does not depend on the choice of x' (namely, we are going to ascertain the well-definedness of $\sigma \cdot x$). Let $y' \in X$ with $\omega_1(y') = x$. The covering map ω_1 is regular by (b); therefore we can find $\mu \in G(\omega_1)$ such that $y' = \mu x'$. By the assumption that $G(\omega_1)$ is normal in $G(\omega)$, we have $\sigma\mu\sigma^{-1} \in G(\omega_1)$, and

$$\omega_1(\sigma y') = \omega_1(\sigma\mu x') = \omega_1(\sigma\mu\sigma^{-1}\sigma x') = \omega_1(\sigma x').$$

Thus $\sigma \cdot x$ is well defined. In much the same way, we can show that $\sigma \cdot e$ is well defined. One can check that $\sigma \cdot : X_1 \longrightarrow X_1$ is a morphism. In order to show that the correspondence $\sigma \longmapsto \sigma \cdot$ yields an action of $G(\omega)$ on X_1, let $\sigma_1, \sigma_2 \in G(\omega)$. Take $x' \in X_1$ with $\omega_1(x') = x$ and put $y' = \sigma_2 x'$. Since $\sigma_2 \cdot x = \omega_1(\sigma_2 x') = \omega_1(y')$, we get

$$\sigma_1 \cdot (\sigma_2 \cdot x) = \omega_1(\sigma_1 y') = \omega_1(\sigma_1\sigma_2 x') = (\sigma_1\sigma_2) \cdot x.$$

Evidently $I_V \cdot x = x$. The same is true for edges. Consequently, we obtain a well-defined action of $G(\omega)$ on X_1.

Looking at in (5.1) and (5.2), we get

$$\omega_2(\sigma \cdot x) = \omega_2(\omega_1(\sigma x')) = \omega(\sigma x') = \omega(x') = \omega_2(\omega_1(x')) = \omega_2(x)$$

and $\omega_2(\sigma \cdot e) = \omega_2(e)$ likewise, so $\sigma \cdot$ is a covering transformation for the covering map ω_2. Therefore the correspondence $\sigma \longmapsto \sigma \cdot$ yields a homomorphism $v : G(\omega) \longrightarrow G(\omega_2)$ with which (5.1) is expressed as

$$\omega_1(\sigma x') = v(\sigma)\omega_1(x') \quad (x' \in V, \ \sigma \in G(\omega)). \tag{5.3}$$

We shall show that through the homomorphism v, the group $G(\omega)$ acts transitively on each fiber of ω_2. Suppose $\omega_2(x) = \omega_2(y)$ for vertices x, y of X_1. For vertices x', y' of X with $\omega_1(x') = x$ and $\omega_1(y') = y$, we have $\omega(x') = \omega(y')$; so there exists $\sigma \in G(\omega)$ with $\sigma x' = y'$. We then have $\sigma \cdot x = \omega_1(\sigma x') = \omega_1(y') = y$ as desired.

The homomorphism v is surjective (and hence ω_1 turns out to be regular). For, given $x \in V_1$ and $\mu \in G(\omega_2)$, from the transitivity of $G(\omega)$-action on V_1, it follows immediately that there is $\sigma \in G(\omega)$ such that $v(\sigma)x(= \sigma \cdot x) = \mu x$, and hence $v(\sigma) = \mu$.

Evidently Ker v coincides with $G(\omega_1)$. We therefore conclude that v induces an isomorphism of $G(\omega)/G(\omega_1)$ onto $G(\omega_2)$ (the first isomorphism theorem).

Conversely, suppose that ω_2 is regular. Applying Theorem 5.4 to the case $Y = X_1$, $\omega_Y = \omega_2$, $f = I_{X_0}$, $F = \omega_1$, we find a homomorphism

$$\rho : G(\omega) \longrightarrow G(\omega_2)$$

such that $\omega_1 \circ \sigma = \rho(\sigma) \circ \omega_1$. The kernel of ρ is $G(\omega_1)$. Consequently $G(\omega_1)$ is a normal subgroup of $G(\omega)$. □

It is straightforward to see that for a regular covering map $\omega : X \longrightarrow X_0$ and a subgroup G of $G(\omega)$, the canonical map of X/G onto $X/G(\omega) = X_0$ is a covering map (and hence the canonical map of X onto X/G is a subcovering map of ω).

We may rephrase what we mentioned at the end of Sect. 3.6 in terms of regular covering maps.

Theorem 5.6. *Let $\omega : X \longrightarrow X_0$ be a regular covering map. The covering graph X is bipartite if and only if*

1. *X_0 is bipartite, or*
2. *There is a subcovering map $\omega_1 : X \longrightarrow X_1$ of ω such that $\omega_2 : X_1 \longrightarrow X_0$ is twofold, and X_1 is bipartite.*

5.3 Fundamental Groups

The "fundamental group", a concept introduced by Poincaré, is another algebraic system constructed from geometric figures (paths in this turn) in a topological space. This concept is much easier to understand compared with homology groups, because artificial expressions like formal sums over vertices or edges are not involved in the definition. The major distinction from homology groups lies in the observation that fundamental groups need not be abelian. Actually as we will see, they are isomorphic to *free groups* which are highly non-abelian except for the free group generated by a single element.

Let $X = (V, E)$ be a graph and $x \in V$. The equivalence class of a loop $c \in C(x,x)$ with respect to the equivalence relation introduced in Sect. 3.5 is denoted by $[c]$, and is called the *homotopy class* of c. We will use the notation $\pi_1(X,x)$ for the set of all homotopy classes of loops with base point x.

We may define multiplication of two homotopy classes $[c_1]$ and $[c_2]$ by $[c_1] \cdot [c_2] = [c_1 \cdot c_2]$. This is well defined because if $c_1 \sim c_1'$ and $c_2 \sim c_2'$, then $c_1 \cdot c_2 \sim c_1' \cdot c_2'$.

With this multiplication, the set $\pi_1(X,x)$ becomes a group. The identity element is given by the equivalence class of the point path $\emptyset_x$. The inverse of $[c]$ is $[\overline{c}]$ because $c \cdot \overline{c} \sim \emptyset_x$. The group $\pi_1(X,x)$ is called the *fundamental group*[4] of X with base point x.

What happens when we change the base point x? The answer is given in the following theorem.

Theorem 5.7 (Base Point Change). *For $x, y \in V$, $\pi_1(X,x)$ is isomorphic to the group $\pi_1(X,y)$.*

The crux of the proof is to use a path c' with $o(c') = y$ and $t(c') = x$ (Fig. 5.3). The correspondence $c \in C(x,x) \longmapsto c' \cdot c \cdot \overline{c'} \in C(y,y)$ yields an isomorphism $f_{x \mapsto y}$ of $\pi_1(X,x)$ onto $\pi_1(X,y)$ (the inverse of $f_{x \mapsto y}$ is given by $f_{y \mapsto x}$ which is defined in

---

[4]Actually $\pi_1(X,x)$ is isomorphic to the fundamental group of X as a topological space.

5.3 Fundamental Groups

Fig. 5.3 Base point change

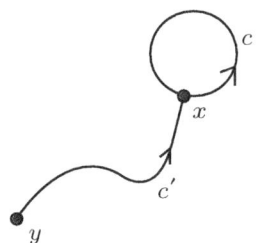

the same way as $f_{x \mapsto y}$ by using $\overline{c'}$). In view of this fact, we occasionally omit the base point and write $\pi_1(X)$ for the fundamental group of X.

A tree X is characterized as a graph with the trivial fundamental group,[5] that is, a graph X with $\pi_1(X) = \{1\}$ is a tree, and vice versa. In fact, if X is a tree, then every loop with base point x is homotopic to the point path $\emptyset_x$. Therefore $\pi_1(X)$ consists only of $[\emptyset_x]$. The converse is obvious.

Let $f : X_1 \longrightarrow X_2$ be a morphism. For a path $c = (e_1, e_2, \ldots, e_n)$ in X_1, the path $f(c)$ in X_2 is defined by $f(c) = (f(e_1), f(e_2), \ldots, f(e_n))$. It is evident that $f(c_1 \cdot c_2) = f(c_1) \cdot f(c_2)$ and $f(\overline{c}) = \overline{f(c)}$. Moreover, if c_1, c_2 are paths in X_1 such that $c_1 \sim c_2$, then $f(c_1) \sim f(c_2)$. Hence the correspondence $c \in C(x,x) \longmapsto f(c) \in C(f(x), f(x))$ yields a homomorphism

$$f_* : \pi_1(X_1, x) \longrightarrow \pi_1(X_2, f(x))$$

by passing to the quotient, that is, $f_*([c]) = [f(c)]$. We call f_* the *induced homomorphism*. For another morphism $g : X_2 \longrightarrow X_3$, we have $(g \circ f)_* = g_* \circ f_*$.

Theorem 5.8 (Covering Homotopy Theorem). *Suppose that two paths $c_1, c_2 \in C(x,y)$ in X_0 are homotopic, and let $\hat{c}_1$ and $\hat{c}_2$ be liftings of c_1 and c_2 for a covering map $\omega : X \longrightarrow X_0$ with $o(\hat{c}_1) = o(\hat{c}_2)$, respectively. Then $t(\hat{c}_1) = t(\hat{c}_2)$ and $\hat{c}_1, \hat{c}_2$ are homotopic.*

Proof. It suffices to prove that if $c_1 \mapsto c_2$, then $\hat{c}_1 \mapsto \hat{c}_2$. But this follows at once from the observation that a back-tracking part of c_1 is lifted to a back tracking part of $\hat{c}_1$ (use condition (ii) in the definition of covering maps). □

The covering homotopy theorem enables us to prove that the induced homomorphism $\omega_* : \pi_1(X) \longrightarrow \pi_1(X_0)$ is injective. The proof is left as an exercise.

We now give a criterion for the existence of a lifting which occupies a very powerful role in the theory of covering graphs.

Theorem 5.9 (Lifting Criterion). *Let $\omega : X \longrightarrow X_0$ be a covering map, and $f_0 : Y = (V_Y, E_Y) \longrightarrow X_0$ be a morphism. Let $x_0 \in X_0, \hat{x}_0 \in X, y_0 \in Y$ with $\omega(\hat{x}_0) = x_0$*

---

[5] In general, an arcwise connected topological space X with a trivial fundamental group is called *simply connected*.

Fig. 5.4 Lifting criterion

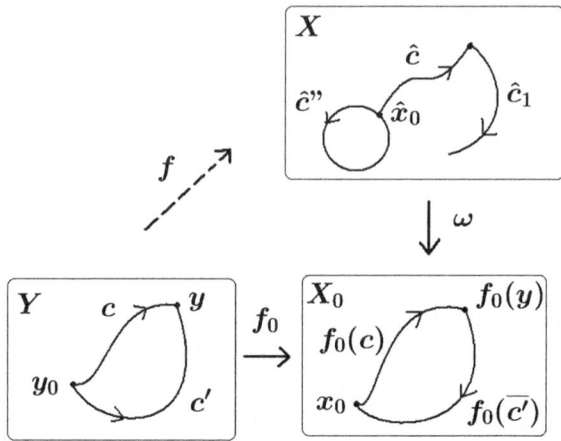

and $f_0(y_0) = x_0$. Then there is a lifting $f : Y \longrightarrow X$ of f_0 with $f(y_0) = \hat{x}_0$ if and only if $f_{0*}(\pi_1(Y, y_0)) \subset \omega_*(\pi_1(X, \hat{x}_0))$.

Proof. If there exists f with $\omega \circ f = f_0$, then $\omega_* \circ f_* = f_{0*}$ so that $f_{0*}(\pi_1(Y, y_0)) \subset \omega_*(\pi_1(X, \hat{x}_0))$. Conversely, suppose $f_{0*}(\pi_1(Y, y_0)) \subset \omega_*(\pi_1(X, \hat{x}_0))$. Given $y \in V_Y$, take a path c in Y with $o(c) = y_0, t(c) = y$, and let $\hat{c}$ be the lifting of $f_0(c)$ in X with $o(\hat{c}) = \hat{x}_0$ (Fig. 5.4).

We then put $f_V(y) = t(\hat{c})$. To show that $f_V(y)$ does not depend on the choice of c, let c' be another path in Y with $o(c') = y_0, t(c') = y$. Take a lifting $\hat{c}_1$ of $f_0(\overline{c'})$ with $o(\hat{c}_1) = t(\hat{c})$. By hypothesis, there exists a loop $\hat{c}'' \in C(\hat{x}_0, \hat{x}_0)$ such that $\omega_*([\hat{c}'']) = f_{0*}([c \cdot \overline{c'}])$, and hence $\omega(\hat{c}'')$ is homotopic to $f_0(c \cdot \overline{c'})$. The product $\hat{c} \cdot \hat{c}_1$ is the lifting of the path $f_0(c \cdot \overline{c'})$ with the origin at $\hat{x}_0$ so that by Theorem 5.8, $\hat{c}''$ is homotopic to $\hat{c} \cdot \hat{c}_1$ and $t(\hat{c} \cdot \hat{c}_1) = \hat{x}_0$. Therefore $\hat{c} \cdot \hat{c}_1$ is a loop with base point $\hat{x}_0$, and the lifting of $f_0(c')$ turns out to have the same terminus as that of $f_0(c)$. In view of the definition of f_V, we see that $\omega_V \circ f_V = f_{0V}$.

What remains to be done is to define a map $f_E : E_Y \longrightarrow E$ such that $f = (f_V, f_E)$ forms a morphism and $\omega_E \circ f_E = f_{0E}$. To this end, let $e \in E_Y$, and take a path c with $t(c) = o(e)$. Using the lifting $\hat{c}$ of the path $f_0(c \cdot e)$ with the origin at $\hat{x}_0$, define $f_E(e)$ as the last edge in the path $\hat{c}$. The map f_E is the desired one. □

As an application of Theorem 5.9, we observe that a covering graph X over X_0 is produced by a "patchwork" of copies of the same tree (see Fig. 5.5). More precisely, take a spanning tree T of X_0, and let $i : T \longrightarrow X_0$ be the inclusion map. Since $\pi_1(T) = \{1\}$, i has a lifting $\hat{i} : T \longrightarrow X$. Clearly $\hat{i}$ is injective[6] so that $\hat{i}(T)$ is a tree isomorphic to T. Then $\{\hat{i}(T) |$ for all liftings $\hat{i}\}$ constitutes a family of disjoint subtrees in X such

---

[6]Precisely speaking, both $\hat{i}_V$ and $\hat{i}_E$ are injective for $\hat{i} = (\hat{i}_V, \hat{i}_V)$.

5.4 Universal Covering Graphs

Fig. 5.5 Patchwork by spanning trees

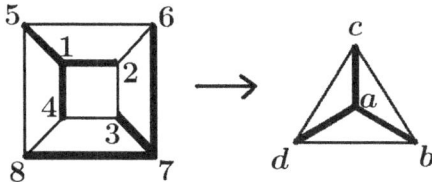

that every vertex is in one of them. This observation gives a useful suggestion when we construct the "universal covering graph" (Sect. 5.4).

It is an easy exercise to prove the following.

Let $\pi_1^{\text{even}}(X)$ be the subset of $\pi_1(X)$ consisting of homotopy classes $[c]$ with even length $|c|$.

1. $\pi_1^{\text{even}}(X)$ is a normal subgroup of $\pi_1(X)$ of index at most two.
2. X is non-bipartite if and only if $\pi_1^{\text{even}}(X)$ is of index-two in $\pi_1(X)$.

Finally we give a relation between the fundamental groups and homology groups. Define a homomorphism

$$h = h_x : \pi_1(X,x) \longrightarrow H_1(X,\mathbb{Z})$$

by putting $h([c]) = \langle c \rangle$. This is well defined since a back-tracking part in c vanishes in $H_1(X,\mathbb{Z})$. For the isomorphism $f_{x \mapsto y} : \pi_1(X,x) \longrightarrow \pi_1(X,y)$ in the base point change, we have $h_y \circ f_{x \mapsto y} = h_y$. The surjective homomorphism h is called the *Hurewicz homomorphism*, which will play a significant role in topological crystallography.

For a morphism $f : X \longrightarrow Y$, the following diagram is commutative.

$$\begin{array}{ccc} \pi_1(X,x) & \xrightarrow{f_*} & \pi_1(Y,f(x)) \\ h \downarrow & & \downarrow h \\ H_1(X,\mathbb{Z}) & \xrightarrow{f_*} & H_1(Y,\mathbb{Z}). \end{array}$$

5.4 Universal Covering Graphs

What we will see in this and the next sections is that among all covering graphs over a given graph X_0, there is the "maximal" one in the sense that every covering graph over X_0 is its subcovering graph.

A *universal covering map* onto X_0 is a $\omega^{\text{uni}} : X_0^{\text{uni}} \longrightarrow X_0$ for which the graph $X_0^{\text{uni}} = (E_0^{\text{uni}}, V_0^{\text{uni}})$ is a tree. The graph X_0^{uni}, which turns out to be uniquely determined (see (ii) below), is called the *universal covering graph* over X_0 (Fig. 5.6).

Fig. 5.6 Universal covering map

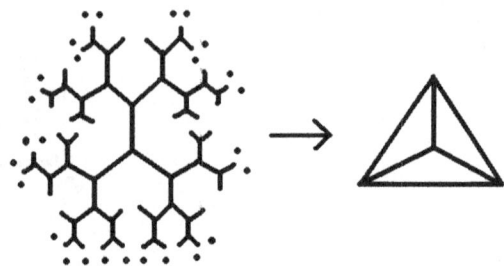

Postponing the proof of the existence of X_0^{uni} until the next section, we shall state several basic properties of X_0^{uni}.

Theorem 5.10. (i) *A universal covering map $\omega^{\text{uni}} : X_0^{\text{uni}} \longrightarrow X_0$ is regular and its covering transformation group is isomorphic to the fundamental group $\pi_1(X_0)$.*
(ii) *A universal covering map is unique in the following sense; If $\omega : X \longrightarrow X_0$ is another universal covering map, then there exists an isomorphism $f : X_0^{\text{uni}} \longrightarrow X$ such that $\omega \circ f = \omega^{\text{uni}}$.*
(iii) *For any covering map $\omega : X \longrightarrow X_0$, there exists a covering map $\omega_1 : X_0^{\text{uni}} \longrightarrow X$ such that $\omega \circ \omega_1 = \omega^{\text{uni}}$. This implies that the universal covering graph X_0^{uni} is maximal among all covering graphs over X_0.*

Proof. To prove (i), we fix a base point x_0 in X_0. Let $\hat{x}_0$ be a vertex in X_0^{uni} with $\omega^{\text{uni}}(\hat{x}_0) = x_0$. First we shall define a homomorphism $\psi = \psi_{x_0, \hat{x}_0} : G(\omega^{\text{uni}}) \longrightarrow \pi_1(X_0, x_0)$. For $g \in G(\omega^{\text{uni}})$, take a path $\hat{c}$ in X_0^{uni} with $o(\hat{c}) = \hat{x}_0$ and $t(\hat{c}) = g\hat{x}_0$. Then the homotopy class $[\omega^{\text{uni}}(\hat{c})] \in \pi_1(X_0, x_0)$ does not depend on the choice of $\hat{c}$ since every path in the tree X_0^{uni} with fixed end points is homotopic to each other. Putting $\psi(g) = [\omega^{\text{uni}}(\hat{c})]$, we obtain the map $\psi : G(\omega^{\text{uni}}) \longrightarrow \pi_1(X_0, x_0)$.

To check that ψ is a homomorphism, let $\hat{c}_1$ be a path with $o(\hat{c}_1) = \hat{x}_0$ and $t(\hat{c}_1) = h\hat{x}_0$ ($h \in G(\omega^{\text{uni}})$). Then $\hat{c}_1 \cdot h\hat{c}$ is a path joining $\hat{x}_0$ and $hg\hat{x}_0$, and

$$\psi(hg) = [\omega^{\text{uni}}(\hat{c}_1 \cdot h\hat{c})] = [\omega^{\text{uni}}(\hat{c}_1) \cdot \omega^{\text{uni}}(h\hat{c})] = [\omega^{\text{uni}}(\hat{c}_1)][\omega^{\text{uni}}(\hat{c})]$$
$$= \psi(h)\psi(g).$$

Clearly $\psi(1) = 1$ so that ψ is a homomorphism.

To show that ψ is injective, suppose that $\omega^{\text{uni}}(\hat{c})$ is homotopic to the point path $\emptyset_{x_0}$. In view of the covering homotopy theorem, $\hat{c}$ must be homotopic to the point path $\emptyset_{\hat{x}_0}$ so that $g\hat{x}_0 = t(\hat{c}) = \hat{x}_0$ and $g = 1$. To prove that ψ is surjective, take $\hat{x} \in X_0^{\text{uni}}$ with $\omega^{\text{uni}}(\hat{x}) = x_0$ and apply Theorem 5.3 in the preceding section to the case that $(Y, y_0) = (X_0^{\text{uni}}, \hat{x}_0), f_0 = \omega^{\text{uni}}$ to obtain a morphism $\sigma : X_0^{\text{uni}} \longrightarrow X_0^{\text{uni}}$ with $\sigma(\hat{x}_0) = \hat{x}$ and $\omega^{\text{uni}} \circ \sigma = \omega^{\text{uni}}$, where we use $\pi_1(X_0^{\text{uni}}, \hat{x}_0) = \{1\}$. In the same way we obtain a morphism $\sigma' : X_0^{\text{uni}} \longrightarrow X_0^{\text{uni}}$ with $\sigma'(\hat{x}) = \hat{x}_0$ and $\omega^{\text{uni}} \circ \sigma' = \omega^{\text{uni}}$. By the unique lifting theorem, we have $\sigma \circ \sigma' = \sigma' \circ \sigma = I$ so that $\sigma \in G(\omega^{\text{uni}})$. Hence ψ is surjective, and ω^{uni} is regular.

5.4 Universal Covering Graphs

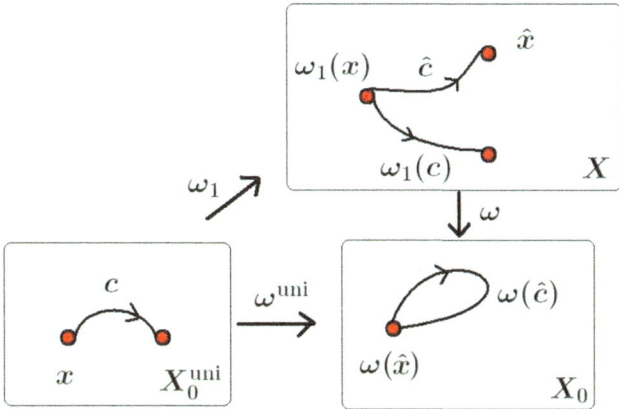

Fig. 5.7 Maximality of the universal covering map

Uniqueness (ii) of a universal covering map is a consequence of the unique lifting theorem and Theorem 5.3. Indeed, using $\pi_1(X_0^{uni}) = \{1\}$, we obtain a morphism $f : X_0^{uni} \longrightarrow X$ with $f(\hat{x}_0) = \hat{x}$ as a lifting of $\omega^{uni} : X_0^{uni} \longrightarrow X_0$ where $\omega^{uni}(\hat{x}_0) = \omega(\hat{x}) = x_0$. Changing the role of X_0^{uni} and X, we also obtain $g : X \longrightarrow X_0^{uni}$ with $g(\hat{x}) = \hat{x}_0$ as a lifting of $\omega : X \longrightarrow X_0$ ($\omega(\hat{x}) = x_0$). We then observe that $f \circ g = I$ and $g \circ f = I$ since $f \circ g$ and $g \circ f$ are liftings of the identity map of X_0 which preserve the base points.

We finally prove (iii). Applying the lifting criterion to $\omega^{uni} : X_0^{uni} \longrightarrow X_0$, we find a morphism $\omega_1 : X_0^{uni} \longrightarrow X$ such that $\omega \circ \omega_1 = \omega^{uni}$.

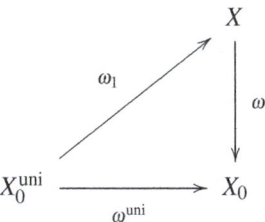

We shall show that ω_1 is a covering map. To check that ω_{1V} is surjective, take any vertex $\hat{x}$ of X, and pick up a vertex x in X_0^{uni} such that $\omega^{uni}(x) = \omega(\hat{x})$. When $\omega_1(x) \neq \hat{x}$, we take a path $\hat{c}$ in X with $o(\hat{c}) = \omega_1(x)$ and $t(\hat{c}) = \hat{x}$ (Fig. 5.7). Then $\omega(\hat{c})$ is a closed path in X_0 since $\omega(\omega_1(x)) = \omega^{uni}(x) = \omega(\hat{x})$. Let c be the lifting of $\omega(\hat{c})$ in X_0^{uni} for the covering map ω^{uni} with $o(c) = x$. Then $\omega_1(c)$ is the lifting of $\omega(\hat{c})$ for ω with $o(\omega_1(c)) = \omega_1(x)$. Thus we have two liftings $\hat{c}$ and $\omega_1(c)$ of $\omega(\hat{c})$ for ω with the same origin $\omega_1(x)$. By the unique path-lifting property, we conclude that $\hat{c} = \omega_1(c)$ and $\hat{x} = t(\hat{c}) = t(\omega_1(c)) = \omega_1(t(c))$. Consequently, ω_{1V} is surjective.

One can check that $\omega_1 : E_{0x}^{uni} \longrightarrow E_{\hat{x}}$ ($\hat{x} = \omega_1(x)$) is a bijection (use the general fact that if $g : B \longrightarrow C$ and $g \circ f : A \xrightarrow{f} B \xrightarrow{g} C$ are bijections, then so is f). This completes the proof of the assertion that ω_1 is a covering map. □

Before proceeding further, we make a remark. In the assertion (iii), $\omega_1 : X_0^{uni} \longrightarrow X$ is the universal covering map with the base graph X; thus $\pi_1(X, \omega_1(\hat{x}_0))$ acts on X_0^{uni} as the covering transformation group. On the other hand, $G(\omega_1) \subset G(\omega^{uni})$ [Theorem 5.5(a)]. Identifying $G(\omega_1)$ with $\pi_1(X, \omega_1(\hat{x}_0))$, and $G(\omega^{uni})$ with $\pi_1(X_0, x_0)$, we have an inclusion map $\pi_1(X, \omega_1(\hat{x}_0)) \longrightarrow \pi_1(X_0, x_0)$. This map coincides with the induced homomorphism $\omega_* : \pi_1(X, \omega_1(\hat{x}_0)) \longrightarrow \pi_1(X_0, x_0)$.

Theorem 5.11. *Let $f : X_0 \longrightarrow Y_0$ be a morphism with $f(x_0) = y_0$, and let $\hat{x}_0 \in X_0^{uni}$ and $\hat{y}_0 \in Y_0^{uni}$ be vertices such that $\omega^{uni}(\hat{x}_0) = x_0$ and $\omega^{uni}(\hat{y}_0) = y_0$ (for simplicity we use the same symbol ω^{uni} for every universal covering map).*

1. *There exists a unique morphism $F : X_0^{uni} \longrightarrow Y_0^{uni}$ such that $F(\hat{x}_0) = \hat{y}_0$ and $\omega^{uni} \circ F = f \circ \omega^{uni}$.*

$$\begin{array}{ccc} X_0^{uni} & \xrightarrow{F} & Y_0^{uni} \\ \omega^{uni} \downarrow & & \downarrow \omega^{uni} \\ X_0 & \xrightarrow{f} & Y_0 \end{array}$$

2. $F^{uni} \circ \sigma = f_*(\sigma) \circ F^{uni}$ *for $\sigma \in \pi_1(X_0, x_0)$.*

Proof. (1) is a consequence of the lifting criterion (F is to be a lifting of $f \circ \omega^{uni}$ for the covering map $\omega^{uni} : Y_0^{uni} \longrightarrow Y_0$).

To prove (2), we apply Theorem 5.4, and find a homomorphism $\rho : \pi_1(X_0, x_0) \longrightarrow \pi_1(Y_0, y_0)$ such that $F^{uni}(\sigma x) = \rho(\sigma) F^{uni}(x)$. We shall show $\rho = f_*$. Let $c \in C(x_0, x_0)$ with $\sigma = [c]$, and $\hat{c}$ be the lifting of c with $o(\hat{c}) = \hat{x}_0$. Recalling how the fundamental group acts on the universal covering graph, we observe that $\sigma \hat{x}_0 = t(\hat{c})$. Notice that $F(\hat{c})$ is the lifting of the loop $f(c) \in C(y_0, y_0)$ with $o(F(\hat{c})) = \hat{y}_0$. Thus setting $\mu = [f(c)] = f_*([c]) \in \pi_1(Y_0, y_0)$, we have $t(F(\hat{c})) = \mu \hat{y}_0$, and

$$\mu \hat{y}_0 = t(F(\hat{c})) = F(t(\hat{c})) = F(\sigma \hat{x}_0) = \rho(\sigma) F(\hat{x}_0) = \rho(\sigma) \hat{y}_0,$$

from which we conclude that $f_*(\sigma) = \mu = \rho(\sigma)$. □

5.5 Construction of Universal Covering Graphs

We now proceed to the proof of the assertion that *for every graph X_0, there exists a universal covering map onto X_0*. The idea suggested in the preceding section underlies the proof. That is, we prepare many copies of a spanning tree in X_0 and

5.5 Construction of Universal Covering Graphs

Fig. 5.8 Construction of the universal covering graph

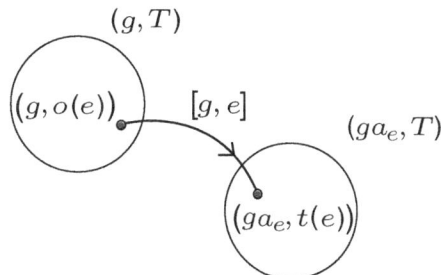

make a "patchwork" by connecting these copies in an appropriate way. To perform this work, we need the notion of *free group*.

Here in general, a free group G with basis A means a group such that any element of G, except for the identity element 1, can be written in one and only one way as a *reduced product* of finitely many elements of A and their inverses, where the term "reduced" means that the array of the form aa^{-1} and $a^{-1}a$ in the product are excluded. Thus, any reduced product $a_1^{\pm 1} \cdots a_n^{\pm 1}$ with $n \geq 1$ cannot be the identity element 1. What we need in our discussion is the fact that starting from any set A, one may construct a free group G with basis A (see Appendix 4).

Fix an orientation E_0^o of $X_0 = (E_0, V_0)$. Choose a spanning tree $T = (V_T, E_T)$ in X_0, and let G be the free group with basis $E_0^o \backslash E_T$.

Though $E_0^o \backslash E_T$ is a subset of G, to avoid the typographical confusion, we shall use the symbol a_e instead of $e \in E_0^o \backslash E_T$ when we regard e as an element of G.

We consider the collection of pairs (g, T) of $g \in G$ and T where each (g, T) is understood to be a copy of the tree T. Vertices and edges in (g, T) are expressed as (g, x) and (g, e), respectively. We shall provide the disjoint union $\bigcup_{g \in G} (g, T)$ with new edges in the following way. For each $g \in G$ and $e \in E_0^o \backslash E_T$, we create a directed edge $[g, e]$ joining $(g, o(e))$ and $(ga_e, t(e))$ (Fig. 5.8). Observe that if $[g, e]$ joins (g_1, T) and (g_2, T), then $g_1 = g$ and $g_2 = ga_e$ so that there is at most one edge joining (g_1, T) and (g_2, T). By forgetting the direction of each new edge and adding to $\bigcup_{g \in G}(g, T)$ these new undirected edges, we obtain a graph $X = (V, E)$. The group G acts freely on the graph X by $g(h, x) = (gh, x)$ ($x \in V_T$), $g(h, e) = (gh, e)$ ($e \in E_T$), and $g[h, e] = [gh, e]$ ($e \in E_0^o \backslash E_T$). The quotient graph X/G is obviously identified with X_0.

We shall show that X is connected. Let $g = a_{e_1}^{\pm 1} \cdots a_{e_n}^{\pm 1}$ ($e_i \in E_0^o \backslash E_T$), and put $g_i = a_{e_1}^{\pm 1} \cdots a_{e_i}^{\pm 1}$. Since (g_i, T) and (g_{i+1}, T) are joined by the edge (g_i, e_{i+1}) (in the case $g_{i+1} = g_i a_{e_{i+1}}$) or by (g_{i+1}, e_{i+1}) (in the case $g_{i+1} = g_i a_{e_{i+1}}^{-1}$), the trees $(1, T)$, $(g_1, T), \ldots, (g_n, T)$ are successively joined by edges in X so that X is connected.

Finally we prove that X is a tree. If there is a geodesic loop c in X with a base point in $(1, T)$ (1 being the identity element of G), then one can find a sequence $(g_0, T), (g_1, T), \ldots, (g_n, T)$ such that $g_0 = g_n = 1$, $g_i \neq g_{i-2}$ ($2 \leq i \leq n$), and (g_i, T) and (g_{i+1}, T) are joined by an edge (g_i, e_{i+1}) or by (g_{i+1}, e_{i+1}) ($0 \leq i \leq n-1$). This implies that $g_{i+1} = g_i a_{e_{i+1}}^{\pm 1}$, and $a_{e_1}^{\pm} \cdots a_{e_n}^{\pm 1} = 1$ since $g_n = 1$. On the other hand

$g_i \neq g_{i-2}$ ($2 \leq i \leq n$) implies that $a_{e_1}^{\pm 1} \cdots a_{e_n}^{\pm 1}$ is a reduced product, and hence it cannot be the identity element in the free group G, thereby leading to a contradiction. Thus X is a tree, and the canonical morphism $\omega : X \longrightarrow X/G = X_0$ is the universal covering map.

As a byproduct of the discussion, we conclude that $\pi_1(X_0)$ is a free group because $\pi_1(X_0)$ is isomorphic to the covering transformation group G of the universal covering map $\omega : X \longrightarrow X_0 = X/G$ which we have just constructed. The basis of the free group $\pi_1(X_0, x)$ corresponding to $E_0^o \backslash E_T$ is described in the following way. For each $e \in E_0^o \backslash E_T$, take two geodesics c_1, c_2 in T with $o(c_1) = x, o(c_2) = x$, $t(c_1) = o(e), t(c_2) = t(e)$, and create the path $c_e = c_1 \cdot e \cdot \bar{c}_2$. The path c_e is a geodesic loop with base point x. Then $\{[c_e] \mid e \in E_0^o \backslash E_T\}$ is a basis of the free group $\pi_1(X_0, x)$. In fact, for the isomorphism $\psi : G \longrightarrow \pi_1(X_0, x)$ defined in the proof of (i) in the previous section, we easily observe that $\psi(a_e) = [c_e]$.

What we observed is paraphrased as: *A group G acting freely on a tree is a free group*. In fact, G is the covering transformation group of the universal covering map $X \longrightarrow X/G$, and hence $G = \pi_1(X/G)$.

5.6 Notes

(I) A primitive idea of covering maps for general spaces is seen in Riemann's epoch-making study of algebraic functions and their integrals (1851). His idea was made rigorous by H. Weyl in the book *Die Idee der Riemannschen Fläche* published in 1913, in which for the first time the notion of covering maps between two-dimensional differentiable manifolds (together with the duality between differential forms and 1-cycles) was introduced. The book *Lehrbuch der Topologie* by H. Seifert and W. Threlfall published in 1934 gave a general set-up and popularized the notion of covering maps based on a theory of topological spaces created by M. F. Fréchet and F. Hausdorff.

We shall content ourselves with the definition of covering spaces. A covering space over a topological space X_0 is a space X and a continuous map $\omega : X \longrightarrow X_0$ such that

(a) ω is surjective, and
(b) Given any $x \in X_0$ there is a connected open set U about x such that ω maps each component of $\omega^{-1}(U)$ homeomorphically onto U.

The unique path-lifting property, the unique lifting theorem, and the covering homotopy theorem are valid in the general set-up.

Fundamental groups and universal covering maps (spaces) are also defined for general topological spaces (with a certain mild condition). Their structures could be quite complicated in general, compared with the case of graphs. For instance,

5.6 Notes

any *finitely presented group*[7] can be the fundamental group of a compact smooth manifold.

(II) The covering transformation group of a regular covering graph over a finite graph X_0 is *finitely generated* since it is a homomorphic image of $\pi_1(X_0)$ (recall that $\pi_1(X_0)$ is a free group with a finite basis). We shall show the converse by introducing the notion of *Cayley–Serre graphs.*[8]

Let G be a group, and let $i : A \longrightarrow G$ be a map of a finite set A into G such that $i(A)$ generates G. We define the graph $X(G,A)$ in the following way (cf. [84]). The set V of vertices is just G. Directed edges are the pairs (g,a), $g \in G$, $a \in A$. The origin and terminus of the edge (g,a) are g and $gi(a)$, respectively. Forgetting orientation, we get a connected regular graph $X(G,A)$ of degree $2|A|$, which is what we call the Cayley–Serre graph. The graph $X(G,A)$ has the natural free G-action given by $g(g',a) = (gg',a)$.

The quotient graph $X(G,A)/G$ is the $|A|$-bouquet graph. In other words, the Cayley–Serre graph $X(G,A)$ is a regular covering graph over a bouquet graph. Conversely, a regular covering graph over a bouquet graph is identified with a Cayley–Serre graph.

One can verify that $X(G,A)$ is a tree if and only if G is the free group with basis A.

(III) As a slight variant of covering map, we introduce the notion of *branched (ramified) covering map*, an analogue of non-constant holomorphic maps between two closed Riemann surfaces.

A morphism $\omega : X = (V,E) \longrightarrow X_0 = (V_0, E_0)$ is said to be a branched covering map[9] if

1. Both $\omega : V \longrightarrow V_0$ and $\omega : E \longrightarrow E_0$ are surjective.
2. There is a positive integer $v(x)$ for each $x \in V$ such that the map $\omega : E_x \longrightarrow E_{0\omega(x)}$ is $v(x)$-to-one map, i.e., $|\{e \in E_x|\ \omega(e) = e_0\}| = v(x)$ for any $e_0 \in E_{0\omega(x)}$.

The number $v(x)$ is the *ramification index* of ω at x. By definition, ω is a covering map if and only if $v(x) = 1$ for every x. If we put

$$n(y) = \sum_{x \in V; \omega(x) = y} v(x),$$

then, for $e \in E_0$, we get $n(o(e)) = |\omega^{-1}(e)| = n(t(e))$. Thus the summation

$$n = \sum_{x \in V; \omega(x) = y} v(x)$$

---

[7] A finitely presented group is a group with a finite set of generators subject to finitely many relations.

[8] The definition of Cayley–Serre graphs is slightly different from that of Cayley graphs (cf. [66]).

[9] This notion is identical with that of *harmonic morphism* introduced by Urakawa [104] and Baker and Norine [6].

does not depend on $y \in V_0$ when every fiber $\omega^{-1}(y)$ is finite. We call n the *degree* of the branched covering map ω.

An example of branched covering map is provided by the canonical morphism onto a quotient graph associated with a group action. To be exact, given an action of a finite group G on X without inversion, if G acts freely on E, then the canonical morphism $\omega : X \longrightarrow X/G$ is a branched covering map. This being the case, $n = |G|$ and $v(x) = |G_x|$ where $G_x = \{g \in G|\ gx = x\}$.

The following theorem is a generalization of the equality $\chi(X) = n\chi(X_0)$ holding for an n-fold covering map.

Theorem 5.12 (cf. [6]). *If X and X_0 are finite graphs, then*

$$\chi(X) = n\chi(X_0) - \sum_{x \in V} \bigl(v(x) - 1\bigr). \tag{5.4}$$

Proof. First note that $\omega : E \longrightarrow E_0$ is n-to-one, and

$$\sum_{x \in V} v(x) = \sum_{y \in V_0} \sum_{x \in V; \omega(x)=y} v(x) = n|V_0|.$$

So the left-hand side of (5.4) becomes

$$n\bigl(|V_0| - |E_0|/2\bigr) - \sum_{x \in V} v(x) + |V| = n|V_0| - |E|/2 - n|V_0| + |V| = \chi(X).$$

□

Equality (5.4) is an analogue of the *Riemann–Hurwitz formula*, and may be considered a result at the threshold of *discrete algebraic geometry* which, in connection with standard realizations, will be discussed intensively in Chap. 10.

Part II
Geometry of Crystal Structures

Chapter 6
Topological Crystals

Crystals, the word derived from the Greek meaning "ice", are the most stable form of all solids. They are found throughout the natural world, and have been always recognized as being distinct from other forms of matter.

It is often said that geometry in Ancient Greece started from the curiosity regarding the shapes of crystals.[1] Actually legend has it that Pythagoras (about 569 BC to about 475 BC) derived the notion of regular polyhedra from the shape of a crystal. This legend is not entirely baseless because the southern part of Italy where Pythagoras dwelled and established his school[2] produces the pyrite crystal (an iron sulfide with the formula FeS_2) whose shape is roughly cubic, octahedral, or dodecahedral.

Euclid's *Elements* (Στοιχεια in Greek), compiled around 300 BC, is one of the most beautiful and influential works of science in the history of humankind. Some sources (such as Proclus) say that Elements contains many results established by mathematicians in the Pythagorean school. It consists of thirteen volumes, and ends up with the classification of regular convex polyhedra; that is, the final volume contains the proof that there are exactly five regular convex polyhedra, namely, the tetrahedron, cube (or regular hexahedron), octahedron, dodecahedron, and icosahedron (Fig. 6.1). According to Plato's dialogue *Timaeus* written around 360 BC, the first mathematician who established the classification is Theaetetus,

---

[1] Even before the Greek period, various properties of figures in the plane and space had been "known" by ancient people as empirical knowledge, some of which were gained from their experiences in land surveys (it is worthwhile to recall that the word "geometry" was derived from the Greek word "geometria" which is the composition of "geo" meaning land and "metria" meaning measurement).

[2] The Pythagorean school in Crotona, a Greek colonial town on the southeastern coast of Italy, was a religious order in which Pythagoras was a sort of cult figure.

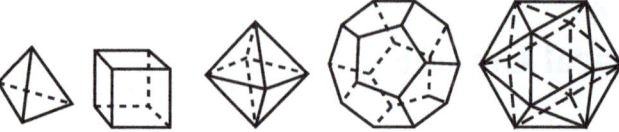

Fig. 6.1 Regular convex polyhedra

Fig. 6.2 Covalent bonding

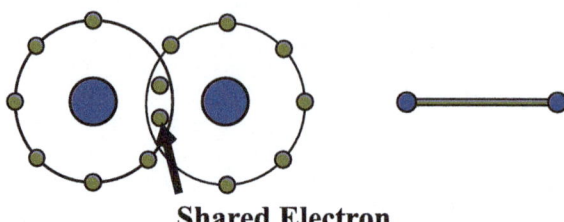

a contemporary of Plato.[3] From today's viewpoint, the classification of regular polyhedra is closely related to that of finite subgroups of the orthogonal group $O(3)$, *point groups* in crystallography [72].

With the exception of the pioneer work by Kepler on crystal structures [see Notes (III) in Chap. 8], it is only in the nineteenth century, more than 2,000 years after ancient Greece, that mathematics started again to play a vital role in crystallography; that is, group theory matured enough to be applied to the morphology of crystals. The study of morphology led early crystallographers to the concepts of *space groups* (crystallographic groups) and points groups which are employed in the classification of all crystals in terms of symmetry (Appendix 5).

On the other hand, as mentioned in the introduction, Max von Laue discovered the microscopic structures of crystals by the diffraction of X-rays (1912). The X-ray technique is based on observing the scattered intensity of an X-ray beam hitting a sample. What was found by von Laue is that crystals are solids composed of atoms arranged in an orderly repetitive array,[4] and hence confirmed the anticipation that crystallographers had conceived long before. Just after von Laue's discovery, W. L. Bragg and his father W. H. Bragg greatly simplified von Laue's description of X-ray interference, and solved the structures of diamond, sodium chloride (NaCl) and Zinc sulfide (ZnS) in 1913.

Today, crystallographers can produce a 3D picture having the density of electrons within the crystal from which the mean positions of the atoms in the crystal can be determined, as well as their mutual interactions.

---

[3]Regular convex polyhedra, also called the Platonic solids, are characterized by the property that their faces are congruent regular polygons, with the same number of faces meeting at each vertex [23, 24].

[4]More precisely, in the crystalline state, atoms oscillate around the positions of equilibrium. We are dealing with the positions of equilibrium when we talk about crystal structures.

Among several ways of interactions, a typical one is *covalent bonding* (Fig. 6.2), a form of chemical bonding that is characterized by the sharing of pairs of electrons between atoms (the term "covalent bond" dates back to 1939). When we draw a picture of a crystal structure (and any molecule), we join two atoms by a line called a *bond* [5] if they interact with each other, no matter which kind of interaction takes place. What we get is a network in space where we ignore the physical characters of atoms and bonds that may be different from one another.

After confirmation that atoms and molecules have physical existence, network representations of molecules also broke away from any fictitious entity.[6] Therefore it is a matter of course for crystallographers to use graph theory for studying crystal structures. Rather surprisingly, however, the systematic approach by means of graph theory was not done until 1950s although, by this time, plenty of crystal structures were understood and represented by networks [108]. The reason may be due to the self-explanatory fact that the ideal model of a crystal is an *infinite* graph, which is not an object that the ordinary graph theory handles.

Well, what can one say about the network associated with a crystal? The salient nature of the network associated with a crystal is the *periodicity* with respect to a lattice group action as indicated in the introduction and Sect. 3.3; that is, there exists a lattice group L of $\mathbb{R}^3$ such that the crystal net[7] is invariant under the translational action of L.

Now forgetting how the network is placed in space, we obtain an abstract graph, say X. The graph X retains, in an abstract way, all the information about the connectivity of atoms in the crystal. In addition, the periodic nature of the crystal net is inherited by X because the lattice group (*period lattice*) L, originally acting on the network in space by translations, acts *freely* on X as automorphisms, and hence yields the canonical morphism ω of X onto the finite quotient graph $X_0 = X/L$ (Theorem 3.1). This passage (in conjunction with the idea of vector assignments mentioned in Example 3.4) brings us to modern crystallography developed since 1980s.

From a mathematical view, one can say more about ω. Namely ω is a covering map whose covering transformation group is L (Theorem 5.2). This simple observation motivates the definition of topological crystals of general dimension. That is,

---

[5]A primitive idea of "bond" had already been conceived as early as the twelfth century. It hypothesizes that certain types of *chemical species* are joined by a type of *chemical affinity*. By the mid-nineteenth century, it became clear that chemical compounds are formed from constituent elements, and the idea of bond had been developed as the theory of *valency* based on the theory of radicals. In particular, the theory of chemical structures by the German chemist August Kekulé, in which he took into account the specific combining power (or valences) of specific atoms, provided a dramatic new clarity of understanding on chemical compounds (1858).

[6]French chemist L. Pasteur observed that the relative spatial arrangement of atoms within molecules is vital in understanding their chemical properties (1861).

[7]In reality, a crystal net has finite size. But we think that the net extends to the whole space when we talk about crystals.

a topological crystal of dimension d is an infinite-fold regular covering graph over a finite graph whose covering transformation group is isomorphic to $\mathbb{Z}^d$.

The most significant feature of topological crystals is that homology groups of their base graphs come up when we describe their structures. Indeed, for a given finite graph X_0, a topological crystal of d-dimension with base graph X_0 corresponds to a subgroup H of the homology group $H_1(X_0,\mathbb{Z})$ whose factor group $H_1(X_0,\mathbb{Z})/H$ is a free abelian group of rank d.

This chapter will exclusively engage in interpreting these abstract aspects of topological crystals.

6.1 Generalities on Abelian Covering Graphs

We set the scene in a bit more generality than needed for crystallography. A regular covering map $\omega : X \longrightarrow X_0$ with abelian covering transformation group is called an *abelian covering map* (over X_0), and the graph X is called an *abelian covering graph* over X_0.

From now on we shall adopt multiplicative notation for additive groups (for example, homology groups and their factor groups) when they acts on a graph as a covering transformation group. Namely for elements α, β in an additive group L acting on a graph X, we write $\alpha\beta x$ and $\alpha^{-1}x$ ($x \in V$) instead of writing $(\alpha+\beta)x$ and $(-\alpha)x$, respectively.

In dealing with the structure of abelian covering maps, we need the notion of commutator subgroups (see Appendix 2 for more details). For a group G, the subgroup $[G,G]$ generated by the set

$$\{aba^{-1}b^{-1}|\, a,b \in G\}$$

is normal and called the *commutator subgroup* of G. The factor group $G/[G,G]$ is abelian, and $[G,G]$ is characterized as the smallest one among all normal subgroups of G with abelian factor groups; that is, if N is a normal subgroup of G such that G/N is abelian, then $[G,G] \subset N$. Therefore if $\phi : G \longrightarrow H$ is a homomorphism into an abelian group H, then ϕ descends to a homomorphism of $G/[G,G]$ into H.

For the free group $F(\Lambda)$ with basis Λ, the factor group $F(\Lambda)/[F(\Lambda),F(\Lambda)]$ is isomorphic to the free abelian group $A(\Lambda)$ with $\mathbb{Z}$-basis Λ (see Appendix 4). Recall that $\pi_1(X_0,x_0)$ is the free group with basis $\Lambda = \{[c_e]|\ e \in E_0^o\backslash E_T\}$ where E_0^o is an orientation and $T = (V_T, E_T)$ is a spanning tree in X_0 (Sect. 5.5). Therefore, $\pi_1(X_0)/[\pi_1(X_0), \pi_1(X_0)]$ is isomorphic to $A(\Lambda)$. On the other hand, in view of the argument in Sect. 4.3, the set $\{\langle c_e \rangle | e \in E_0^o\backslash E_T\}$ is a $\mathbb{Z}$-basis of $H_1(X_0,\mathbb{Z})$. Since $H_1(X_0,\mathbb{Z})$ is an abelian group, the Hurewicz homomorphism $h : \pi_1(X_0,x_0) \longrightarrow H_1(X_0,\mathbb{Z})$ descends to a surjective homomorphism ϕ of the factor group $\pi_1(X_0)/[\pi_1(X_0), \pi_1(X_0)]$ onto $H_1(X_0,\mathbb{Z})$. One can check that $h([c_e]) = \langle c_e \rangle$, and hence ϕ bijectively maps the basis $\Lambda = \{[c_e]|\ e \in E_0^o\backslash E_T\}$ of $\pi_1(X_0)/[\pi_1(X_0), \pi_1(X_0)]$ onto the $\mathbb{Z}$-basis $\{\langle c_e \rangle | e \in E_0^o\backslash E_T\}$ of $H_1(X_0,\mathbb{Z})$. Therefore we conclude that ϕ is an isomorphism.

6.1 Generalities on Abelian Covering Graphs

Let $\omega^{\mathrm{uni}} : X_0^{\mathrm{uni}} \longrightarrow X_0$ be the universal covering map, and consider the quotient graph
$$X_0^{\mathrm{ab}} = X_0^{\mathrm{uni}}/\bigl[\pi_1(X_0), \pi_1(X_0)\bigr].$$
We thus have a regular covering map $\omega^{\mathrm{ab}} : X_0^{\mathrm{ab}} \longrightarrow X_0$.

Theorem 6.1. *1. X_0^{ab} is an abelian covering graph over X_0 whose covering transformation group is (isomorphic to) $H_1(X_0, \mathbb{Z})$.*
2. Let X be an abelian covering graph over X_0. Then there exists a subcovering map $\omega_1 : X_0^{\mathrm{ab}} \longrightarrow X$ of $\omega^{\mathrm{ab}} : X_0^{\mathrm{ab}} \longrightarrow X_0$.
In this sense, X_0^{ab} is the maximal abelian covering graph[8] over X_0.

(1) is immediate from the argument we made so far. For (2), we note that the factor group $\pi_1(X_0)/\pi_1(X)$ is abelian so that $[\pi_1(X_0), \pi_1(X_0)] \subset \pi_1(X)$. Thus we may apply the lifting criterion (Theorem 5.9) to obtain ω_1.

If $\omega : X \longrightarrow X_0$ is an abelian covering map, then, identifying $\pi_1(X_0, x_0)$ with the covering transformation group $G(\omega^{\mathrm{uni}})$ of the universal covering map $\omega^{\mathrm{uni}} : X_0^{\mathrm{uni}} \longrightarrow X_0$, we have the homomorphism $\nu : \pi_1(X_0, x_0) \longrightarrow G(\omega)$. Since $G(\omega)$ is abelian, ν descends to a homomorphism $\mu : H_1(X_0, \mathbb{Z}) \longrightarrow G(\omega)$, i.e., $\mu \circ h = \nu$ (remember that h is the Hurewicz homomorphism):

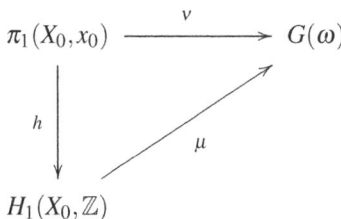

The subgroup $\mathrm{Ker}\,\mu$ of $H_1(X_0, \mathbb{Z})$ is identified with the covering transformation group of the subcovering map $\omega_1 : X_0^{\mathrm{ab}} \longrightarrow X$. Conversely, given a subgroup H of $H_1(X_0, \mathbb{Z})$, we obtain an abelian covering map $\omega : X \longrightarrow X_0$ with $\mathrm{Ker}\,\mu = H$ and $G(\omega) = H_1(X_0, \mathbb{Z})/H$ by putting $X = X_0^{\mathrm{ab}}/H$.

We look at the homomorphism μ more closely. Let $\alpha \in H_1(X_0, \mathbb{Z})$, and represent it by a closed path c in X_0 (this is possible as seen in Sect. 4.2). For a lifting $\tilde{c}$ of c in X, there exists a unique $\sigma \in G(\omega)$ such that $t(\tilde{c}) = \sigma o(\tilde{c})$. Here σ does not depend on the choice of a lifting $\tilde{c}$. Indeed, any lifting of c is expressed as $\delta\tilde{c}$ with $\delta \in G(\omega)$, and hence, thanks to commutativity of $G(\omega)$,

$$t(\delta\tilde{c}) = \delta t(\tilde{c}) = \delta\sigma o(\tilde{c}) = \sigma\delta o(\tilde{c}) = \sigma o(\delta\tilde{c}).$$

Theorem 6.2. *For a closed path c in X_0 and its lifting $\tilde{c}$ in X, we have*

$$t(\tilde{c}) = \mu(\langle c \rangle) o(\tilde{c}), \qquad (6.1)$$

---

[8] X_0^{ab} is also called the *homology universal covering graph*.

Fig. 6.3 Lifting of a closed path

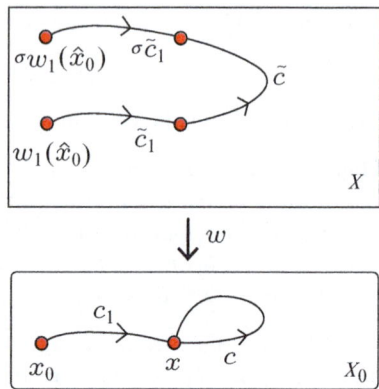

or equivalently, $\sigma = \mu(\alpha)$. In particular, a lifting of a closed path c in X_0 is closed if and only if $\langle c \rangle \in H$. Thus in the case $X = X_0^{ab}$, a lifting of a closed path c in X_0 is closed if and only if $\langle c \rangle = 0$.

Proof. To derive (6.1), we first assume that $o(c) = x_0$. Remember that when we identified $\pi_1(X_0, x_0)$ with $G(\omega^{uni})$ (see the proof of (i) in Sect. 5.4), we chose $\hat{x}_0 \in V_0^{uni}$ with $\omega^{uni}(\hat{x}_0) = x_0$, and had the relation

$$[c]o(\hat{c}) = t(\hat{c}) \tag{6.2}$$

for the lifting $\hat{c}$ of c in X_0^{uni} with $o(\hat{c}) = \hat{x}_0$, where, on the left-hand side, we think of $[c]$ as a covering transformation under the identification of $\pi_1(X_0, x_0)$ with $G(\omega^{uni})$. From (6.2), it follows that

$$\mu(\langle c \rangle)o(\omega_1(\hat{c})) = \mu(h([c])\omega_1(o(\hat{c})) = \nu([c])\omega_1(o(\hat{c})) = \omega_1([c]o(\hat{c}))$$
$$= \omega_1(t(\hat{c})) = t(\omega_1(\hat{c})),$$

where we have used (5.3). The path $\omega_1(\hat{c})$ is the lifting of c in X whose origin is $\omega_1(\hat{x}_0)$.

Next consider an arbitrary closed path c in X_0. We take a path c_1 in X_0 such that $o(c_1) = x_0$ and $t(c_1) = o(c)$ (Fig. 6.3).

We apply the above argument to the closed path $c' = c_1 \cdot c \cdot \overline{c_1}$ with $o(c') = x_0$ and its lifting $\tilde{c}'$ in X with $o(\tilde{c}') = \omega_1(\hat{x}_0)$ to get $\mu(\langle c' \rangle)o(\tilde{c}') = t(\tilde{c}')$. Put $\sigma = \mu(\langle c' \rangle)$, and note that $\tilde{c}'$ can be expressed as $\tilde{c}_1 \cdot \tilde{c} \cdot \sigma \overline{\tilde{c}_1}$ where $\tilde{c}_1$ is the lifting of c_1 with $o(\tilde{c}_1) = \omega_1(\hat{x}_0)$, and $\tilde{c}$ is the lifting of c with $o(\tilde{c}) = t(\tilde{c}_1)$. This implies that $t(\tilde{c}) = \sigma o(\tilde{c})$ (see Fig. 6.3). Thus for any lifting $\tilde{c}$ of c, we have $t(\tilde{c}) = \sigma o(\tilde{c})$. Using $\sigma = \mu(\langle c' \rangle) = \mu(\langle c \rangle)$, we get $t(\tilde{c}) = \mu(\langle c \rangle)o(\tilde{c})$, as desired. □

Theorem 6.3. *Let $\omega : X \longrightarrow X_0$ be an abelian covering map. Then*

$$\omega_*(H_1(X, \mathbb{Z})) = \operatorname{Ker} \mu.$$

6.1 Generalities on Abelian Covering Graphs

Fig. 6.4 Petersen graph

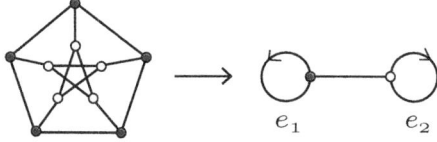

This is a corollary of the above theorem. Indeed, representing a homology class $\alpha \in H_1(X, \mathbb{Z})$ by a closed path c in X and noting that c is a lifting of $\omega(c)$, we have

$$\mu(\omega_*(\alpha)) = \mu(\langle \omega(c) \rangle) = 0.$$

Hence $\omega_*\big(H_1(X,\mathbb{Z})\big) \subset \operatorname{Ker}\mu$. On the other hand, a lifting c in X of any closed path c_0 in X_0 with $\langle c_0 \rangle \in \operatorname{Ker}\mu$ is closed so that $\langle c_0 \rangle = \omega_*(\langle c \rangle)$, and $\omega_*\big(H_1(X,\mathbb{Z})\big) \supset \operatorname{Ker}\mu$.

Example 6.1. The *Petersen graph* X drawn in Fig. 6.4 is an abelian covering graph over the graph X_0 with two loop edges joined by a separating edge whose covering transformation group is isomorphic to $\mathbb{Z}_5$. Indeed, X has a free rotational action of $\mathbb{Z}_5$ whose quotient graph is X_0.

We may check that $\operatorname{Ker}\mu = \mathbb{Z}(2e_1 - e_2) + \mathbb{Z}(2e_2 + e_1)$ (recall $H_1(X_0,\mathbb{Z}) = \mathbb{Z}e_1 + \mathbb{Z}e_2$). The correspondence $me_1 + ne_2 \longmapsto m + 2n$ yields an isomorphism of $H_1(X_0,\mathbb{Z})/\operatorname{Ker}\mu$ onto $\mathbb{Z}_5$. □

The following theorem is an abelian-covering version of Theorem 5.11.

Theorem 6.4. *Let $f : X_0 \longrightarrow Y_0$ be a morphism with $f(x_0) = y_0$, and let $\tilde{x}_0 \in X_0^{ab}$ and $\tilde{y}_0 \in Y_0^{ab}$ be vertices such that $\omega^{ab}(\tilde{x}_0) = x_0$ and $\omega^{ab}(\tilde{y}_0) = y_0$. Then there exists a unique morphism $F : X_0^{ab} \longrightarrow Y_0^{ab}$ such that $F(\tilde{x}_0) = \tilde{y}_0$ and $\omega^{ab} \circ F = f \circ \omega^{ab}$.*

$$\begin{array}{ccc} X_0^{ab} & \xrightarrow{F} & Y_0^{ab} \\ \omega^{ab} \downarrow & & \downarrow \omega^{ab} \\ X_0 & \xrightarrow{f} & Y_0 \end{array}$$

Moreover we have $F \circ \sigma = f_*(\sigma) \circ F$ for $\sigma \in H_1(X_0, \mathbb{Z})$.

Proof. Let $\hat{x}_0$ and $\hat{y}_0$ be vertices in X_0^{uni} and Y_0^{uni} lying on the fibers of x_0 and y_0 under the subcovering maps $\omega_1 : X_0^{uni} \longrightarrow X_0^{ab}$ and $\omega_1 : Y_0^{uni} \longrightarrow Y_0^{ab}$, respectively. We use Theorem 5.11 to find $\widehat{F} : X_0^{uni} \longrightarrow Y_0^{uni}$ such that $\widehat{F}(\hat{x}_0) = \hat{y}_0$ and $\omega^{uni} \circ \widehat{F} = f \circ \omega^{uni}$.

Since $\widehat{F} \circ \sigma = f_*(\sigma) \circ \widehat{F}$ for $\sigma \in \pi_1(X_0, x_0)$ and

$$f_*\big([\pi_1(X_0,x_0), \pi_1(X_0,x_0)]\big) \subset [\pi_1(Y_0,y_0), \pi_1(Y_0,y_0)],$$

we find, in view of Theorem 3.2, a morphism $F : X_0^{ab} \longrightarrow Y_0^{ab}$ such that $\omega_1 \circ \widehat{F} = F \circ \omega_1$. Evidently this F has the properties in the theorem. □

Based on the above theorem one can establish the following theorem whose proof is carried out in much the same way as that of the above theorem.

Theorem 6.5. *Let X, Y be abelian covering graphs over X_0, Y_0 corresponding to subgroups $H_X \subset H_1(X_0, \mathbb{Z})$, $H_Y \subset H_1(Y_0, \mathbb{Z})$ (so that $X = X_0^{ab}/H_X$, $Y = Y_0^{ab}/H_Y$). Let $f : X_0 \longrightarrow Y_0$ be a morphism with $f(x_0) = y_0$, and let $\hat{x}_0, \hat{y}_0$ be vertices in X, Y lying on the fibers of x_0 and y_0 under the covering maps $\omega_X : X \longrightarrow X_0$ and $\omega_Y : Y \longrightarrow Y_0$, respectively. If $f_*(H_X) \subset H_Y$, then there exists a unique morphism $F : X \longrightarrow Y$ such that $F(\hat{x}_0) = \hat{y}_0$ and $\omega_Y \circ F = f \circ \omega_X$.*

Let us examine the bipartiteness of an abelian covering graph. The following gives a criterion for the bipartiteness of $X = X_0^{ab}/H$ in terms of the homomorphism $\varepsilon_1 : H_1(X_0, \mathbb{Z}) \longrightarrow \mathbb{Z}_2$ (see Sect. 4.3):

X is bipartite if and only if $\varepsilon_1(\alpha) = 0$ for every $\alpha \in H$. In particular, the maximal abelian covering graph X_0^{ab} is always bipartite.

In order to check this it suffices to recall that any closed path in X is obtained by lifting a closed path in X_0 whose homology class is in H.

We close this section with the result concerning loop edges and parallel edges in abelian covering graphs.

Theorem 6.6. *Let $X = X_0^{ab}/H$ be an abelian covering graph over X_0.*

(1) X has a loop edge if and only if X_0 has a loop edge e such that $e \in H$ as a 1-cycle.
(2) X has parallel edges if and only if X_0 has parallel edges e_1, e_2 such that $e_1 - e_2 \in H$.

In particular, X_0^{ab} is a combinatorial graph, i.e., X_0^{ab} has neither loop edges nor parallel edges.

Indeed, the image of a loop edge e' (respectively parallel edges e_1', e_2') in X by the covering map $\omega : X \longrightarrow X_0$ is a loop edge (respectively parallel edges) in X_0. Theorem 6.2 applied to the closed path e' [respectively $(e_1', \overline{e_2'})$] proves the "only if" part. The converse is also shown by using Theorem 6.2.

6.2 Topological Crystals

A *topological crystal* X is an infinite-fold regular covering graph (possibly with loop edges or parallel edges) over a *finite graph*,[9] X_0, whose covering transformation group is a *free* abelian group. We call X a topological crystal *over* the *base graph*[10] X_0. The covering transformation group L will be called an *abstract period lattice*. The *dimension*[11] of X, symbolically $\dim X$, is defined to be the rank of L. The maximal abelian covering graph over X_0 is a topological crystal whose abstract period lattice is $H_1(X_0, \mathbb{Z})$, and will be called the *maximal topological crystal* over X_0.

From what we have mentioned in the preceding section, it is deduced that every topological crystal X over a fixed finite graph X_0 is obtained by choosing a subgroup H of $H_1(X_0, \mathbb{Z})$ such that the factor group $L = H_1(X_0, \mathbb{Z})/H$ is free abelian (or equivalently H is a direct summand of $H_1(X_0, \mathbb{Z})$; see Appendix 2). We shall call H a *vanishing subgroup*. An important fact we will use later is that one can choose a $\mathbb{Z}$-basis $\{\alpha_1, \ldots, \alpha_b\}$ ($b = b_1(X_0)$) of $H_1(X_0, \mathbb{Z})$ such that $\{\alpha_{d+1}, \ldots, \alpha_b\}$ ($d = \dim X$) is a $\mathbb{Z}$-basis of the vanishing subgroup H. Indeed, if we take homology classes $\alpha_1, \ldots, \alpha_d$ such that $\{\mu(\alpha_1), \ldots, \mu(\alpha_d)\}$ is a $\mathbb{Z}$-basis of L and also take a $\mathbb{Z}$-basis $\{\alpha_{d+1}, \ldots, \alpha_b\}$ of H, then $\{\alpha_1, \ldots, \alpha_b\}$ is a $\mathbb{Z}$-basis of $H_1(X_0, \mathbb{Z})$ (Appendix 2).

In particular, in oder to obtain nets of 3D (hypothetical) crystals from a finite graph X_0, we select a vanishing subgroup H of $H_1(X_0, \mathbb{Z})$ such that $H_1(X_0, \mathbb{Z})/H$ is of rank three. Although, in some cases, this construction may yield "pathological" crystal structures, we do not exclude this possibility (see Sect. 7.1).

In this connection, it should be pointed out that given an integer d with $1 \leq d < b_1(X_0)$, there are infinitely many vanishing groups H with rank $H_1(X_0, \mathbb{Z})/H = d$. This tells us that there are infinitely many d-dimensional topological crystals over X_0. Thus in order to enumerate crystal structures in a systematic way, we need to introduce a sort of "magnitude" of vanishing groups. Towards this end, we first define the norm[12] $\|\alpha\|_1$ of a 1-chain $\alpha = \sum_{e \in E_0^o} a_e e$ by setting

$$\|\alpha\|_1 = \sum_{e \in E_0^o} |a_e|,$$

where it should be noted that $\|\alpha\|_1$ does not depend on the choice of an orientation E_0^o. For a subset $S = \{\alpha_{d+1}, \ldots, \alpha_b\}$ of $H_1(X_0, \mathbb{Z})$ which forms a $\mathbb{Z}$-basis of a vanishing subgroup, we put

---

[9] From now on we tacitly assume that $\deg x \geq 2$ for every vertex x.

[10] A base graph is also called a *fundamental finite graph*. In crystallography, it is called a *quotient graph* (of a periodic graph).

[11] $\dim X$ does not depend on the choice of an abstract period lattice.

[12] Actually one may take any norm. The norm defined here is used to describe the large deviation property of random walks on topological crystals [Notes (III) in Chap. 9].

$$h(S) = \max(\|\alpha_{d+1}\|_1, \ldots, \|\alpha_b\|_1),$$

and define the *height* of H by $h(H) = \min_S h(S)$, where S runs over all $\mathbb{Z}$-bases of H. Consider two sets

$$R_1 = \{S|\, h(S) \leq h\}, \quad R_2 = \{H|\, h(H) \leq h\}.$$

Clearly R_1 is a finite set. The correspondence $S \mapsto H$ (generated by S) yields a surjective map of R_1 onto R_2. Thus we get:

Theorem 6.7. *For any positive number h and positive integer d with $1 \leq d < b_1(X_0)$, there are only a finite number of vanishing subgroups H of the homology group $H_1(X_0, \mathbb{Z})$ satisfying*

1. $\operatorname{rank} H_1(X_0, \mathbb{Z})/H = d$,
2. $h(H) \leq h$.

This theorem enables us to enumerate topological crystals in a systemetic way, at least in theory. More precisely, we first list finite graphs X_0 with $b_1(X_0) \leq b$ (Theorem 4.4), and then enumerate vanishing groups $H \subset H_1(X_0, \mathbb{Z})$ with $h(H) \leq h$.

We have an easy criterion for a topological crystal X over X_0 to be maximal.

Theorem 6.8. *X is maximal if and only if $\dim X = b_1(X_0)$.*

Indeed, for the subgroup $H = G(\omega_1)$ of $H_1(X_0, \mathbb{Z})$ corresponding to the sub-covering map $\omega_1 : X_0^{\mathrm{ab}} \longrightarrow X$ of $\omega^{\mathrm{ab}} : X_0^{\mathrm{ab}} \longrightarrow X_0$, if the rank of the factor group $H_1(X_0, \mathbb{Z})/H$ is $b_1(X_0)$, then $\operatorname{rank} H = 0$. Since H is free, we conclude $H = \{0\}$. For example, the *topological diamond* (the crystal structure of diamond) is a covering graph over the graph X_0 with two vertices joined by four parallel edges, and $b_1(X_0) = 3$. Hence it is the maximal topological crystal over X_0.

An obvious question that must be considered is what about the case when the covering transformation group L is not free. This being the case, we take the *torsion part*

$$\operatorname{Tor}(L) = \{\sigma \in L|\, n\sigma = 0 \text{ for a non-zero integer } n\},$$

which is known to be a finite subgroup of L such that $L/\operatorname{Tor}(L)$ is free abelian. Thus $\operatorname{Tor}(L)$ is a direct summand of L. Taking a subgroup L_1 of L such that $L = L_1 \oplus \operatorname{Tor}(L)$, we obtain an abelian covering map $\omega_1 : X \longrightarrow X_1 = X/L_1$ with the free abelian covering transformation group L_1. In short, changing the base graph, we can make X a topological graph in our sense.

We shall cite several examples of topological crystals. More examples together with their "canonical" realizations in space will be given in Sect. 8.3.

Example 6.2. The maximal topological crystal over the 3-bouquet graph is realized as the *cubic lattice*, the net associated with the 3-dimensional standard lattice $\mathbb{Z}^3$. Using the notation in Example 4.2 with $n = 3$, we find that the homology group

$$H_1(B_3, \mathbb{Z}) = \{a_1 e_1 + a_2 e_2 + a_3 e_3|\, a_i \in \mathbb{Z}\}$$

6.2 Topological Crystals

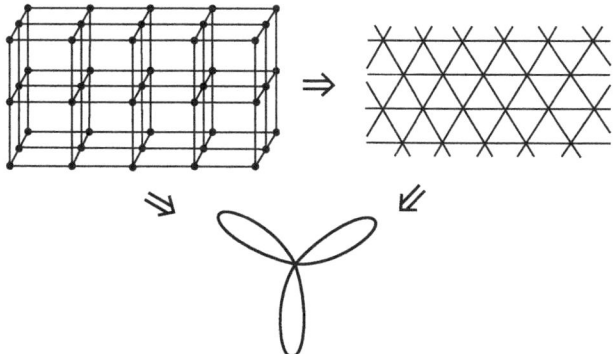

Fig. 6.5 The cubic lattice and the triangular lattice

acts on the set of vertices $\mathbb{Z}^3$ of the cubic lattice[13] by

$$(x_1, x_2, x_3) \mapsto (x_1 + a_1, x_2 + a_2, x_3 + a_3).$$

On the other hand, the *triangular lattice* is its subcovering graph. More precisely, we let H be the vanishing subgroup of $H_1(B_3, \mathbb{Z})$ generated by $e_1 + e_2 + e_3$, that is,

$$H = \mathbb{Z}(e_1 + e_2 + e_3)(= \{a(e_1 + e_2 + e_3)| a \in \mathbb{Z}\}).$$

The triangular lattice as a 2-dimensional topological crystal is the quotient graph of the cubic lattice by the H-action (Fig. 6.5).

More generally, the maximal topological crystal over the d-bouquet graph B_d is realized as the net associated with the d-dimensional standard lattice, a generalization of the square and cubic lattices,[14] which is also called the *hypercubic lattice*. To be exact, the set of vertices of the d-dimensional hypercubic lattice is $\mathbb{Z}^d$, and vertices adjacent to $(n_1, \ldots, n_d)$ are given by

$$(n_1 \pm 1, n_2, \ldots, n_d), (n_1, n_2 \pm 1, \ldots, n_d), \ldots, (n_1, n_2, \ldots, n_d \pm 1).$$

□

Example 6.3. Two 1-dimensional topological crystals in Fig. 6.6 are obtained from the 2-bouquet graph with loop edges e_1, e_2 by taking $\mathbb{Z}e_1$ and $\mathbb{Z}(e_1 + e_2)$ as a vanishing subgroup H, respectively. □

---

[13] By abuse of language, the term "cubic lattice" is often used to express the 3-dimensional standard lattice group. In crystallography, this, as a lattice group in $\mathbb{R}^3$, is called the *primitive cubic lattice*. A general cubic lattice is a lattice *commensurable* with the primitive cubic lattice. Here two lattice groups L_1 and L_2 of $\mathbb{R}^d$ are called commensurable if $L_1 \cap L_2$ are of finite index in both L_1 and L_2. See Appendix 2 for the definition of *subgroups of finite index*.

[14] The square lattice as a topological crystal is called the *quadrangle lattice*.

Fig. 6.6 One-dimensional topological crystals

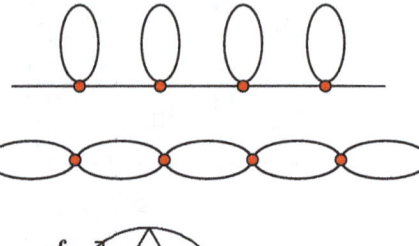

Fig. 6.7 The kagome lattice

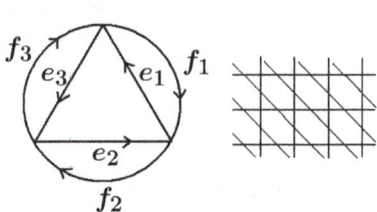

Example 6.4. A single-walled *carbon nanotube*[15] is obtained by rolling the graphene sheet in an appropriate way. As a topological crystal, it is one-dimensional, and is constructed from the hexagonal lattice X by choosing a non-zero element **a** of an abstract period lattice L of X; that is, $X_1 = X/\mathbb{Z}\mathbf{a}$ is the topological crystal associated with a carbon nanotube. □

Example 6.5. Consider the graph X_0 depicted in Fig. 6.7 (left). We set

$$\alpha_1 = e_1 + f_1, \quad \alpha_2 = e_2 + f_2, \quad \alpha_3 = e_1 + e_2 + e_3, \quad \alpha_4 = f_1 + f_2 + f_3.$$

Then $\{\alpha_1, \ldots, \alpha_4\}$ is a $\mathbb{Z}$-basis of $H_1(X_0, \mathbb{Z})$. We let H be the vanishing subgroup of $H_1(X_0, \mathbb{Z})$ generated by α_3 and α_4. The quotient graph $X = X_0^{\text{ab}}/H$ is what we call the (topological) *kagome lattice* (see also Fig. 8.1 in Sect. 8.3 giving the "standard" shape of the kagome lattice).

"Kagome" is a Japanese word meaning "woven-bamboo pattern". As the term expressing a lattice, it was first coined by Japanese physicist Koji Fushimi (1951). It is known that some minerals, for instance jarosites ($KFe_3^{3+}(OH)_6(SO_4)_2$) and herbertsmithite ($ZnCu_3(OH)_6Cl_2$), contain layers with the kagome lattice arrangement of atoms in their crystal structure. □

Example 6.6. Look at the graph X_0 depicted in Fig. 6.8 (left). We set $\alpha = f_1 + f_2 + g_1 + g_2$, and let H be the (cyclic) subgroup generated by α. Then X_0^{ab}/H is the two-dimensional topological crystal depicted on the right. The shape of the maximal topological crystal X_0^{ab} is given in Fig. 8.17 in Sect. 8.3. □

In the above examples, the realization of each covering graph is displayed without any explanation. The reader may wonder how we get it. As a matter of

---

[15] Carbon nanotubes are allotropes of carbon discovered by Sumio Iijima in 1991 and have remarkable properties, making them potentially useful in many applications in nanotechnology.

6.2 Topological Crystals

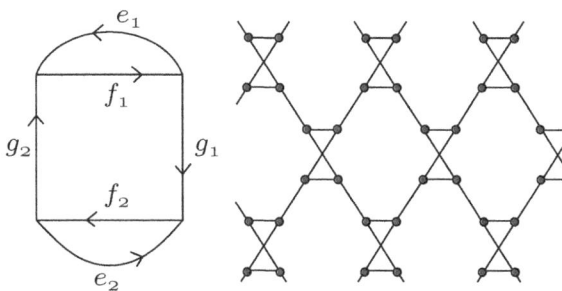

Fig. 6.8 A two-dimensional topological crystal

fact, it is not easy in general to derive the shape of a realization directly from the data of a topological crystal (for instance, the reader should ponder how to get the topological crystal in Fig. 6.8 out of the information of its vanishing group). This motivates us to look for a unified way to realize topological crystals in space, which is the main subject in the subsequent chapters.

The examples of crystal structures we have given so far have "small" base graphs (look back at diamond and Lonsdaleite as real crystals for instance). There are several classes of real crystals whose base graphs are considerably large. A typical one is the class of *zeolites*.[16] *Cage compounds* (polycyclic compounds having the form of a cage) with periodic structures also belong to such a class.

Finally we deal with the bipartiteness of topological crystals (see the previous section and Sect. 4.3). *A topological crystal $X = X_0^{\text{ab}}/H$ is bipartite if and only if the vanishing subgroup H is generated by $\alpha_1, \ldots, \alpha_k$ such that $\varepsilon_1(\alpha_i) = 0$*. In fact, expressing $\alpha \in H$ as $\sum_{i=1}^{k} n_i \alpha_i$, we get

$$\varepsilon_1(\alpha) = \sum_{i=1}^{k} n_i \varepsilon_1(\alpha_i) = 0.$$

The reader may look at the examples above to make sure that this criterion actually works.

Remark. It is not necessarily true that the composition $\omega = \omega_2 \circ \omega_1 : X \longrightarrow X_0$ of two abelian covering maps $\omega_1 : X \longrightarrow X_1$ and $\omega_2 : X_1 \longrightarrow X_0$ is an abelian covering map. For instance, the topological diamond is an abelian covering graph over the square lattice (this fact corresponds to the observation that when we take a look at a ready-made model of the diamond crystal, we can find a direction in which we see the square lattice; see Sect. 7.2), and the square lattice is the maximal topological crystal over the 2-bouquet graph, but the topological diamond cannot be a topological crystal over the 2-bouquet graph.

---

[16]Zeolites are microporous, aluminosilicate minerals commonly used as commercial adsorbents.

One more remark: a subcovering map $\omega_1 : X_0^{ab} \longrightarrow X_1$ of the maximal covering map $\omega : X_0^{ab} \longrightarrow X_0$ is not necessarily the maximal abelian covering map over X_1. What we can say is that there is a covering map $\omega_{10} : X_1^{ab} \longrightarrow X_0^{ab}$ such that the composition $\omega_1 \circ \omega_{10} : X_1^{ab} \longrightarrow X_1$ is the maximal abelian covering map over X_1. □

> **Summary.** A d-dimensional *topological crystal* is a regular covering graph over a finite graph, say X_0, with a free abelian covering transformation group L isomorphic to $\mathbb{Z}^d$. The finite graph X_0 is said to be the *base graph*. The group L is said to be the *abstract period lattice*. Any topological crystal over X_0 is obtained from a (vanishing) subgroup H of the homology group $H_1(X_0, \mathbb{Z})$ such that $H_1(X_0, \mathbb{Z})/H$ is free abelian. The factor group $H_1(X_0, \mathbb{Z})/H$ is identified with the abstract period lattice. The dimension of the topological crystal corresponding to H is equal to $\operatorname{rank} H_1(X_0, \mathbb{Z})/H$.

6.3 Symmetry of Topological Crystals

As Hermann Weyl (1885–1955) said, a guiding principle in modern mathematics is the lesson: *Whenever you have to do with a structure-endowed entity Σ, try to determine its group of automorphisms.*

According to this guiding principle, we shall pay attention to automorphisms of a topological crystal; especially we look at the relationship between them and automorphisms of its base graph.

Let X be a topological crystal over X_0, and let $H \subset H_1(X_0, \mathbb{Z})$ be the vanishing subgroup corresponding to X. We define the subgroup $\operatorname{Aut}_H(X_0)$ of $\operatorname{Aut}(X_0)$ by setting

$$\operatorname{Aut}_H(X_0) = \{g_0 \in \operatorname{Aut}(X_0) | \ g_{0*}(H) \subset H\}.$$

We also define the subgroup $\operatorname{Aut}(X/X_0)$ of $\operatorname{Aut}(X)$ by

$$\operatorname{Aut}(X/X_0) = \{g \in \operatorname{Aut}(X) | \text{ there is } g_0 \in \operatorname{Aut}(X_0)$$
$$\text{such that } \omega \circ g = g_0 \circ \omega\},$$

where $\omega : X \longrightarrow X_0$ is the covering map. The correspondence $g \mapsto g_0$ yields a homomorphism

$$p : \operatorname{Aut}(X/X_0) \longrightarrow \operatorname{Aut}(X_0).$$

The abstract period lattice L corresponding to the vanishing group H is a subgroup of $\operatorname{Aut}(X/X_0)$ (remember that L is the covering transformation group of $\omega : X \longrightarrow X_0$). Let $i : L \longrightarrow \operatorname{Aut}(X/X_0)$ be the inclusion map.

Theorem 6.9. *1. Image $p = \operatorname{Aut}_H(X_0)$, and $\operatorname{Ker} p = \operatorname{Image} i$.*

6.3 Symmetry of Topological Crystals

2. $\mathrm{Aut}(X/X_0)$ coincides with the normalizer[17] of L, i.e.,

$$\mathrm{Aut}(X/X_0) = \{g \in \mathrm{Aut}(X) | \, gLg^{-1} = L\}.$$

Proof. 1. Since $\sigma \in \mathrm{Ker}\, p$ is a covering transformation, the proof of the assertion $\mathrm{Ker}\, p = \mathrm{Image}\, i$ is trivial. What we have to prove is that $\mathrm{Image}\, p = \mathrm{Aut}_H(X_0)$. But this is a consequence of Theorems 6.3 and 6.5. Indeed, for any $g \in \mathrm{Aut}(X/X_0)$, if we put $p(g) = g_0$, then in view of Theorem 6.3, we have

$$g_{0*}(H) = g_{0*}(\omega_*(H_1(X,\mathbb{Z}))) = \omega_* g_*(H_1(X,\mathbb{Z})) = \omega_*(H_1(X,\mathbb{Z})) \subset H,$$

which implies $g_0 \in \mathrm{Aut}_H(X_0)$, whence $\mathrm{Image}\, p \subset \mathrm{Aut}_H(X_0)$. Conversely, for $g_0 \in \mathrm{Aut}_H(X_0)$, one can take two morphisms $g, h : X \longrightarrow X$ such that $\omega \circ g = g_0 \circ \omega$ and $\omega \circ h = g_0^{-1} \circ \omega$ (Theorem 6.5). Then

$$\omega \circ h \circ g = g_0^{-1} \circ \omega \circ g = g_0^{-1} \circ g_0 \circ \omega = \omega.$$

Likewise $\omega \circ g \circ h = \omega$; therefore both $g \circ h$ and $h \circ g$ are covering transformations. So g is an automorphism and belongs to $\mathrm{Aut}(X/X_0)$. We thus conclude $g_0 = p(g) \in \mathrm{Image}\, p$.

2. Since $L = \mathrm{Ker}\, p$ is a normal subgroup of $\mathrm{Aut}(X/X_0)$, it is enough to prove that any $g \in \mathrm{Aut}(X)$ with $gLg^{-1} = L$ belongs to $\mathrm{Aut}(X/X_0)$. If $\omega(x) = \omega(x')$, there exists $\sigma \in L$ such that $x' = \sigma x$. Since $g\sigma g^{-1}$ is in L, putting $\sigma' = g\sigma g^{-1}$, we have $gx' = g\sigma x = \sigma' gx$ and $\omega(gx') = \omega(gx)$. Likewise $\omega(ge') = \omega(ge)$ for $e, e' \in E$ with $\omega(e') = \omega(e)$. Thus the correspondences $\omega(x) \mapsto \omega(gx)$ and $\omega(e) \mapsto \omega(ge)$ are well defined, and yield a morphism g_0 of X_0 into itself. It is checked that $g_0 \in \mathrm{Aut}(X_0)$. From the definition of g_0, it follows that $g_0 \circ \omega = \omega \circ g$, and hence $g \in \mathrm{Aut}(X/X_0)$. □

In general, given two homomorphisms $i : H \longrightarrow G$ and $p : G \longrightarrow N$ of groups, if

(a) i is injective,
(b) p is suejective,
(c) $\mathrm{Ker}\, p = \mathrm{Image}\, i$,

then we say that

$$1 \longrightarrow H \xrightarrow{i} G \xrightarrow{p} N \longrightarrow 1$$

is an *exact sequence*. Therefore what the theorem asserts is that

$$0 \longrightarrow L \xrightarrow{i} \mathrm{Aut}(X/X_0) \xrightarrow{p} \mathrm{Aut}_H(X_0) \longrightarrow 1. \tag{6.3}$$

is an exact sequence.

---

[17] See Appendix 2.

From (6.3), it follows that the factor group $\mathrm{Aut}(X/X_0)/L$ is isomorphic to the finite group $\mathrm{Aut}_H(X_0)$, and hence $\mathrm{Aut}(X/X_0)$ is *virtually abelian*. Here in general, a group G is said to be virtually abelian if there is a normal abelian subgroup H of finite index.[18]

This fact raises the question "Is $\mathrm{Aut}(X/X_0)$ isomorphic to a d-dimensional *crystallographic group*[19]?" This question is natural because we are dealing with an abstract model of a crystal, and the concept of crystallographic group was derived from studying the (macroscopic) symmetry of crystals.

The characterization of crystallographic groups due to L. Auslander and M. Kuranishi says that $\mathrm{Aut}(X/X_0)$ is isomorphic to a crystallographic group if the abstract period lattice L is *maximal* among all abelian subgroups of $\mathrm{Aut}(X/X_0)$ in the sense that there is no abelian subgroup of $\mathrm{Aut}(X/X_0)$ properly containing L (see Appendix 5). Thus we shall ask whether L is maximal or not.

For each $g \in \mathrm{Aut}(X/X_0)$, define $g_L \in \mathrm{Aut}(L)$ by setting $g_L(\sigma) = g\sigma g^{-1}$. Since $g_L = I$ for $g \in L$, the homomorphism $g \mapsto g_L$ induces a homomorphism φ of $\mathrm{Aut}_H(X_0) = \mathrm{Aut}(X/X_0)/L$ into $\mathrm{Aut}(L)$ (the first isomorphism theorem).

Theorem 6.10. *If the homomorphism $\varphi : \mathrm{Aut}_H(X_0) \longrightarrow \mathrm{Aut}(L)$ is injective, then L is maximal, and hence $\mathrm{Aut}(X/X_0)$ is isomorphic to a crystallographic group. Conversely, if L is maximal, then φ is injective.*

Proof. Suppose that $g \in \mathrm{Aut}(X/X_0)$ commutes with every element of L. Then $g_L = I$. By injectivity of φ, we conclude that $g \in L$. Thus an abelian subgroup of $\mathrm{Aut}(X/X_0)$ containing L must coincide with L.

Conversely suppose that L is maximal, and that there is an element $g \in \mathrm{Aut}(X/X_0)\backslash L$ with $g_L = 1$. Then $g\sigma = \sigma g$ for every $\sigma \in L$ so that the subgroup L' generated by g and L is an abelian group containing properly L. This is a contradiction. □

We shall show
$$g_L(\mu(\alpha)) = \mu(g_{0*}(\alpha)) \quad (g \in \mathrm{Aut}(X/X_0),\ g_0 = p(g),\ \alpha \in H_1(X_0,\mathbb{Z})).$$

Towards this end, take a closed path c representing α, and let $\hat{c}$ be a lifting of c in X. Since $t(\hat{c}) = \mu(\alpha)o(\hat{c})$, we find that
$$t(g(\hat{c})) = gt(\hat{c}) = g\mu(\alpha)o(\hat{c}).$$

Noting that $g(\hat{c})$ is a lifting of $g_0(c)$, and that $g_{0*}(\alpha)$ is represented by the closed path $g_0(c)$, we obtain

---

[18] See Appendix 2.

[19] A d-dimensional crystallographic group is a discrete subgroup, say Γ, of the motion group $\mathbf{M}(d)$ of $\mathbb{R}^d$ with compact quotient $\mathbf{M}(d)/\Gamma$ (see Appendix 5). Crystallographic groups are virtually abelian, but not vice versa.

$$t(g(\hat{c})) = \mu(g_{0*}(\alpha))o(g(\hat{c})) = \mu(g_{0*}(\alpha))go(\hat{c}).$$

Hence $g\mu(\alpha)o(\hat{c}) = \mu(g_{0*}(\alpha))go(\hat{c})$. This implies that $g\mu(\alpha)x = \mu(g_{0*}(\alpha))gx$ for every $x \in V$, and $g\mu(\alpha)g^{-1}x = \mu(g_{0*}(\alpha))x$ ($x \in V$). Both $g\mu(\alpha)g^{-1}$ and $\mu(g_{0*}(\alpha))$ belong to L, and L acts freely on V; thereby $g\mu(\alpha)g^{-1} = \mu(g_{0*}(\alpha))$.

Let $X_0 = (V_0, E_0)$ be a finite graph with no separating edges. Assume that X_0 is not isomorphic to the circuit graph. As proved in Sect. 4.5, the correspondence

$$g \in \mathrm{Aut}(X_0) \longmapsto g_* \in \mathrm{Aut}(H_1(X_0, \mathbb{Z}))$$

is an injective homomorphism (Theorem 4.5). What we have seen above says that this homomorphism coincides with $\varphi : \mathrm{Aut}_H(X_0) \longrightarrow \mathrm{Aut}(L)$ for $H = \{0\}$ and $L = H_1(X_0, \mathbb{Z})$. This implies that $H_1(X_0, \mathbb{Z})$ is maximal as an abelian subgroup of $\mathrm{Aut}(X_0^{\mathrm{ab}}/X_0)$, and hence $\mathrm{Aut}(X_0^{\mathrm{ab}}/X_0)$ is isomorphic to a crystallographic group.

In Sect. 7.6, we will see that $\mathrm{Aut}(X/X_0)$ is related to a crystallographic group in a different manner. The structure of the full automorphism group $\mathrm{Aut}(X)$, which could be much more subtle than $\mathrm{Aut}(X/X_0)$, will be discussed in Sect. 9.6 in connection with a random walk on X.

Remark. If we know beforehand that $\mathrm{Aut}(X)$ is a crystallographic group, and that L is its maximal abelian subgroup of finite index, then we find that $\mathrm{Aut}(X/X_0) = \mathrm{Aut}(X)$. This is so because L is normal in $\mathrm{Aut}(X)$. Using this fact, we will show in Sect. 10.3 that if X_0 has no separating edges, then $\mathrm{Aut}(X_0^{\mathrm{ab}}) = \mathrm{Aut}(X_0^{\mathrm{ab}}/X_0)$ (Theorem 10.5).

6.4 Notes

(I) It is fun to look into a ready-made crystal model. Its tricky structure stimulates our thought and imagination, and may arouse a new issue. For instance, a glance at the model of the diamond crystal[20] reveals that it has infinitely many closed geodesics with different lengths. A question related to this observation is whether this is true for a general topological crystal, especially for the maximal topological crystal X_0^{ab} over a finite graph X_0.

Note that a closed geodesic in X_0^{ab} is a lifting of a closed geodesic in X_0 whose homology class is zero (Theorem 6.2). Thus if we regard a closed geodesic in X_0^{ab} in the same light as another when they are transformed into each other by a covering transformation, then counting closed geodesics in X_0^{ab} reduces to counting closed geodesics c in X_0 with $\langle c \rangle = 0$.

To be exact, we consider the set $\mathcal{P}$ consisting of *prime cycles* in a finite graph X_0 with $b_1(X_0) \geq 2$. Here a prime cycle is an equivalence class of a closed geodesic

---

[20]The diamond crystal is a join of the same hexagonal rings (called also the chair conformation in chemical terminology).

which is not a power[21] of another one, and two closed paths are *equivalent* if one is obtained by a cyclic permutation of edges in another. The answer to the above question is simply stated as $\left|\{\mathfrak{p} \in \mathcal{P}|\ [\mathfrak{p}] = 0\}\right| = \infty$, where $[\mathfrak{p}]$ stands for the homology class of $\mathfrak{p}$. This, of course, does not surprise us.

More intriguing is the fact that any homology class $\alpha \in H_1(X_0, \mathbb{Z})$ contains infinitely many prime cycles. In this fact the reader might see an analogy with the Dirichlet theorem for arithmetic progression, which asserts that given coprime positive integers a, d there are infinitely many primes in the arithmetic progression $a, a+d, a+2d, \ldots$. More precisely, in terms of the *Euler function* $\varphi(d)$, i.e., the number of positive integers not greater than and coprime to d, one has

$$\left|\{\text{prime } p \leq x|\ p = a+kd \text{ for some } k\}\right| \sim \varphi(d)^{-1} \frac{x}{\log x} \quad (x \uparrow \infty). \tag{6.4}$$

The proof of this density theorem relies on the properties of the *L*-function

$$L(s,\chi) = \prod_p (1-\chi(p)p^{-s})^{-1},$$

where χ is a unitary character of the multiplicative group $(\mathbb{Z}/d\mathbb{Z})^\times$, which is the Galois group of the cyclotomic field $\mathbb{Q}(e^{2\pi\sqrt{-1}/d})$ (note that $|(\mathbb{Z}/d\mathbb{Z})^\times| = \varphi(d)$).

One can consider an analogue of the *L*-function in the graph theory. Indeed, for a unitary character χ of the homology group $H_1(X_0, \mathbb{Z})$, the graph-theoretic *L*-function is defined by

$$L(u,\chi) = \prod_{\mathfrak{p} \in \mathcal{P}} \left(1-\chi([\mathfrak{p}])u^{|\mathfrak{p}|}\right)^{-1}. \tag{6.5}$$

Using analytic properties of this function, one may prove that there exist positive constants $h > 1$ and C not depending on on α such that

$$\left|\{\mathfrak{p}|\ [\mathfrak{p}] = \alpha,\ |\mathfrak{p}| \leq n\}\right| \sim C\frac{h^n}{n^{b/2+1}} \quad (n \uparrow \infty),$$

where $b = b_1(X_0)$ (cf. [54, 57]). This is an analogue of (6.4). Thus, what we said above is not a joke (yet it might seem a surprise that totally different disciplines are linked via a similar underlying idea).

The graph-theoretic *L*-function is what we call the *Ihara zeta function*, which was originally introduced by Ihara [51] in 1966 in his study of co-compact torsion-free discrete subgroups of the *p*-adic special linear group $SL_2(\mathbb{Q}_p)$, and was interpreted in terms of finite graphs by the author [89] following the suggestion stated in the preface of Serre [84]. See Bass [8], Sunada [90, 95], Terras [101] for more about Ihara zeta functions.

---

[21] For a closed path c, the product $c \cdots c$ (*m* times) is called the *m*-th *power* of c.

6.4 Notes

(II) Theorem 6.3 says, in paticular, that for an n-fold abelian covering map $\omega : X \longrightarrow X_0$ over a finite graph X_0, the following equality holds:

$$\left| H_1(X_0, \mathbb{Z}) / \omega_*(H_1(X, \mathbb{Z})) \right| = n.$$

Related to this fact is:

Theorem 6.11. *Let $\omega : X \longrightarrow X_0$ be an n-fold covering map over a finite graph X_0. Then*

$$\left| H_1(X_0, \mathbb{Z}) / \omega_*(H_1(X, \mathbb{Z})) \right| \leq n.$$

The equality holds if and only if ω is an abelian covering map.

This theorem is a consequence of the following commutative diagram.

$$\begin{array}{ccccc}
\pi_1(X) & \xrightarrow{\omega_*} & \pi_1(X_0) & \longrightarrow & \pi_1(X_0)/\omega_*(\pi_1(X)) \\
{\scriptstyle h}\downarrow & & {\scriptstyle h}\downarrow & & {\scriptstyle h}\downarrow \\
H_1(X,\mathbb{Z}) & \xrightarrow{\omega_*} & H_1(X_0,\mathbb{Z}) & \longrightarrow & H_1(X_0,\mathbb{Z})/\omega_*(H_1(X,\mathbb{Z})).
\end{array}$$

In this diagram the map h represented by the rightmost vertical arrow is the one induced from the Hurewicz homomorphism, and is surjective.

Theorem 6.11 can be interpreted as a covering-theoretic analogue of a statement in the fundamental theorem of *class field theory*,[22] which somehow offers a "moral support" for the Ihara zeta function introduced in (I). To explain this analogy briefly, we introduce the free abelian group I_X generated by all prime cycles in a finite graph X. If we use the multiplicative notation, then each element $\mathfrak{a}$ in I_X is uniquely expressed as $\mathfrak{a} = \mathfrak{p}_1^{a_1} \cdots \mathfrak{p}_g^{a_g}$, where $\mathfrak{p}_i$ is a prime cycle and $a_i \in \mathbb{Z}$ (what we have in mind here is Dedekind's theorem on prime ideal factorizations). The subgroup I_X^0 consisting of $\mathfrak{a} \in I_X$ with $\mathfrak{a} = 0$ as a 1-cycles in X is an analogue of the group of principal ideals. Obviously I_X/I_X^0 is isomorphic to $H_1(X,\mathbb{Z})$. The abelian groups I_X and I_X/I_X^0 are the graph-theoretic analogues of *ideal groups* and *ideal class groups*, respectively.

Let $\mathfrak{P}$ be a prime cycle in X represented by a prime closed geodesic $\hat{c}$. For an n-fold covering map $\omega : X \longrightarrow X_0$, the image $\omega(\hat{c})$ is a power of a prime closed geodesic, say $\omega(\hat{c}) = c^m$. The equivalence class $\mathfrak{p}$ of c and the positive integer m do not depend on the choice of $\hat{c}$ representing $\mathfrak{P}$. We define $N_\omega(\mathfrak{P})$ to be $\mathfrak{p}^m$, and

---

[22] The fundamental theorem characterizes abelian extensions of number fields in terms of ideal class groups (see [71]). The idea of class field theory in a geometric setting was stated in [87].

extend N_ω to a homomorphism of I_X into I_{X_0}. Evidently the following diagram is commutative:

$$\begin{array}{ccc} I_X & \xrightarrow{N_\omega} & I_{X_0} \\ \downarrow & & \downarrow \\ H_1(X,\mathbb{Z}) & \xrightarrow[\omega_*]{} & H_1(X_0,\mathbb{Z}), \end{array}$$

where the vertical arrows represent the canonical homomorphisms. Furthermore we can show that $I_{X_0}/\bigl(I_{X_0}^0 \cdot N_\omega(I_X)\bigr)$ is isomorphic to

$$H_1(X_0,\mathbb{Z})/\omega_*\bigl(H_1(X,\mathbb{Z})\bigr).$$

Thus we have the following statement in which one can see a direct analogy with the one in the fundamental theorem of class field theory:

For any covering map $\omega : X \longrightarrow X_0$, we have $\bigl|I_{X_0}/\bigl(I_{X_0}^0 \cdot N_\omega(I_X)\bigr)\bigr| \leq n$. The equality holds if and only if ω is abelian.

The reader might wonder why such a topic irrelevant to crystallography was presented here. The reason is that author's view of crystallography is, to a large extent, motivated by an analogy with number theory. Actually this analogy gives rise to the inducement to look for a canonical way to realize topological crystals in a space. See Notes (VI) in Chap. 8 for more about author's way of thinking.

Chapter 7
Standard Realizations

We have so far interpreted how elementary algebraic topology can be used for a set-up of topological crystallography. One can say that our discussion is perfect on the basis of pure reflection. Namely, the idea of topological crystals as a product of conceptual thought satisfies enough the people working in pure mathematics. For practical purposes, however, this situation is unsatisfactory because a topological crystal does not inhabit real space, so it is not visible even for the 3D case if we would leave it intact.[1] For this reason, it is a natural desire to find an "actual shape" of a topological crystal.

The issue of topological crystals becomes more interesting if we look for the most natural way to place them in space. For instance, the crystal net of diamond is one of the realizations of the "topological diamond" in space. Figure 7.1 illustrates another realization.[2] Thus a natural question is: "what are the characteristics possessed by the diamond crystal net that differ from all other realizations of the topological diamond?"

The crystal net of diamond is more symmetric than the graphite-like realization. This observation motivates us to seek the most symmetric realization in general. This chapter will give, in an accessible manner, a detailed account of the concept of *standard realization* proposed in [58] as a candidate for the most symmetric and canonical realization. For this purpose, we shall define "energy" of a crystal net (per unit cell) with which we set up a minimal principle. Standard realizations are characterized as energy-minimizers. The energy we introduce in Sect. 7.4 is unrealistic from a physical point of view, but very amenable to mathematical treatment. Thus our approach is, as mentioned in the introduction, rather unconventional

---

[1] Hardy says in *A Mathematician's Apology* (Cambridge University Press, 1967) "I believe that mathematical reality lies outside us, that our function is to discover or observe it, and that the theorems which we prove, and which we describe grandiloquently as our 'creations', are simply our notes of our observations".

[2] It is interesting to compare this realization with the graphite-like realization of Lonsdaleite in Fig. 3.10". The reader may be aware of a subtle difference between them.

Fig. 7.1 The graphite-like realization of the topological diamond

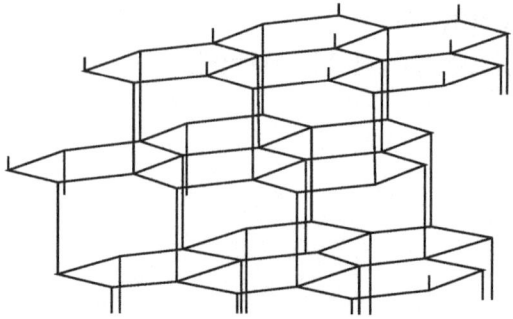

in that we focus only on the geometric shapes of crystal nets (namely, we are not going to explain the physical and chemical reason "why diamond is diamond?"). In this sense we follow the Greek tradition in seeking beautiful shapes such as regular convex polyhedra, away from the curiosity about crystals themselves.

The reader might think that we are going to digress largely from the topological theory because what we shall deal with are the "metric" aspects of crystal structures. However, our argument relies heavily on homology theory, and hence is within the framework of topological crystallography.

From now on the reader is expected to have a mastery of matrix theory. But a few notations are explained here for convenience: We express a matrix A by (a_{ij}) for simplicity when the $(i,j)^{\text{th}}$ entry of A is a_{ij}. The transpose of $A = (a_{ij})$ expressed by ${}^t A$ is the matrix whose $(i,j)^{\text{th}}$ entry is a_{ji}. The *trace* of a square matrix A, the sum of the diagonal entries of A, is denoted by tr A. Remember the basic facts about trace and determinant: tr $AB =$ tr BA, det$AB =$ det A det B, det${}^t A =$ det$A$.

We also freely use abstract linear algebra in the discussion. See Appendix 3 for the terminology such as vector spaces, linear operators, and inner products. When we think of $\mathbb{R}^d$ as a vector space, we express $\mathbf{x} \in \mathbb{R}^d$ by a column vector, say $\mathbf{x} = {}^t(x_1,\ldots,x_d)$. The (standard) inner product $\langle \mathbf{x}, \mathbf{y} \rangle$ of two vectors $\mathbf{x} = {}^t(x_1,\ldots,x_d)$, $\mathbf{y} = {}^t(y_1,\ldots,y_d)$ is ${}^t\mathbf{x}\mathbf{y} = x_1 y_1 + \cdots + x_n y_n$. The norm $\|\mathbf{x}\|$ is $\langle \mathbf{x}, \mathbf{x} \rangle^{1/2}$.

7.1 Periodic Realizations

Suppose we are given a crystal in space. A crystal as a network in space is thought of as the image of a map Φ from an abstract graph X into $\mathbb{R}^3$, where the image of each edge is a line segment in $\mathbb{R}^3$. As mentioned at the beginning of the previous chapter, the periodic nature of the crystal is reflected in the fact that X is a topological crystal on which the period lattice L acts as automorphisms, and this action is identified with the translational action on the crystal net via Φ. In short, the map Φ satisfies $\Phi(\sigma x) = \Phi(x) + \sigma$ for every $\sigma \in L$ and vertex x.

7.1 Periodic Realizations

In this section, motivated by this observation, we shall make the process the other way around. Namely, we start with a topological crystal, and then realize it in space as a crystal net.

Recall that a general graph $X = (V,E)$ is identified with a one-dimensional cell complex; therefore each edge e is identified with the interval $[0,1]$ (Sect. 3.1). A map Φ of X into $\mathbb{R}^d$ is said to be *piecewise linear* provided that the restriction $(\Phi|e)(t)$ $(0 \leq t \leq 1)$ is a linear function in the variable t with values in $\mathbb{R}^d$ for every $e \in E$ and $(\Phi|\bar{e})(t) = (\Phi|e)(1-t)$ $((\Phi|e)(t)$ is possibly constant).

Let $X = (V,E)$ be a d-dimensional topological crystal with an abstract period lattice L, and let $X_0 = (V_0, E_0)$ be its base graph. We also let $\omega : X \longrightarrow X_0$ be the covering map, and let $\mu : H_1(X_0, \mathbb{Z}) \longrightarrow L$ be the homomorphism introduced in Sect. 6.1. A piecewise linear map $\Phi : X \longrightarrow \mathbb{R}^d$ is said to be a *periodic realization*[3] if there exists an injective homomorphism $\rho : L \longrightarrow \mathbb{R}^d$ satisfying

1. $\Phi(\sigma x) = \Phi(x) + \rho(\sigma)$ $(x \in V,\ \sigma \in L)$.
2. $\rho(L)$ is a lattice subgroup of $\mathbb{R}^d$.

Condition (1) means that Φ is L-equivariant. We call ρ and $\rho(L)$ the *period homomorphism* and *period lattice* for Φ, respectively. We occasionally write (Φ, ρ) to express a periodic realization even though ρ is uniquely determined by Φ. By abuse of language, we call the image $\Phi(X)$ a crystal net or simply a crystal.

For a periodic realization Φ, the vector

$$\mathbf{v}(e) = \Phi\bigl(t(e)\bigr) - \Phi\bigl(o(e)\bigr) \quad (e \in E),$$

satisfies $\mathbf{v}(\sigma e) = \mathbf{v}(e)$, which implies that $\mathbf{v}$ is identified with an $\mathbb{R}^d$-valued function $\mathbf{v}_0$ on E_0 with $\mathbf{v}_0(\bar{e}) = -\mathbf{v}_0(e)$, i.e., $\mathbf{v}_0(\omega(e)) = \mathbf{v}(e)$ for $e \in E$ (occasionally we use the symbol $\mathbf{v}(e)$ for $\mathbf{v}_0(\omega(e))$). The system of vectors $\{\mathbf{v}_0(e)\}_{e \in E_0}$, which is the precise formulation of vector assignments stated in Example 3.4, is a *building block*[4] of the realization Φ in the sense that it completely determines Φ if we fix the image $\Phi(x_0)$ of a reference point $x_0 \in V$. Indeed, for any vertex $x \in V$, by joining x_0 and x by a path $c = (e_1, \ldots, e_n)$, we have

$$\begin{aligned}\Phi(x) &= \Phi(x_0) + \mathbf{v}(e_1) + \cdots + \mathbf{v}(e_n) \\ &= \Phi(x_0) + \mathbf{v}_0\bigl(\omega(e_1)\bigr) + \cdots + \mathbf{v}_0\bigl(\omega(e_n)\bigr).\end{aligned}$$

Figure 7.2 exhibits the building block of the dice lattice.

The period homomorphism ρ is expressed in terms of the building block $\{\mathbf{v}_0(e)\}_{e \in E_0}$ as follows. For a 1-chain $\alpha = \sum_{e \in E_0} a_e e \in C_1(X_0, \mathbb{Z})$, we set

---

[3] In crystallography, the term "placement" (or "drawing") is used.

[4] In terms of cohomology theory, the function $\mathbf{v}_0$ on E_0 is an $\mathbb{R}^d$-valued 1-cochain of X_0 [see Notes (I) in Chap. 4]. In crystallography, a finite graph together with a building block is referred to as a (vector) *labeled quotient graph*.

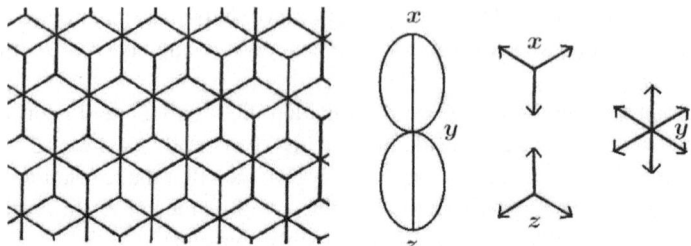

Fig. 7.2 Building block of the dice lattice

$$\widehat{\mathbf{v}}_0(\alpha) = \sum_{e \in E_0} a_e \mathbf{v}_0(e), \tag{7.1}$$

which gives a well-defined homomorphism of $C_1(X_0, \mathbb{Z})$ into $\mathbb{R}^d$ because of the equality $\mathbf{v}_0(\bar{e}) = -\mathbf{v}_0(e)$. For $\alpha \in H_1(X_0, \mathbb{Z})$, we find that

$$\rho(\mu(\alpha)) = \widehat{\mathbf{v}}_0(\alpha), \tag{7.2}$$

that is, the following diagram is commutative.

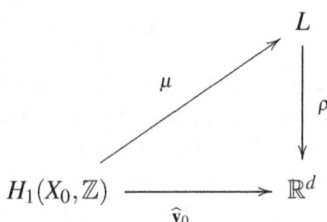

Actually, joining x_0 and σx_0 ($\sigma \in L$) by a path $c' = (f_1, \ldots, f_m)$ and also joining x_0 and x by a path $c = (e_1, \ldots, e_n)$, we have

$$\begin{aligned}
\Phi(\sigma x) &= \Phi(x_0) + \mathbf{v}_0(\omega(f_1)) + \cdots + \mathbf{v}_0(\omega(f_m)) \\
&\quad + \mathbf{v}_0(\omega(\sigma e_1)) + \cdots + \mathbf{v}_0(\omega(\sigma e_n)) \\
&= \Phi(x) + \mathbf{v}_0(\omega(f_1)) + \cdots + \mathbf{v}_0(\omega(f_m)),
\end{aligned}$$

where we note that $\mathbf{v}_0(\omega(\sigma e_i)) = \mathbf{v}_0(\omega(e_i))$, and $c' \cdot \sigma c$ is a path joining x_0 and σx. Thus $\rho(\sigma) = \widehat{\mathbf{v}}_0(\langle \omega(c') \rangle)$. Recalling that $\mu(\langle \omega(c') \rangle) = \sigma$ [see (6.1)], we get (7.2).

Equation (7.2) tells us that

$$\mathrm{Ker}\,(\widehat{\mathbf{v}}_0 : H_1(X_0, \mathbb{Z}) \longrightarrow \mathbb{R}^d) = \mathrm{Ker}\,\mu$$

7.1 Periodic Realizations

because ρ is injective, and that Image $\widehat{\mathbf{v}}_0 = \{\widehat{\mathbf{v}}_0(\alpha) | \ \alpha \in H_1(X_0, \mathbb{Z})\}$ coincides with the period lattice $\rho(L)$ because μ is surjective. Moreover, $\{\mathbf{v}_0(e)\}_{e \in E_0}$ spans $\mathbb{R}^d$; that is, every vector of $\mathbb{R}^d$ is expressed as a linear combination of $\{\mathbf{v}_0(e)\}_{e \in E_0}$.

One may ask the question: Can a general system of vectors $\{\mathbf{v}_0(e)\}_{e \in E_0}$ in $\mathbb{R}^d$ satisfying $\mathbf{v}_0(\bar{e}) = -\mathbf{v}_0(e)$ ($e \in E_0$) be the building block for a periodic realization Φ of a certain topological crystal over X_0?

To settle this question we consider the homomorphism

$$\widehat{\mathbf{v}}_0 : H_1(X_0, \mathbb{Z}) \longrightarrow \mathbb{R}^d$$

defined by (7.1), and put $H = \text{Ker}\,\widehat{\mathbf{v}}_0$.

Theorem 7.1. *If* Image $\widehat{\mathbf{v}}_0$ *is a lattice group in* $\mathbb{R}^d$, *then* $\{\mathbf{v}_0(e)\}_{e \in E_0}$ *is the building block for a periodic realization* Φ *of the topological crystal* $X_0^{\text{ab}}/H = (V, E)$.

Proof. Recall that the abstract period lattice for X_0^{ab}/H is $L = H_1(X_0, \mathbb{Z})/H$. The homomorphism $\mu : H_1(X_0, \mathbb{Z}) \longrightarrow L$ associated with the covering map $\omega : X_0^{\text{ab}} \longrightarrow X_0$ coincides with the canonical homomorphism. Since $H = \text{Ker}\,\widehat{\mathbf{v}}_0$, the homomorphism $\widehat{\mathbf{v}}_0 : H_1(X_0, \mathbb{Z}) \longrightarrow \mathbb{R}^d$ induces an injective homomorphism $\rho : L \longrightarrow \mathbb{R}^d$ such that $\rho \circ \mu = \widehat{\mathbf{v}}_0$. Here, by assumption, $\rho(L) = \text{Image}\,\widehat{\mathbf{v}}_0$ is a lattice group in $\mathbb{R}^d$ (which is going to be the period lattice). A candidate of Φ is, of course, given by

$$\Phi(x_0) = \mathbf{0},$$
$$\Phi(x) = \mathbf{v}_0(\omega(e_1)) + \cdots + \mathbf{v}_0(\omega(e_n)) \qquad (x \neq x_0), \qquad (7.3)$$

where $c = (e_1, \ldots, e_n)$ is a path in X joining x_0 and x. What we have to prove is that this is well defined in the sense that if $c' = (e'_1, \ldots, e'_m)$ is another path joining x_0 and x, then

$$\mathbf{v}_0(\omega(e_1)) + \cdots + \mathbf{v}_0(\omega(e_n)) = \mathbf{v}_0(\omega(e'_1)) + \cdots + \mathbf{v}_0(\omega(e'_m)).$$

This is checked by observing that $c \cdot \overline{c'}$ is a closed path in X and

$$[\mathbf{v}_0(\omega(e_1)) + \cdots + \mathbf{v}_0(\omega(e_n))] - [\mathbf{v}_0(\omega(e'_1)) + \cdots + \mathbf{v}_0(\omega(e'_m))]$$
$$= \widehat{\mathbf{v}}_0(\langle \omega(c \cdot \overline{c'}) \rangle) = \mathbf{0},$$

where we should note that $\langle \omega(c \cdot \overline{c'}) \rangle \in \text{Ker}\,\mu = \text{Ker}\,\widehat{\mathbf{v}}_0 = H$. According to the definition (7.3), Φ is evidently a periodic realization whose period homomorphism is $\sigma \in L \mapsto \widehat{\mathbf{v}}_0(\langle \omega(c) \rangle)$ where c is a path with $t(c) = \sigma o(c)$. Since $\mu(\langle \omega(c) \rangle) = \sigma$, we find that $\widehat{\mathbf{v}}_0(\langle \omega(c) \rangle) = \rho(\sigma)$. Thus the period homomorphism is ρ. □

Remark. The discussion above might remind the reader of the idea used when we show that a voltage obeying *Kirchhoff's voltage law* is expressed as a *potential difference*. In terms of a finite graph $X = (V, E)$, the voltage law asserts that a function v (*voltage*) on E with $v(\overline{e}) = -v(e)$, $e \in E$, must satisfy $v(e_1) + \cdots + v(e_n) = 0$ for every closed path $(e_1, \ldots, e_n)$.

An electric potential f (a function on V) is constructed as follows. Fixing a reference vertex x_0 and taking a path $c = (e_1, \ldots, e_n)$ joining x_0 and x, we put $f(x) = v(c) = v(e_1) + \cdots + v(e_n)$. By virtue of the voltage law, for another path $c' = (e'_1, \ldots, e'_m)$ joining x_0 and x, we have

$$\bigl(v(e_1) + \cdots + v(e_n)\bigr) - \bigl(v(e'_1) + \cdots + v(e'_m)\bigr) = v(c \cdot \overline{c'}) = 0.$$

Therefore $f(x)$ does not depend on the choice of c. The function f clearly satisfies $v(e) = f\bigl(t(e)\bigr) - f\bigl(o(e)\bigr)$ so that the voltage is expressed as a potential difference. □

Closely related to the discussion above and useful afterwards is:

Theorem 7.2. *Let X be a topological crystal over X_0 whose abstract period lattice is L, and let $\rho : L \longrightarrow \mathbb{R}^d$ be an injective homomorphism such that $\rho(L)$ is a lattice group of $\mathbb{R}^d$. Then there exists a periodic realization Φ whose period homomorphism is ρ.*

Proof. Let H be the vanishing group so that $H_1(X_0, \mathbb{Z})/H = L$. In view of Theorem 4.1 in Chap. 4, the homomorphism $\rho \circ \mu : H_1(X_0, \mathbb{Z}) \longrightarrow \rho(L)$ extends to a homomorphism $f : C_1(X_0, \mathbb{Z}) \longrightarrow \rho(L)$. If we put $\mathbf{v}(e) = f(e)$, then $\widehat{\mathbf{v}}(\alpha) = f(\alpha) = \rho\bigl(\mu(\alpha)\bigr)$ for $\alpha \in H_1(X_0, \mathbb{Z})$, and $\mathrm{Ker}\, \mathbf{v}|H_1(X_0, \mathbb{Z}) = \mathrm{Ker}\, \mu = H$. Therefore $\{\mathbf{v}(e)\}_{e \in E_0}$ is a building block of a periodic realization of X whose period homomorphism is ρ. □

Our definition of periodic realizations does not exclude pathological ones. Namely, different vertices in X are allowed to be mapped to the same point, the image of an edge may collapse, or the image of two edges may overlap. For instance, any periodic realization of a topological crystal with loop edges or parallel edges is pathological in this sense.

In the 3-dimensional case, the image of a pathological realization (except for parallel edges which may be allowed as "multiple bonds") cannot be a model of a "realistic" crystal. In this connection, it is natural to ask what kind of hypothesis we should impose on a periodic realization Φ so that it represents the network of a realistic crystal. An easy-going idea is to assume the condition that $\Phi : X \longrightarrow \mathbb{R}^d$ is injective. It seems also reasonable to assume, in addition to the injectivity, that for a fixed c with $0 < c \leq 1$,

$$\|\Phi(y) - \Phi(x)\| \leq c \max_{e \in E_x} \|\mathbf{v}(e)\| \implies y \text{ is adjacent to } x, \text{ or } y = x.$$

This roughly means that two atoms close enough to each other must be joined by a bond (see Notes (I) in this chapter). But these conditions may not be enough (or may

7.1 Periodic Realizations

be too strong) to characterize the "reality" of a crystal model.[5] If we want to be more serious about this problem, we need to go over the physical and chemical aspect of crystals, which is beyond the scope of topological crystallography. We are not going to be involved in this question.

In the mathematical framework it is still interesting to ask whether a periodic realization constructed in a certain way is pathological or not. Among many possible conditions, easy to check is the following one:

A periodic realization Φ of a topological crystal $X = (V, E)$ is said to be *non-degenerate* if the following two conditions are satisfied (otherwise said to be *degenerate*):

1. Φ restricted to V is injective.
2. $\mathbf{v}(e) \neq \mathbf{0}$ for every $e \in E$, and the map $e \in E_x \longmapsto \mathbf{v}(e)/\|\mathbf{v}(e)\|$ into the unit sphere is injective for every $x \in V$.

Condition (2) is self-explanatory; it merely asserts that edges with the same origin do not overlap (but it is still possible to have overlapping edges which do not share an end point). A topological crystal with a non-degenerate realization must be combinatorial.

Before concluding this section, we make a short digression. Given a periodic realization (Φ, ρ), we obtain the flat torus $\mathbb{R}^d/\rho(L)$ associated with the period lattice $\rho(L)$, which is, as mentioned in Example 2.5, a substitute for a unit cell. We also have a map $\Phi_0 : X_0 \longrightarrow \mathbb{R}^d/\rho(L)$ such that the following diagram is commutative (i.e., $\pi \circ \Phi = \Phi_0 \circ \omega$), where the image of each edge is a straight line segment[6] in $\mathbb{R}^d/\rho(L)$.

$$\begin{array}{ccc} X & \xrightarrow{\Phi} & \mathbb{R}^d \\ \omega \downarrow & & \downarrow \pi \\ X_0 & \xrightarrow{\Phi_0} & \mathbb{R}^d/\rho(L). \end{array}$$

The base graph X_0 is thought of as a graph version of a "unit cell". This implies that the matter of periodic realizations reduces to that of maps between two kinds of unit cells. This view is of considerable use when we relate our problem to a differential-geometric one, and motivates us to study maps from a finite graph into a general Riemannian manifold [see Notes (IV) in this chapter and also Chap. 10].

Example 7.1. As an illustration, we look at the honeycomb lattice. In Fig. 7.3, two arrows represent a basis for the period lattice. This being the case, X_0 is the graph with two vertices joined by three parallel edges (Example 3.4), and $\Phi_0 : X_0 \longrightarrow \mathbb{R}^2/L$ is illustrated on the right of Fig. 7.3, where the torus pretends to have a flat metric (as noticed in Example 2.5, a flat torus cannot be realized in $\mathbb{R}^3$). □

---

[5] An alternative idea to characterize real crystal structures is to use Voronoi diagrams.
[6] Recall that a flat torus locally possesses a Euclidean structure, and so the term "straight line segment" makes sense.

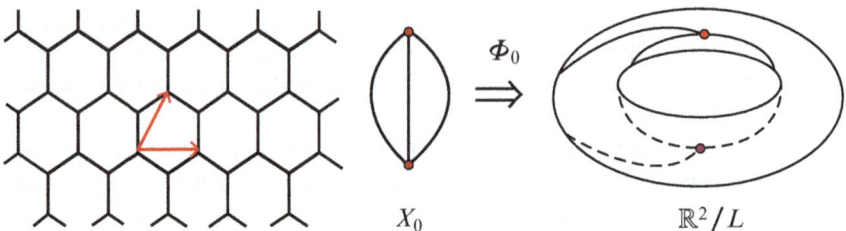

Fig. 7.3 Map from a finite graph into a flat torus

Remark. In Example 6.4, a topological carbon nanotube is defined as the quotient graph $X/\mathbb{Z}\mathbf{a}$ of the honeycomb lattice X, where $\mathbf{a}$ is a non-zero element of the period lattice L. The quotient graph $X/\mathbb{Z}\mathbf{a}$ is realized in $\mathbb{R}^2/\mathbb{Z}\mathbf{a}$ by the inclusion map $i : X/\mathbb{Z}\mathbf{a} \longrightarrow \mathbb{R}^2/\mathbb{Z}\mathbf{a}$. This realization is compatible with the natural actions of L on both $X/\mathbb{Z}\mathbf{a}$ and $\mathbb{R}^2/\mathbb{Z}\mathbf{a}$; namely the inclusion map i is an L-equivariant map. By identifying $\mathbb{R}^2/\mathbb{Z}\mathbf{a}$ with a circular cylinder C in $\mathbb{R}^3$, we obtain a network model of a carbon nanotube. The induced action of L on C is given by helical transformations.[7] □

In some materials, two or more than two crystal nets interpenetrate each other without any chemical bonds connecting them. Mathematically, this situation is described by the term "linking" or "entanglement" of disjoint finite graphs realized in the 3-dimensional torus.

7.2 Projection and Reduction

Let us look at a ready-made model of a crystal. What we find out when we turn the model around is that there are some specific directions toward which we may see two-dimensional crystal structures (ignoring the effect of perspective). For instance, in the case of the diamond crystal, one can see the square lattice or a distorted honeycomb lattice (the distortion is caused by parallel edges). For Lonsdaleite, the genuine honeycomb lattice can be seen in a suitable direction.

Mathematically, this phenomenon is interpreted in the following way. Let $\Phi : X \longrightarrow \mathbb{R}^d$ be a periodic realization of a d-dimensional topological crystal X with a period homomorphism $\rho : L \longrightarrow \mathbb{R}^d$. An orthogonal projection[8] $P : \mathbb{R}^d \longrightarrow W$ onto a d'-dimensional subspace W of $\mathbb{R}^d$ is said to be a *rational projection* (with respect

---

[7] A helical transformation means a transformation $(\theta, z) \mapsto (\theta + a, z + b)$ where (θ, z) is the cylindrical polar coordinate representing a point on C.

[8] In general, an *orthogonal projection* (or simply a *projection*) is a linear map P of a vector space W with an inner product $\langle \cdot, \cdot \rangle$ into itself such that $P^2 = P$ and $\langle P\mathbf{x}, \mathbf{y} \rangle = \langle \mathbf{x}, P\mathbf{y} \rangle$ ($\mathbf{x}, \mathbf{y} \in W$). See Appendix 3.

7.2 Projection and Reduction

to the period lattice $\rho(L)$) if $P(\rho(L))$ is a lattice group of W. What we shall observe is that if P is a rational projection, then $P(\Phi(X))$ coincides with the image of a periodic realization of a d'-dimensional topological crystal X'.

Choosing an orthonormal basis of W, we identify W with $\mathbb{R}^{d'}$. If put $L' = L/\mathrm{Ker}(P\circ\rho)$, then the quotient graph $X' = (V',E') = X/\mathrm{Ker}(P\circ\rho)$ is an abelian covering graph over $X_0 = X/L$ whose covering transformation group is L' (see Theorem 5.5 (c) in Sect. 5.2). The covering map $\omega : X \longrightarrow X_0$ is the composition of two covering maps: $\omega : X \xrightarrow{\omega_1} X' \xrightarrow{\omega_2} X_0$. Furthermore $P\circ\rho : L \longrightarrow \mathbb{R}^{d'} = W$ induces a homomorphism $\rho' : L' \longrightarrow \mathbb{R}^{d'}$. We also observe that if $\omega_1(x) = \omega_1(y)$, then $P(\Phi(x)) = P(\Phi(y))$. This is because there is $\sigma \in \mathrm{Ker}(P\circ\rho)$ such that $y = \sigma x$, and

$$P(\Phi(y)) = P(\Phi(\sigma x)) = P(\Phi(x)) + P(\rho(\sigma)) = P(\Phi(x)).$$

From this observation one can find a map $\Phi' : V' \longrightarrow \mathbb{R}^{d'}$ such that $P\circ\Phi = \Phi'\circ\omega_1$.

$$\begin{array}{ccc} X & \xrightarrow{\Phi} & \mathbb{R}^d \\ \omega_1 \downarrow & & \downarrow P \\ X' & \xrightarrow{\Phi'} & \mathbb{R}^{d'} \end{array}$$

To show that Φ' is a periodic realization whose period homomorphism is ρ', we again use Theorem 5.5 (c) in Sect. 5.2. Let $p : L \longrightarrow L'$ be the canonical homomorphism. For $x' \in V'$, $\sigma' \in L'$, we take $x \in V$ and $\sigma \in L$ such that $\omega_1(x) = x'$ and $p(\sigma) = \sigma'$. Then $\sigma'x' = \omega_1(\sigma x)$, whence

$$\Phi'(\sigma'x') = \Phi'(\omega_1(\sigma x)) = P(\Phi(\sigma x)) = P(\Phi(x)) + P(\rho(\sigma))$$
$$= \Phi'(x') + \rho'(\sigma').$$

If we denote by $\{\mathbf{v}'_0(e)\}_{e\in E_0}$ the building block of Φ', then

$$\mathbf{v}'_0(e) = P(\mathbf{v}_0(e)). \tag{7.4}$$

By abuse of language, we call Φ' a rational projection of Φ.

If P is not rational, then $P(\rho(L))$ is not discrete in W, and hence $P(\Phi(X))$ cannot be the image of a periodic realization. One should notice that almost all projections are not rational.

In the definition of periodic realizations we did not exclude the possibility that $\mathbf{v}(e)$ vanishes. This often happens when we consider a rational projection. The existence of such an edge makes the realization degenerate. Namely, an edge e in X with $\mathbf{v}(e) = \mathbf{0}$ is contracted to a point in $\mathbb{R}^d$. This suggests to us to consider the graph $X^{\mathrm{red}} = (V^{\mathrm{red}}, E^{\mathrm{red}})$ derived from X by contracting every edge e with $\mathbf{v}(e) = \mathbf{0}$

to a vertex, and to construct its realization Φ^{red} whose image coincides with that of Φ. Here $E^{\text{red}} = \{e \in E |\ \mathbf{v}(e) \neq \mathbf{0}\}$. The graph X^{red} will be a topological crystal, and the system of vectors $\{\mathbf{v}^{\text{red}}(e)\}_{e \in E^{\text{red}}}$ obtained by restricting $\mathbf{v}$ to E^{red} will be the building block of Φ^{red} of X^{red}. To be exact, using the covering map $\omega : X \longrightarrow X_0$, we set

$$E_Y = \{e \in E |\ \mathbf{v}(e)(= \mathbf{v}_0(\omega(e))) = \mathbf{0}\}.$$

Let $Y = (V_Y, E_Y)$ be the graph (not necessarily connected) obtained by collecting edges in E_Y together with their end points. The graph Y is invariant under the action of the abstract period lattice L. Let $\{Y_\alpha\}_{\alpha \in A}$, $Y_\alpha = (V_{Y\alpha}, E_{Y\alpha})$, be the family of connected components of Y. Contracting each Y_α to a single point, we obtain a graph X^{red}, which we call the *reduction* of X. The group L acts freely on X^{red} in a natural manner.

By doing the same for X_0 using $E_{Y_0} = \{e \in E_0 |\ \mathbf{v}_0(e) = \mathbf{0}\}$, we obtain a graph Y_0 and the reduction X_0^{red} of X_0. The inverse image $\omega^{-1}(E_{Y_0})$ coincides with E_Y. One can check that $\omega(Y_\alpha)$ is a connected component of Y_0.

We show that $\omega : Y_\alpha \longrightarrow \omega(Y_\alpha)$ is an isomorphism. For this, it is enough to prove that ω is injective on the set of vertices. Suppose $\omega(x) = \omega(y)$ for distinct vertices x, y in Y_α. Then $y = \sigma x$ with $\sigma \neq 0$ in L. Joining x, y by a path $c = (e_1, \ldots, e_n)$, we have

$$\mathbf{0} \neq \rho(\sigma) = \Phi(y) - \Phi(x) = \mathbf{v}(e_1) + \cdots + \mathbf{v}(e_n) = \mathbf{0},$$

whence a contradiction.

A less pedestrian argument shows that if $\omega(Y_\alpha) = \omega(Y_\beta)$, then $Y_\alpha = \sigma Y_\beta$ for some $\sigma \in L$. The implication is that the quotient graph X^{red}/L is identified with X_0^{red}, and hence X^{red} is a topological crystal over X_0^{red}. Clearly $\{\mathbf{v}^{\text{red}}(e)\}_{e \in E^{\text{red}}}$ is the building block of a periodic realization of X^{red}.

7.3 Idea

From the next section we start to seek the most canonical periodic realization of a topological crystal, which is to be called the *standard realization*. Towards this end, we shall appeal to a certain minimal principle, having in mind Euler's wise remark "Since the fabric of the Universe is most perfect and the work of a most wise creator, nothing at all takes place in the Universe in which some rule of maximum or minimum does not appear".

Our proto-idea is set forth in two geometric inequalities for the tetrahedron, which turns out to be closely related to the standard realization of the topological diamond. Figure 7.4 represents a fundamental parallelotope and part of a periodic

7.3 Idea

Fig. 7.4 A fundamental parallelotope of a periodic realization of the topological diamond

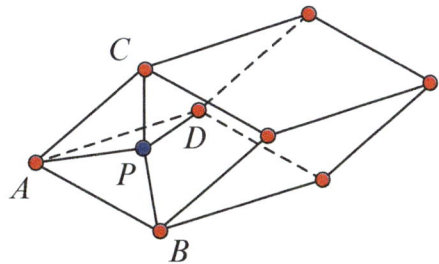

realization of the topological diamond (A, B, C, D, P are vertices in it). Note that the volume of the tetrahedron $ABCD$ is equal to one sixth of that of the fundamental parallelotope.

1. *For the tetrahedron ABCD, the point P that minimizes*

$$(PA)^2 + (PB)^2 + (PC)^2 + (PD)^2$$

   is the barycenter of ABCD.
   In general, the barycenter of n points $\mathbf{x}_1, \ldots, \mathbf{x}_n$ in $\mathbb{R}^d$ is given by

$$\frac{1}{n}(\mathbf{x}_1 + \cdots + \mathbf{x}_n).$$

   The statement in (1) is a special case of the fact that the barycenter is characterized as the point that minimizes the value

$$f(\mathbf{x}) = \|\mathbf{x} - \mathbf{x}_1\|^2 + \cdots + \|\mathbf{x} - \mathbf{x}_n\|^2,$$

   which is a consequence of the following equality.

$$\sum_{i=1}^{n} \|\mathbf{x} - \mathbf{x}_i\|^2 = n \left\| \mathbf{x} - \frac{1}{n} \sum_{i=1}^{n} \mathbf{x}_i \right\|^2 + \frac{1}{2n} \sum_{i,j=1}^{n} \|\mathbf{x}_i - \mathbf{x}_j\|^2.$$

2. *Let* $\mathrm{vol}(ABCD)$ *denote the volume of the tetrahedron ABCD. For any point P, we have the inequality*

$$(PA)^2 + (PB)^2 + (PC)^2 + (PD)^2 \geq 3^{5/3} \mathrm{vol}(ABCD)^{2/3}.$$

   The equality holds if and only if ABCD is equilateral, and P is its barycenter.

   We entrust the reader with the proof. In effect the idea of proof is almost the same as the one we are going to explain in characterizing standard realizations (Sect. 7.5).

7.4 Harmonic Realizations

Let us think of a crystal net as a system of *harmonic oscillators*, that is, each edge is supposed to represent a harmonic oscillator whose energy is the square of its length (Fig. 7.5). This simple model leads to the notion of *energy* functional defined by[9]

$$\mathcal{E}(\Phi,\rho) = \text{vol}(D_{\rho(L)})^{-2/d} \sum_{e \in E_0} \|\mathbf{v}_0(e)\|^2, \tag{7.5}$$

where $\{\mathbf{v}_0(e)\}_{e \in E_0}$ is the building block of a periodic realization (Φ,ρ), and $\text{vol}(D_{\rho(L)})$ stands for the volume of a fundamental parallelotope $D_{\rho(L)}$ associated with the period lattice $\rho(L)$ (which is also equal to $\text{vol}(\mathbb{R}^d/\rho(L))$, the volume of the flat torus $\mathbb{R}^d/\rho(L)$). The quantity $\mathcal{E}(\Phi,\rho)$ may be understood as the "normalized" total potential energy of the system *per unit cell*.

Before going further, it is appropriate to explain how to compute the volume of $D_{\rho(L)}$. In general, given a $\mathbb{Z}$-basis $\{\mathbf{a}_1,\ldots,\mathbf{a}_d\}$ of a lattice group W in $\mathbb{R}^d$, the volume of the fundamental parallelotope D_W associated with $\{\mathbf{a}_1,\ldots,\mathbf{a}_d\}$ is equal to $|\det(\mathbf{a}_1,\ldots,\mathbf{a}_d)|$, where $(\mathbf{a}_1,\ldots,\mathbf{a}_d)$ is the matrix obtained by lining up the column vectors $\mathbf{a}_1,\ldots,\mathbf{a}_d$. This is also equal to $\det(\langle \mathbf{a}_i, \mathbf{a}_j \rangle)^{1/2}$ because

$$^t(\mathbf{a}_1,\ldots,\mathbf{a}_d)(\mathbf{a}_1,\ldots,\mathbf{a}_d) = (\langle \mathbf{a}_i,\mathbf{a}_j \rangle),$$

and

$$\det(^t(\mathbf{a}_1,\ldots,\mathbf{a}_d)(\mathbf{a}_1,\ldots,\mathbf{a}_d)) = (\det(\mathbf{a}_1,\ldots,\mathbf{a}_d))^2.$$

This remark will be useful in a later discussion.

The reader might wonder why we put the term $\text{vol}(D_{\rho(L)})^{-2/d}$ in the definition (7.5) of the "normalized" energy. The reason is that we want to have a similarity-free

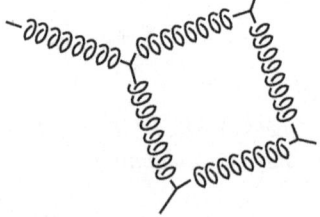

Fig. 7.5 A system of harmonic oscillators

---

[9] A real crystal (crystalline solid) is also physically regarded as a system of harmonic oscillators under an appropriate approximation of the equation of motion, but the shape of energy is much more complicated (cf. [85]). See Radin [79, 80] for a physical treatment, in which crystalline symmetry is related to physical energy.

7.4 Harmonic Realizations

quantity. To be precise, let M be a motion[10] (congruent transformation) of $\mathbb{R}^d$ whose linear part is an orthogonal matrix $U \in O(d)$, i.e.,

$$M(\mathbf{x}) = U\mathbf{x} + \mathbf{a} \qquad (\mathbf{a} \in \mathbb{R}^d).$$

The transformation of the form cM ($c > 0$) is called a *similar transformation*. Clearly $(cM \circ \Phi, cU \circ \rho)$ is also a periodic realization, and $\mathcal{E}(\Phi, \rho)$ is *similarity invariant* in the following sense:

$$\mathcal{E}(cM \circ \Phi, cU\rho) = \mathcal{E}(\Phi, \rho), \tag{7.6}$$

where we use $\mathrm{vol}(D_{cU \circ \rho(L)}) = c^d \mathrm{vol}(D_{\rho(L)})$.

For clarity, we introduce the set of periodic realizations with a fixed period homomorphism ρ:

$$\mathrm{Per}(X, \rho) = \{(\Phi, \rho) |\ \text{periodic realizations with a fixed } \rho\}.$$

Theorem 7.3. *A minimizer[11] (Φ, ρ) of the energy $\mathcal{E}$ restricted to $\mathrm{Per}(X, \rho)$ is characterized by*

$$\sum_{e \in E_{0x}} \mathbf{v}_0(e) = \mathbf{0} \qquad (x \in V_0). \tag{7.7}$$

Equality (7.7) says that $\Phi(x)$ is the barycenter of $\Phi(y_1), \ldots, \Phi(y_k)$ for vertices $y_1, \ldots, y_k \in V$ adjacent to $x \in V$ ($k = \deg x$), where $y_1, \ldots, y_k$ are counted with multiplicity (the multiplicity of y_i is the number of edges joining x and y_i).

A realization Φ satisfying (7.7) turns out to be unique up to translations. We call it a *harmonic realization*, following the notion of harmonic map in Riemannian geometry [see Notes (IV) in this chapter]. Crystallographers call it an *equilibrium placement*[12] [30].

We are now going to give a proof of Theorem 7.3. Suppose that Φ satisfies (7.7). One may assume without loss of generality that $\mathrm{vol}(D_{\rho(L)}) = 1$ so that $\mathcal{E}(\Phi, \rho) = \sum_{e \in E_0} \|\mathbf{v}_0(e)\|^2$. For $\Phi' \in \mathrm{Per}(X, \rho)$, we put $\mathbf{f}(x) = \Phi'(x) - \Phi(x)$. Then for every $\sigma \in L$,

$$\mathbf{f}(\sigma x) = \Phi'(\sigma x) - \Phi(\sigma x) = [\Phi'(x) + \rho(\sigma)] - [\Phi(x) + \rho(\sigma)] = \mathbf{f}(x).$$

Namely, $\mathbf{f}$ is a periodic function with values in $\mathbb{R}^d$, and is identified with a function on V_0. We now have

---

[10] A motion is a parallel translation, rotation, reflection, or their composition.

[11] A minimizer in this case is a periodic realization (Φ_0, ρ) such that $\mathcal{E}(\Phi_0, \rho) = \min \mathcal{E}(\Phi, \rho)$.

[12] This name sounds natural because (7.7) says that the crystal as a system of harmonic oscillators is in equilibrium in the sense that the total force acting on any "atom" from its nearest neighbors vanishes (we regard $\mathbf{v}(e)$ as the force generated by the harmonic oscillator corresponding to e).

$$\mathcal{E}(\Phi') = \mathcal{E}(\Phi + \mathbf{f}) = \sum_{e \in E_0} \Big(\|\mathbf{v}_0(e)\|^2 + 2\langle \mathbf{v}_0(e), [\mathbf{f}(t(e)) - \mathbf{f}(o(e))] \rangle$$
$$+ \|\mathbf{f}(t(e)) - \mathbf{f}(o(e))\|^2 \Big)$$
$$\geq \mathcal{E}(\Phi) + 2 \sum_{e \in E_0} \langle \mathbf{v}_0(e), [\mathbf{f}(t(e)) - \mathbf{f}(o(e))] \rangle$$
$$= \mathcal{E}(\Phi) + 2 \sum_{e \in E_0} \langle \mathbf{v}_0(e), \mathbf{f}(t(e)) \rangle - 2 \sum_{e \in E_0} \langle \mathbf{v}(e), \mathbf{f}(o(e)) \rangle.$$

Noting that the correspondence $e \mapsto \overline{e}$ is a bijection, we have

$$\mathcal{E}(\Phi') \geq \mathcal{E}(\Phi) + 2 \sum_{e \in E_0} \langle \mathbf{v}_0(\overline{e}), \mathbf{f}(t(\overline{e})) \rangle - 2 \sum_{e \in E_0} \langle \mathbf{v}_0(e), \mathbf{f}(o(e)) \rangle$$
$$= \mathcal{E}(\Phi) - 4 \sum_{e \in E_0} \langle \mathbf{v}_0(e), \mathbf{f}(o(e)) \rangle$$
$$= \mathcal{E}(\Phi) - 4 \sum_{x \in V_0} \Big\langle \sum_{e \in E_{0x}} \mathbf{v}_0(e), \mathbf{f}(x) \Big\rangle = \mathcal{E}(\Phi).$$

In this computation, we used $\mathbf{v}_0(\overline{e}) = -\mathbf{v}_0(e)$ and $t(\overline{e}) = o(e)$. Reflecting on the calculation, we conclude that the equality $\mathcal{E}(\Phi') = \mathcal{E}(\Phi)$ holds only if

$$\sum_{e \in E_0} \|\mathbf{f}(t(e)) - \mathbf{f}(o(e))\|^2 = 0,$$

or equivalently, $\mathbf{f}(t(e)) = \mathbf{f}(o(e))$ holds for every e. Therefore $\mathcal{E}(\Phi') = \mathcal{E}(\Phi)$ if and only if $\mathbf{f}$ is constant; that is, Φ' is a translation of Φ. This completes the proof of Theorem 7.3.

Theorem 7.3 does not include the claim that for a given injective homomorphism $\rho : L \longrightarrow \mathbb{R}^d$ such that $\rho(L)$ is a lattice group of $\mathbb{R}^d$, there exists a harmonic realization of X whose period homomorphism is ρ. This claim is actually true, and is proved in the following way [Notes (IV) in this chapter gives an alternative proof in a more general set-up].

We begin with an arbitrary periodic realization Φ_0 whose period homomorphism is ρ (see Theorem 7.2 for the existence of such Φ_0). Recall that any periodic realization Φ with the same period homomorphism ρ can be written as $\Phi = \Phi_0 + \mathbf{f} \circ \omega$ with an $\mathbb{R}^d$-valued function $\mathbf{f}$ on V_0 (where $\omega : X \longrightarrow X_0$ is the covering map). We thus wish to look for a function $\mathbf{f}$ such that $\Phi_0 + \mathbf{f} \circ \omega$ is a harmonic realization.

If $\{\mathbf{v}_0(e)\}_{e \in E_0}$ is the building block of Φ_0, then the building block $\{\mathbf{v}(e)\}_{e \in E_0}$ of $\Phi_0 + \mathbf{f} \circ \omega$ is given by $\mathbf{v}(e) = \mathbf{v}_0(e) + \mathbf{f}(t(e)) - \mathbf{f}(o(e))$. Thus in order that Φ be harmonic, $\mathbf{f}$ must be a solution of the equation

$$\sum_{e \in E_{0x}} [\mathbf{f}(t(e)) - \mathbf{f}(o(e))] = -\sum_{e \in E_{0x}} \mathbf{v}_0(e).$$

7.4 Harmonic Realizations

To make this equation concise, we define the *discrete Laplacian* Δ on X_0 by setting

$$(\Delta \mathbf{f})(x) = \sum_{e \in E_{0x}} \big[\mathbf{f}(t(e)) - \mathbf{f}(o(e))\big];$$

[the reason why we call Δ the discrete Laplacian is explained in Notes (II) in this chapter]. If we put

$$\mathbf{g}(x) = -\sum_{e \in E_{0x}} \mathbf{v}_0(e),$$

the above equation is expressed as

$$\Delta \mathbf{f} = \mathbf{g}. \tag{7.8}$$

This is a discrete analogue of the *Poisson equation*. Note that $\mathbf{g}$ satisfies $\sum_{x \in V_0} \mathbf{g}(x) = 0$; indeed

$$\sum_{x \in V_0} \mathbf{g}(x) = \sum_{e \in E_0} \mathbf{v}_0(e) = 0$$

because $\mathbf{v}_0(\overline{e}) = -\mathbf{v}_0(e)$. Therefore, in order to prove the existence of $\mathbf{f}$, it is enough to check that the equation $\Delta f = g$ in the scalar case has a solution f under the condition $\sum_{x \in V_0} g(x) = 0$.

What we require for establishing this claim is the fact that the operator Δ is symmetric with respect to the following inner product on the space of functions on V_0:

$$\langle f_1, f_2 \rangle = \sum_{x \in V_0} f_1(x) f_2(x).$$

Indeed

$$\langle \Delta f_1, f_2 \rangle = \sum_{x \in V_0} \sum_{e \in E_{0x}} \big[f_1(t(e)) - f_1(o(e))\big] f_2(x)$$

$$= \sum_{e \in E_0} \big[f_1(t(e)) - f_1(o(e))\big] f_2(o(e))$$

$$= \frac{1}{2}\bigg(\sum_{e \in E_0} \big[f_1(t(e)) - f_1(o(e))\big] f_2(o(e))$$

$$+ \sum_{e \in E_0} \big[f_1(o(e)) - f_1(t(e))\big] f_2(t(e)) \bigg)$$

$$= -\frac{1}{2} \sum_{e \in E_0} \big[f_1(t(e)) - f_1(o(e))\big] \big[f_2(t(e)) - f_2(o(e))\big],$$

and so $\langle \Delta f_1, f_2 \rangle = \langle f_1, \Delta f_2 \rangle$. We also have

$$\langle \Delta f, f \rangle = -\frac{1}{2} \sum_{e \in E_0} \left[f(t(e)) - f(o(e)) \right]^2,$$

from which we conclude that if $\Delta f = 0$, then f is constant. Now using[13] $(\mathrm{Ker}\,\Delta)^\perp =$ Image Δ, we find that Image Δ coincides with the space of functions g with $\sum_{x \in V_0} g(x) = 0$. This finishes the proof of our claim.

There is a large degree of freedom in the choice of a harmonic realization if we do not fix a period homomorphism. Indeed any affine image of a harmonic realization is a harmonic realization. This means that if (Φ, ρ) is harmonic, then so is $(A\Phi + \mathbf{b}, A\rho)$ for any invertible square matrix A of size d and any vector $\mathbf{b}$ (the transformation T of $\mathbb{R}^d$ defined by $T\mathbf{x} = A\mathbf{x} + \mathbf{b}$ is said to be *affine*).

Theorem 7.4. *Harmonic realizations of a topological crystal X over X_0 are unique up to affine transformations.*

Proof. Let (Φ, ρ) and (Φ', ρ') be two harmonic realizations of X. If $\{\sigma_1, \ldots, \sigma_d\}$ is a $\mathbb{Z}$-basis of L, then $\{\rho(\sigma_1), \ldots, \rho(\sigma_d)\}$ and $\{\rho'(\sigma_1), \ldots, \rho'(\sigma_d)\}$ are $\mathbb{Z}$-bases of the period lattices $\rho(L)$ and $\rho'(L)$, respectively, (and hence bases of the vector space $\mathbb{R}^d$). Therefore one can find an invertible square matrix A of size d such that $\rho'(\sigma_k) = A\rho(\sigma_k)$ ($k = 1, \ldots, d$), or equivalently $\rho'(\sigma) = A\rho(\sigma)$ ($\sigma \in L$). Since $A\Phi$ is the harmonic realization associated with the period homomorphism $\rho' = A\rho$, by the uniqueness of a harmonic realization with a fixed period homomorphism, one can find a vector $\mathbf{b}$ such that $\Phi' = A\Phi + \mathbf{b}$. □

7.5 Standard Realizations

In the preceding section, we fixed a period homomorphism $\rho : L \longrightarrow \mathbb{R}^d$. We now remove this constraint (but we still leave the abstract period lattice L untouched). If (Φ, ρ) is a minimizer of $\mathcal{E}$, then Φ must be harmonic so that it suffices to seek a minimizer among harmonic realizations.

First we give a characterization of the minimizers.

Theorem 7.5. *A periodic realization Φ with a building block $\{\mathbf{v}_0(e)\}_{e \in E_0}$ minimizes $\mathcal{E}$ if and only if the following two conditions are satisfied.*

1. $\sum_{e \subset E_{0x}} \mathbf{v}_0(e) = \mathbf{0}$ *for every* $x \in V_0$,
2. *There exists a positive constant c such that*

$$\sum_{e \in E_0} \langle \mathbf{x}, \mathbf{v}_0(e) \rangle \mathbf{v}_0(e) = c\mathbf{x} \quad (\mathbf{x} \in \mathbb{R}^d). \tag{7.9}$$

---

[13] See Appendix 3.

7.5 Standard Realizations

Moreover, a minimizer is unique up to similar transformations of $\mathbb{R}^d$ when we fix the abstract period lattice L. To be precise, if Φ' is another periodic realization with a building block $\{\mathbf{v}'_0(e)\}_{e \in E_0}$ satisfying (1) and (2) with a constant c', then there exists a motion M such that $\Phi' = (c'/c)^{1/2} M \circ \Phi$.

If we define the linear operator $T : \mathbb{R}^d \longrightarrow \mathbb{R}^d$ by

$$T\mathbf{x} = \sum_{e \in E_0} \langle \mathbf{x}, \mathbf{v}_0(e) \rangle \mathbf{v}_0(e),$$

then condition (2) is equivalent to that T coincides with a scalar multiplication cI. This is also equivalent to

$$\sum_{e \in E_0} \langle \mathbf{v}_0(e), \mathbf{x} \rangle^2 = c\|\mathbf{x}\|^2 \quad (\mathbf{x} \in \mathbb{R}^d).$$

A minimizer of $\mathcal{E}$ is called the *standard realization* or *canonical placement* of X (associated with the abstract period lattice L). This notion is, as mentioned in the introduction, identical to what J.-G. Eon calls the *archetypical representation* [43].

Before proceeding to the proof of the theorem, we make some remarks. If we write $\mathbf{v}_0(e) = {}^t(v_1(e), \ldots, v_d(e))$ and use the Kronecker delta

$$\delta_{ij} = \begin{cases} 1 & (i = j) \\ 0 & (i \neq j), \end{cases}$$

then (7.9) is expressed in coordinate form as

$$\sum_{e \in E_0} v_i(e) v_j(e) = c \delta_{ij} \quad (i, j = 1, \ldots, d). \tag{7.10}$$

This implies that if Φ is standard, then

$$\mathrm{vol}(D_{\rho(L)})^{2/d} \mathcal{E}(\Phi, \rho) = \sum_{e \in E_0} \sum_{i=1}^{d} v_i(e)^2 = cd.$$

Suppose that a harmonic realization (Φ, ρ) satisfies (7.9). For another harmonic realization (Φ', ρ'), we put

$$\mathbf{v}'_0(e) = \Phi'(t(e)) - \Phi'(o(e)) = {}^t(v'_1(e), \ldots, v'_d(e)).$$

As proved at the end of the previous section, there is an invertible square matrix $A = (a_{ij})$ of size d and a vector $\mathbf{b} \in \mathbb{R}^d$ such that

$$(\Phi', \rho') = (A\Phi + \mathbf{b}, A\rho).$$

Notice
$$\text{vol}(D_{\rho'(L)}) = |\det A| \cdot \text{vol}(D_{\rho(L)}). \tag{7.11}$$

In fact, choosing a $\mathbb{Z}$-basis $\{\sigma_1, \ldots, \sigma_d\}$ of L, we find that

$$\begin{aligned}
\text{vol}(D_{\rho'(L)}) &= |\det(\rho'(\sigma_1), \ldots, \rho'(\sigma_d))| = |\det A(\rho(\sigma_1), \ldots, \rho(\sigma_d))| \\
&= |\det A| \cdot |\det(\rho(\sigma_1), \ldots, \rho(\sigma_d))| \\
&= |\det A| \cdot \text{vol}(D_{\rho(L)}).
\end{aligned}$$

We are now in the position to establish the inequality $\mathcal{E}(\Phi', \rho') \geq \mathcal{E}(\Phi, \rho)$. An easy way of working this out is to make use of the following inequality:

For a symmetric matrix S of size d with positive eigenvalues,

$$\text{tr}\, S \geq d(\det S)^{1/d}, \tag{7.12}$$

where the equality holds if and only if S is a scalar matrix, i.e., $S = \lambda I_d$ with $\lambda > 0$.

When S is a diagonal matrix, (7.12) reduces to the well-known inequality (*Inequality of arithmetic and geometric means*):

$$\frac{1}{d}(a_1 + \cdots + a_d) \geq (a_1 \cdots a_d)^{1/d} \ (a_i > 0)$$

The equality holds if and only if $a_1 = \cdots = a_d$. In the general case we take an orthogonal matrix T such that ${}^t T S T$ is a diagonal matrix. Using the properties $\text{tr}\, AB = \text{tr}\, BA$, we obtain

$$\text{tr}\, S = \text{tr}\, {}^t T S T \geq d \left(\det {}^t T S T\right)^{1/d} = d \left(\det S\right)^{1/d}.$$

The equality holds if and only if all the diagonal entries of ${}^t T S T$ are the same. Since ${}^t T S T = \lambda I_d \iff S = \lambda I_d$, our claim is proved.

We go back to the claim $\mathcal{E}(\Phi', \rho') \geq \mathcal{E}(\Phi, \rho)$. In coordinate form, the relation between $\{\mathbf{v}(e)\}$ and $\{\mathbf{v}'(e)\}$ is written as $v'_i(e) = \sum_{j=1}^{d} a_{ij} v_j(e)$ so that

$$\text{vol}(D_{\rho'(L)})^{2/d} \mathcal{E}(\Phi') = \sum_{i=1}^{d} \sum_{e \in E_0} v'_i(e)^2 = \sum_{i=1}^{d} \sum_{e \in E_0} \sum_{h,k=1}^{d} a_{ih} a_{ik} v_h(e) v_k(e)$$

$$= c \sum_{i=1}^{d} \sum_{e \in E_0} \sum_{h,k=1}^{d} a_{ih} a_{ik} \delta_{hk} = c \sum_{i=1}^{d} \sum_{h=1}^{d} a_{ih} a_{ih}$$

7.5 Standard Realizations

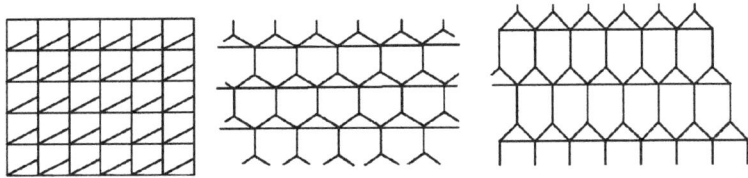

A realization   A harmonic realization   The standard realization

Fig. 7.6 Various realizations of the same crystal structure

$$= c \cdot \mathrm{tr}\,{}^t A A \geq cd \left(\det {}^t A A \right)^{1/d} = cd |\det A|^{2/d}$$

$$= cd \frac{\mathrm{vol}\left(D_{\rho'(L)}\right)^{2/d}}{\mathrm{vol}\left(D_{\rho(L)}\right)^{2/d}} = \mathrm{vol}\left(D_{\rho'(L)}\right)^{2/d} \mathcal{E}(\Phi),$$

and hence $\mathcal{E}(\Phi', \rho') \geq \mathcal{E}(\Phi, \rho)$. Here the equality holds if and only if ${}^t A A = \lambda I_d$ with $\lambda > 0$; that is, $U = \lambda^{-1/2} A$ is an orthogonal matrix. This completes the proof of our claim.

Figure 7.6 illustrates three kinds of realizations of a topological crystal.

Here comes the issue of the existence of standard realizations. Actually it is easy to establish the existence if we do not care about their explicit constructions[14] fitting in with the enumeration of topological crystals explained in Sect. 6.2. We start with a harmonic realization (Ψ, ρ) of X whose building block is $\{\mathbf{u}(e)\}_{e \in E_0}$. Define the linear operator T of $\mathbb{R}^d$ into itself by

$$T\mathbf{x} = \sum_{e \in E_0} \langle \mathbf{x}, \mathbf{u}(e) \rangle \mathbf{u}(e). \tag{7.13}$$

Since

$$\langle T\mathbf{x}, \mathbf{y} \rangle = \sum_{e \in E_0} \langle \mathbf{x}, \mathbf{u}(e) \rangle \cdot \langle \mathbf{y}, \mathbf{u}(e) \rangle,$$

the operator T is symmetric, and positive in the sense that $\langle T\mathbf{x}, \mathbf{x} \rangle > 0$ for $\mathbf{x} \neq \mathbf{0}$. Hence one can find a (unique) symmetric and positive operator $T^{-1/2}$ with $(T^{-1/2})^2 = T^{-1}$. Then

$$\sum_{e \in E_0} \langle \mathbf{x}, T^{-1/2}\mathbf{u}(e) \rangle T^{-1/2}\mathbf{u}(e) = T^{-1/2} \sum_{e \in E_0} \langle T^{-1/2}\mathbf{x}, \mathbf{u}(e) \rangle \mathbf{u}(e)$$

$$= T^{-1/2} T T^{-1/2} \mathbf{x} = \mathbf{x}.$$

---

[14] See the next chapter for an explicit construction.

Fig. 7.7 The hydrogen bond

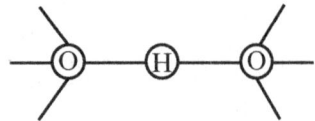

This implies that $\Phi = T^{-1/2}\Psi$ is the standard realization with the period homomorphism $T^{-1/2}\rho$.

Remark. Given a general periodic realization Ψ of X with a building block $\{\mathbf{u}(e)\}_{e \in E_0}$, let $\lambda^{\min}$ (respectively $\lambda^{\max}$) be the minimal (respectively maximal) eigenvalue of the symmetric operator T defined by (7.13). The ratio

$$S(\Psi) = \lambda^{\max}/\lambda^{\min} \geq 1$$

is considered representing the *degree of distorsion* from the standard realization. Indeed, $S(\Psi) = 1$ if and only if Ψ is standard. □

There are a few examples of real crystals whose nets happen to be standard in our sense. The typical ones are, as will be seen in Chap. 8, diamond and Lonsdaleite. Classical 2D crystal lattices such as the square lattice, honeycomb, and kagome lattice are also standard realizations. Needless to say, the nets associated with many real crystals are not realized in the standard way. This is, in particular, the case for crystals composed of more than one kind of atoms. For instance, the crystal structure of ordinary ice[15] is the same as that of Lonsdaleite if we think of *hydrogen bonds* as edges (Fig. 7.7), while its realization in space is slightly distorted from that of Lonsdaleite. Nevertheless, the concept of standard realization is still useful not only for designing crystal nets in a systematic way, but also for describing general realizations in a quantitative way. Indeed, one can use the standard realization as a basis of comparison as noted in the above remark.

7.6 Various Properties of Standard Realizations

We are now going to explore some basic properties of standard realizations, which claim the naturality of the concept as viewed from another angle.

The first one is about the symmetric property of the standard realization (Φ, ρ) with respect to the group $\mathrm{Aut}(X/X_0)$. We have seen, in Sect. 6.3, that $\mathrm{Aut}(X/X_0)$ is virtually abelian. Hence one can say that it is nearly a crystallographic group. What

---

[15] Ice is a crystal phase of water (H_2O). "Ordinary ice" means the most common phase of ice on earth which is called **Ih**. Among several phases, there is a phase **Ic** whose crystal structure is the same as that of diamond. This phase occurs in the atmosphere. See Scott [83] for more information on ice.

7.6 Various Properties of Standard Realizations

we shall observe now is that the standard realization gives a more direct connection between $\mathrm{Aut}(X/X_0)$ and a crystallographic group.

Let $g \in \mathrm{Aut}(X/X_0)$. Recalling $gLg^{-1} = L$, one may define the homomorphism $\rho_g : L \longrightarrow \mathbb{R}^d$ by setting $\rho_g(\sigma) = \rho(g\sigma g^{-1})$, with which $\Phi \circ g$ satisfies

$$(\Phi \circ g)(\sigma x) = \Phi\big((g\sigma g^{-1})gx\big) = \Phi(gx) + \rho(g\sigma g^{-1}) = (\Phi \circ g)(x) + \rho_g(\sigma),$$

so $\Phi \circ g$ is a periodic realization with the period homomorphism ρ_g. If $g_0 \in \mathrm{Aut}(X_0)$ is the automorphism of X_0 induced from g, then $\{\mathbf{v}_0(g_0(e))\}_{e \in E_0}$ is the building block for $\Phi \circ g$. Evidently

$$\sum_{e \in E_{0x}} \mathbf{v}_0(g_0(e)) = \mathbf{0},$$

$$\sum_{e \in E_0} \langle \mathbf{x}, \mathbf{v}_0(g_0(e)) \rangle \mathbf{v}_0(g_0(e)) = c\mathbf{x},$$

where c is the same constant as the one in (7.9), and hence $(\Phi \circ g, \rho_g)$ is a standard realization. From the uniqueness, it follows that there exists a motion $M(g)$ such that $\Phi \circ g = M(g) \circ \Phi$. We may easily check that $g \mapsto M(g)$ is a homomorphism of $\mathrm{Aut}(X/X_0)$ into $\mathbf{M}(d)$ where $\mathbf{M}(d)$ is the *motion group* consisting of all motions of $\mathbb{R}^d$.

In general, the homomorphism M is not injective. We shall show that $\mathrm{Ker}\, M$ is a finite (normal) subgroup of $\mathrm{Aut}(X/X_0)$. Note that

$$\mathrm{Ker}\, M = \{g \in \mathrm{Aut}(X/X_0)|\ \Phi(gx) = \Phi(x),\ (x \in V)\}.$$

For $\sigma \in (\mathrm{Ker}\, M) \cap L$, we have $\Phi(x) = \Phi(\sigma x) = \Phi(x) + \rho(\sigma)$. Thus $\sigma = 1$, and $(\mathrm{Ker}\, M) \cap L = \{1\}$. Therefore the restriction $p|(\mathrm{Ker}\, M)$ of the homomorphism $p : \mathrm{Aut}(X/X_0) \longrightarrow \mathrm{Aut}(X_0)$ introduced in Sect. 6.3 is injective, and hence $\mathrm{Ker}\, M$ is isomorphic to the finite group $p(\mathrm{Ker}\, M)$.

One can show that Image M is a crystallographic group (Sect. 9.6). Therefore, $\mathrm{Aut}(X/X_0)$ is *nearly* a crystallographic group in the sense that its factor group by a finite normal subgroup is crystallographic.

So far we have treated the symmetry relating to $\mathrm{Aut}(X/X_0)$, the group of automorphisms preserving the abstract period lattice L. What about the case of the full automorphism group $\mathrm{Aut}(X)$? Namely we ask whether there exists a homomorphism M of $\mathrm{Aut}(X)$ into the motion group such that $\Phi \circ g = M(g) \circ \Phi$. The answer is "Yes", but the proof is much involved (see Sect. 9.6).

Remark. What we have observed above says somehow that the standard realization has "maximal symmetry" with respect to $\mathrm{Aut}(X/X_0)$. There is an alternative way to construct a periodic realization with maximal symmetry. The idea is to change, if necessary, the standard inner product $\langle \cdot, \cdot \rangle_0$ on $\mathbb{R}^d$ to make a given harmonic realization maximally symmetric (cf. [30]).

Let (Φ,ρ) be a harmonic realization. The composition $\Phi \circ g$ is a harmonic realization whose period homomorphism is ρ_g. Let $A(g)$ be the invertible matrix such that $A(g)\rho(\sigma) = \rho_g(\sigma)$ ($\sigma \in L$). Then $g \mapsto A(g)$ is a homomorphism of $\mathrm{Aut}(X/X_0)$ into the group $GL_d(\mathbb{R})$ of invertible matrices. In view of Theorem 7.4, we find a vector $\mathbf{b}(g)$ with $\Phi \circ g = A(g)\Phi + \mathbf{b}(g)$. What we want is an inner product $\langle \cdot, \cdot \rangle$ on $\mathbb{R}^d$ satisfying

$$\langle A(g)\mathbf{x}, A(g)\mathbf{y} \rangle = \langle \mathbf{x}, \mathbf{y} \rangle \quad (g \in \mathrm{Aut}(X/X_0),\ \mathbf{x}, \mathbf{y} \in \mathbb{R}^d).$$

Since $A(L) = \{I_d\}$, the homomorphism A induces a homomorphism of the finite group $\mathrm{Aut}_H(X_0)$ into $GL_d(\mathbb{R})$ [Theorem 6.9 (2)]. Define the inner product $\langle \cdot, \cdot \rangle$ by setting

$$\langle \mathbf{x}, \mathbf{y} \rangle = \sum_{h \in \mathrm{Aut}_H(X_0)} \langle h\mathbf{x}, h\mathbf{y} \rangle_0.$$

This, "averaging" process makes this new inner product invariant under the action of $\mathrm{Aut}_H(X_0)$ (and hence invariant under the action of $\mathrm{Aut}(X/X_0)$).

The harmonic realization obtained in this way does not necessarily coincide with the standard realization. But one can prove that if the action of $\mathrm{Aut}_H(X_0)$ on $\mathbb{R}^d$ is irreducible, then this realization is standard (cf. Theorem 7.10 in Notes in this chapter). □

In Sect. 7.2 we introduced the notions of rational projection and reduction of a periodic realization. We shall observe that standardness is preserved by these operations.

First, if Φ is standard, then so is the reduction Φ^{red}. In fact, the building block $\{\mathbf{v}_0^{\mathrm{red}}(e)\}_{e \in E_0^{\mathrm{red}}}$ of Φ^{red} is obtained from $\{\mathbf{v}_0(e)\}_{e \in E_0}$ by removing zero vectors. The condition of standardness is not affected by this elimination procedure.

Next let Φ' be a rational projection of the standard realization Φ. According to (7.4), the building block $\{\mathbf{v}_0'(e)\}_{e \in E_0}$ satisfies $\sum_{e \in E_{0x}} \mathbf{v}_0'(e) = \mathbf{0}$. Furthermore, for $\mathbf{x} \in W = \mathbb{R}^{d'}$, we find that

$$\sum_{e \in E_0} \langle \mathbf{x}, \mathbf{v}_0'(e) \rangle \mathbf{v}_0'(e) = \sum_{e \in E_0} \langle \mathbf{x}, P(\mathbf{v}_0(e)) \rangle P(\mathbf{v}_0(e))$$

$$= P\Big(\sum_{e \in E_0} \langle \mathbf{x}, (\mathbf{v}_0(e)) \rangle \mathbf{v}_0(e) \Big) = P(c\mathbf{x}) = c\mathbf{x}.$$

Therefore Φ' is the standard realization of X'.

So far, we have fixed a base graph X_0 of a topological crystal X when we were discussing the uniqueness of the standard realization. What happens if we replace the base graph X_0 by another one, say X_0', while leaving X unchanged? The answer is: *the similarity class of standard realizations of X does not depend on the choice of a base graph*. Actually, this claim is easy to prove if X_0 and X_0' are "commensurable" in the sense that their abstract period lattices are commensurable

7.7 Notes

(we leave the proof for the reader as an exercise). However, this is not obvious in the non-commensurable case, which may happen for pathological topological crystals. The proof will be given in Sect. 9.6.

The positive constant c in (7.9) does not play a significant role in our discussion since only similarity-free concepts are our concern. Having this in mind, we adopt the following definition:

A periodic realization (Φ_0, ρ_0) is said to be the *normalized standard realization* if it satisfies

$$\sum_{e \in E_{0x}} \mathbf{v}_0(e) = \mathbf{0} \quad (x \in V_0), \tag{7.14}$$

$$\sum_{e \in E_0} \langle \mathbf{x}, \mathbf{v}_0(e) \rangle \mathbf{v}_0(e) = 2\mathbf{x} \quad (\mathbf{x} \in \mathbb{R}^d). \tag{7.15}$$

If we choose an orientation E_0^o, then (7.15) is written as

$$\sum_{e \in E_0^o} \langle \mathbf{x}, \mathbf{v}_0(e) \rangle \mathbf{v}_0(e) = \mathbf{x}.$$

If (Φ, ρ) satisfies (7.14), (7.15), then there is a motion M of $\mathbb{R}^d$ with $\Phi = M \circ \Phi_0$. Indeed, if Φ is another minimizer of $\mathcal{E}$, then $\mathcal{E}(\Phi) = \mathcal{E}(\Phi_0)$, and hence $\Phi = \lambda M \circ \Phi_0$. In view of the normalization condition, we have $\lambda = 1$.

Summary.
1. *Uniqueness of the normalized standard realization.* If (Φ, ρ) and (Φ', ρ') satisfies (7.14), (7.15), then Φ' is congruent to Φ in the sense that there is a motion M of $\mathbb{R}^d$ with $\Phi' = M \circ \Phi$.
2. *Energy inequality.* If (Φ_0, ρ_0) is the normalized standard realization, then for any periodic realization (Φ, ρ) with a building block $\{\mathbf{v}(e)\}_{e \in E_0}$, the following inequality holds:

$$\sum_{e \in E_0} \|\mathbf{v}(e)\|^2 \geq 2d \mathrm{vol}(D_{\rho_0(L)})^{-2/d} \cdot \mathrm{vol}(D_{\rho(L)})^{2/d}, \tag{7.16}$$

   where the equality holds if and only if there are a motion M and a positive constant λ such that $\Phi = \lambda M \circ \Phi_0$.

7.7 Notes

(I) Because of the chemical nature of constituent atoms in real crystals, the degree of the corresponding nets cannot be arbitrarily large. For instance, the degree of the net associated with a carbon crystals is less than or equal to 4. We also have a

Fig. 7.8 Kissing number

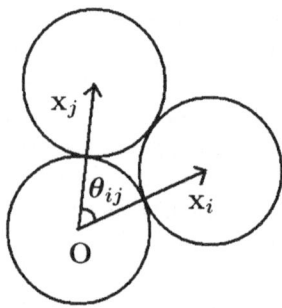

mathematical restriction on degree when we only handle topological crystals with non-pathological realizations. More precisely, we have

Theorem 7.6. *Let $\Phi : X \longrightarrow \mathbb{R}^d$ be a non-degenerate periodic realization satisfying the conditions:*

$$\|\Phi(y) - \Phi(x)\| \leq \max_{e \in E_x} \|\mathbf{v}(e)\| \implies y \text{ is adjacent to } x, \text{ or } y = x.$$

If the girth of X is greater than three, then $\deg x \leq k(d)$, for every vertex x of X, where $k(d)$ is the d-dimensional kissing number.

The d-dimensional *kissing number* is the greatest number of non-overlapping unit spheres in $\mathbb{R}^d$ that touch another given unit sphere. For instance, $k(2) = 6$. Historically the issue of the kissing number arose from the famous controversy[16] in 1694 between Isaac Newton and David Gregory: Newton insisted that $k(3) = 12$, while Gregory thought that $k(3) = 13$. The correct answer is $k(3) = 12$ as Newton believed. In the higher dimensional case, it is known that $k(4) = 24$, $k(8) = 240$, $k(24) = 196,560$; in other cases, we only know lower and upper bounds of $k(d)$.

Given a point $\mathbf{x} \in \mathbb{R}^d$, we denote by $S(\mathbf{x})$ the unit sphere whose center is $\mathbf{x}$. Let $\{\mathbf{x}_1, \ldots, \mathbf{x}_k\}$ be a set of vectors in $\mathbb{R}^d$ such that $\|\mathbf{x}_i\| = 2$ $(i = 1, \ldots, k)$. The spheres $S(\mathbf{x}_i)$ touch $S(\mathbf{0})$. The necessary and sufficient condition in order that $S(\mathbf{x}_1), \ldots, S(\mathbf{x}_k)$ do not overlap is that the angle θ_{ij} of vectors $\mathbf{x}_i$ and $\mathbf{x}_j$ satisfies

$$\theta_{ij} \geq \frac{\pi}{3} \quad (i \neq j)$$

(see Fig. 7.8).

The proof of the theorem above is done as follows. Take any vertex $x \in V$ and distinct vertices $y_1, y_2 \in V$ adjacent to x. Also take $e_i \in E_x$ such that $t(e_i) = y_i$ ($i = 1, 2$). Since the girth is greater than three, y_1 and y_2 are not adjacent, and hence, in view of our assumption, we have

---

[16]This is closely related to the Kepler conjecture about the densest packing of equally sized spheres; see Notes (III) in Chap. 8.

7.7 Notes

$$\|\Phi(y_2) - \Phi(y_1)\| = \|\mathbf{v}(e_2) - \mathbf{v}(e_1)\| > \max\{\mathbf{v}(e_1), \mathbf{v}(e_2)\},$$

from which it follows that

$$2\langle \mathbf{v}(e_1), \mathbf{v}(e_2)\rangle < \min\{\|\mathbf{v}(e_1)\|^2, \|\mathbf{v}(e_2)\|^2\}.$$

Therefore the angle θ between two vectors $\mathbf{v}(e_1)$ and $\mathbf{v}(e_2)$ is greater than $\pi/3$. This implies $|E_x| \leq k(d)$, as desired.

(II) Let us say a few words about *discrete Laplacian*, a notion with many ramifications in discrete geometric analysis.

First, we explain why the operator Δ defined by

$$\Delta f(x) = \sum_{e \in E_x} \big[f(t(e)) - f(o(e))\big]$$

is regarded as a discrete version of the Laplacian. The Laplacian Δ as a differential operator is

$$\Delta f = \sum_{i=1}^{d} \frac{\partial^2 f}{\partial x_i^2}.$$

In particular, $\Delta = \dfrac{d^2}{dx^2}$ in the one-dimensional case. Recall the following expression of $\dfrac{d^2}{dx^2}$ derived from Taylor's theorem:

$$\frac{d^2 f}{dx^2} = \lim_{t \to 0} \frac{1}{t^2}\big[f(x+t) + f(x-t) - 2f(x)\big].$$

On the other hand, the discrete Laplacian on the standard one-dimensional lattice $\mathbb{Z}$ is given by

$$(\Delta f)(n) = f(n+1) + f(n-1) - 2f(n). \tag{7.17}$$

In view of this expression, the reader may agree that (7.17) is a discretization of $\dfrac{d^2 f}{dx^2}$.

A periodic realization $\Phi : X \longrightarrow \mathbb{R}^d$ is harmonic if and only if Φ is a solution of the *discrete Laplace equation* $\Delta \Phi = \mathbf{0}$. Namely, each entry of the vector-valued function Φ is a *harmonic function* in the discrete sense.

The notion of discrete Laplacian is generalized to the case of *weighted graphs*. Here a graph $X = (V, E)$ is *weighted* if we are given positive-valued functions m_V on V and m_E on E such that $m_E(\overline{e}) = m_E(e)$. The discrete Laplacian on the weighted graph X is defined by

$$(\Delta f)(x) = \frac{1}{m_V(x)} \sum_{e \in E_x} m_E(e)[f(t(e)) - f(o(e))].$$

The discrete Laplacian we have considered so far is the one in the case $m_V \equiv 1$, $m_E \equiv 1$.

If $\sup_{x \in V} m_V(x)^{-1} \sum_{e \in E_x} m_E(e) < \infty$, then the discrete Laplacian Δ is a hermitian operator acting on the Hilbert space

$$\ell^2(V, m_V) = \left\{ f : V \longrightarrow \mathbb{C} \mid \sum_{x \in V} |f(x)|^2 m_V(x) < \infty \right\}.$$

As a sample of resemblance between the Laplacian and its discrete analogue, we quote:

Theorem 7.7 (A Discrete Analogue of the Maximum Principle). *Let $X = (V, E)$ be a connected weighted graph, and let f be real-valued.*

1. *If $(\Delta f)(x) \geq 0$ for every $x \in V$, and if f attains its maximum, then f is constant.*[17]
2. *If $\Delta f = 0$ and if f is not constant, then f attains neither maximum nor minimum in V. In particular, if X is a finite graph, then any harmonic function must be constant.*

Proof. 1. Suppose that f attains its maximum at $x \in V$. Then

$$0 \leq \Delta f(x) = m_V(x)^{-1} \sum_{e \in E_x} m_E(e)\{f(t(e)) - f(x)\}.$$

Since $f(t(e)) \leq f(x)$, this implies that $f(x) = f(t(e))$ for all $e \in E$ with $o(e) = x$. Since X is connected, we conclude that f is constant (Lemma 3.4.1).

2. This is an immediate consequence of (1). □

The uniqueness of harmonic realization is a consequence of this theorem. Indeed, for two harmonic realizations (Φ, ρ), (φ', ρ), the function $\mathbf{f} = \Phi' - \Phi$ is identified with a function on V_0, as we have seen in Sect. 7.4. Each entry of the vector-valued function $\mathbf{f}$ is harmonic so that $\mathbf{f}$ is a constant vector.

The following theorem is a slight generalization of what we have mentioned before, and is verified by using the fact that Δ is a hermitian operator of $\ell^2(V, m_V)$.

Theorem 7.8. *Let X be a finite graph. Given an $\mathbb{R}$-valued function g on V, the Poisson equation $\Delta f = g$ has an $\mathbb{R}$-valued solution g if and only if*

$$\sum_{x \in V} g(x) m_V(x) = 0.$$

If f_1 and f_2 are solutions of $\Delta f = g$, then $f_1 - f_2$ is constant.

---

[17] A function f satisfying $(\Delta f)(x) \geq 0$ is called a *subharmonic function*.

7.7 Notes

Looking at things through the notion of discrete Laplacians, we can see lots of conceptual resemblance between the theory of electric circuits and crystallography. For instance, the Poisson equation shows up when we treat a *resistive electric circuit*, an electric network consisting of resistors alone. To explain this, we regard the circuit as a finite weighted graph such that $m_V \equiv 1$ and $m_E(e)$ is the conductance (the reciprocal of the resistance) of the resistor represented by e. A current i and voltage v are identified with real-valued functions on E such that $i(\bar{e}) = -i(e)$ and $v(\bar{e}) = -v(e)$. In view of Ohm's law, they are related to each other by $m_E(e)v(e) = i(e)$. Kirchhoff's current law asserts

$$\sum_{e \in E_x} i(e) - g(x) = 0,$$

while, by Kirchhoff's voltage law, there exists a function f on V with $v(e) = f(t(e)) - f(o(e))$ (see Remark after Theorem 7.1). Thus combining Ohm's law and Kirchhoff's laws, we get the discrete Poisson equation $\Delta f = g$. Theorem 7.8 applied to this case states that *a steady flow of electricity takes place in the circuit if and only if the total amount of current entering the circuit from outside is zero.* Since the potential f is uniquely determined up to additive constants, the current (and hence the voltage) is also uniquely determined. This is what was established by G. Kirchhoff at the age of 21 (1845).

Theorems 7.7 and 7.8 have applications to random walks on topological crystals, an advanced topic discussed in Chap. 9.

(III) Equation (7.9) implies that the system of vectors $\{\mathbf{v}_0(e)\}_{e \in E_0}$ (and $\{\mathbf{v}_0(e)\}_{e \in E_0^o}$ for an orientation E_0^o) is a *tight frame*, a concept introduced in *wavelet analysis*[18] [82].

In general, a countable set of vectors $\{\mathbf{v}_\alpha\}$ in a Hilbert space $\mathcal{H}$ is said to be a tight frame if there exists a positive number c such that

$$\sum_\alpha \langle \mathbf{x}, \mathbf{v}_\alpha \rangle \mathbf{v}_\alpha = c\mathbf{x}$$

for every $\mathbf{x} \in \mathcal{H}$. Obviously any complete orthonormal basis is a tight frame (with $c = 1$) so that tight frames are a generalization of complete orthonormal basis. For many applications of wavelet analysis; say, signal processing, filtering and numerical analysis, tight frames are more important than orthonormal basis.

A set of vectors $\{\mathbf{v}_1, \ldots, \mathbf{v}_k\}$ is a tight frame in $\mathbb{R}^d$ with $c = 1$ if and only if it is the image by an orthogonal projection onto $\mathbb{R}^d$ of an orthonormal basis in the k-dimensional space. More precisely we have the following.

Theorem 7.9. *Let W be a d-dimensional subspace of $\mathbb{R}^k$, and let*

$$P : \mathbb{R}^k \longrightarrow \mathbb{R}^k$$

---

[18] This fact was suggested by Peter Kuchment.

be an orthogonal projection onto W.

1. Given an orthonormal basis $\{\mathbf{f}_1,\ldots,\mathbf{f}_k\}$ of $\mathbb{R}^k$, if we put $P\mathbf{f}_i = \mathbf{v}_i$, then

$$\langle \mathbf{x}, \mathbf{v}_1\rangle \mathbf{v}_1 + \cdots + \langle \mathbf{x}, \mathbf{v}_k\rangle \mathbf{v}_k = \mathbf{x} \quad (7.18)$$

for any $\mathbf{x} \in W$.

2. Conversely, given a system of vectors $\{\mathbf{v}_1,\ldots,\mathbf{v}_k\}$ in W, if (7.18) holds for any $\mathbf{x} \in W$, then there exist an orthonormal basis $\{\mathbf{f}_1,\ldots,\mathbf{f}_k\}$ of $\mathbb{R}^k$ such that $\mathbf{v}_i = P\mathbf{f}_i$ ($i = 1,\ldots,k$).

Proof. (1) is easily shown. To prove (2), define the linear operator $P' : \mathbb{R}^k \longrightarrow \mathbb{R}^k$ by

$$P'\mathbf{x} = \sum_{i=1}^{k} \langle \mathbf{x}, \mathbf{v}_i\rangle \mathbf{v}_i.$$

Then $P' = P$ because writing $\mathbf{x} = \mathbf{x}_1 + \mathbf{x}_2$ ($\mathbf{x}_1 \in W, \mathbf{x}_2 \in W^\perp$), we have $P'\mathbf{x} = P'\mathbf{x}_1 = \mathbf{x}_1 = P\mathbf{x}$.

For the square matrix $A = (\mathbf{v}_1,\ldots,\mathbf{v}_k)$ and $\mathbf{x} = {}^t(x_1,\ldots,x_k)$, we have $A\mathbf{x} = \sum_{i=1}^{k} x_i \mathbf{v}_i$, and hence

$$A^t A \mathbf{x} = (\mathbf{v}_1,\ldots,\mathbf{v}_k) \begin{pmatrix} \langle \mathbf{v}_1, \mathbf{x}\rangle \\ \vdots \\ \langle \mathbf{v}_k, \mathbf{x}\rangle \end{pmatrix} = \sum_{i=1}^{k} \langle \mathbf{x}\cdot\mathbf{v}_i\rangle \mathbf{v}_i = P\mathbf{x}.$$

Therefore $A^t A = P$. We also have

$$PA\mathbf{x} = \sum_{i=1}^{k} x_i P\mathbf{v}_i = \sum_{i=1}^{k} x_i \mathbf{v}_i = A\mathbf{x},$$

so $PA = A$.

We shall look for an orthogonal matrix U such that $U^t A = P$. When ${}^t A\mathbf{x} = {}^t A\mathbf{y}$, we get $P\mathbf{x} = A^t A\mathbf{x} = A^t A\mathbf{y} = P\mathbf{y}$, so by putting $U^t A\mathbf{x} = P\mathbf{x}$, we may define U on Image${}^t A$. Noting

$$\|U^t A\mathbf{x}\|^2 = \|P\mathbf{x}\|^2 = \langle P\mathbf{x}, P\mathbf{x}\rangle = \langle P\mathbf{x}, \mathbf{x}\rangle = \langle A^t A\mathbf{x}, \mathbf{x}\rangle$$
$$= \langle {}^t A\mathbf{x}, {}^t A\mathbf{x}\rangle = \|{}^t A\mathbf{x}\|^2,$$

we may extend U to the orthogonal complement of Image${}^t A$ in such a way that U is an orthogonal transformation of $\mathbb{R}^k$.

Taking the transpose of both sides of $U^t A = P$, we get $A^t U = P$, and hence $A = PU$. Since $A\mathbf{e}_i = \mathbf{v}_i$ for the fundamental basis $\{\mathbf{e}_1,\ldots,\mathbf{e}_k\}$ of $\mathbb{R}^k$, by putting $\mathbf{f}_i = U\mathbf{e}_i$, we get an orthonormal basis $\{\mathbf{f}_1,\ldots,\mathbf{f}_k\}$ with $P\mathbf{f}_i = \mathbf{v}_i$. □

7.7 Notes

Fig. 7.9 A regular polygon

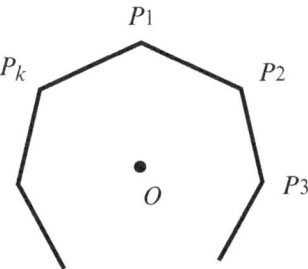

As will be explained in Sect. 8.1, we shall implicitly employ the above idea when we make up a building block of the normalized standard realization.

A tight frame appears in the following situation.

Theorem 7.10. *Suppose that a finite group G acts on $\mathbb{R}^d$ as orthogonal transformations, and let $\mathcal{S} = \{\mathbf{v}_1, \ldots, \mathbf{v}_k\}$ be a set of vectors in $\mathbb{R}^d$ which is invariant under the G-action. If the G-action on $\mathbb{R}^d$ is irreducible,[19] then $\mathcal{S}$ is a tight frame.*

Indeed, if we define the linear operator $T : \mathbb{R}^d \longrightarrow \mathbb{R}^d$ by setting $T\mathbf{x} = \sum_{i=1}^{k} \langle \mathbf{x}, \mathbf{v}_i \rangle \mathbf{v}_i$, then

$$Tg\mathbf{x} = \sum_{i=1}^{k} \langle g\mathbf{x}, \mathbf{v}_i \rangle \mathbf{v}_i = \sum_{i=1}^{k} \langle \mathbf{x}, g^{-1}\mathbf{v}_i \rangle \mathbf{v}_i = g \sum_{i=1}^{k} \langle \mathbf{x}, g^{-1}\mathbf{v}_i \rangle g^{-1}\mathbf{v}_i$$

$$= g \sum_{i=1}^{k} \langle \mathbf{x}, \mathbf{v}_i \rangle \mathbf{v}_i = gT\mathbf{x}.$$

Therefore the G-action commutes with T. By Shur's lemma (or looking at eigenspaces of T), we conclude that $T = cI$ for some positive scalar c, as desired.

Using this theorem, one can prove that if the points $P_1, \ldots, P_k$ ($k \geq 3$) in the plane $\mathbb{R}^2$ form a k-regular polygon with barycenter O (Fig. 7.9), then the vectors $\mathbf{v}_1 = \overrightarrow{OP_1}, \ldots, \mathbf{v}_k = \overrightarrow{OP_k}$ yield a tight frame (of course, one can prove this by a direct computation). For some two-dimensional crystal nets, this fact can be used to check that they are standard realizations. For example, the dice lattice in Example 3.5 is the standard realization. In fact, the building block of the dice lattice yields two regular hexagons (see Fig. 7.2).

By the same reasoning, we obtain a tight frame of $\mathbb{R}^3$ from a regular convex polyhedron.

(IV) The notion of harmonic realization is closely related to *harmonic maps*, critical maps for a certain energy functional defined on the set of maps between Riemannian

---

[19] This means that a G-invariant subspace of $\mathbb{R}^d$ is either $\{\mathbf{0}\}$ or the whole space $\mathbb{R}^d$.

manifolds (possibly with singularities). Actually the differential-geometric interpretation is a great help in seeing what really goes on in our discussion on harmonic and standard realizations. We assume here familiarity with Riemannian geometry (see [68] for the details).

In the case of smooth Riemannian manifolds, a smooth map $f : M \longrightarrow N$ is *harmonic* if it is a critical point of the Dirichlet energy functional

$$\mathcal{E}(f) = \int_M \|df\|^2 dv_M,$$

where $\|df\|^2$ is the squared norm of the differential of f, with respect to the induced metric on the bundle $T^*M \otimes f^{-1}TN$, and dv_M denotes the measure on M induced by its metric. The Euler–Lagrange equation associated to the functional $\mathcal{E}$, which characterizes harmonic maps, reads

$$\operatorname{tr} \nabla df = 0,$$

where ∇ is the connection (covariant differentiation) on $T^*M \otimes f^{-1}TN$ induced by the Levi-Civita connections on M and N.

Now we shall restrict ourselves to the case of maps from a finite graph into a Riemannian manifold (see [58]). Let $X_0 = (V_0, E_0)$ be a finite graph. We think of X_0 as a one-dimensional Riemannian manifold with singularity where each edge is identified with the interval $[0,1]$ having the metric dt^2. For a piecewise smooth map Φ of $X_0 = (V_0, E_0)$ into a complete Riemannian manifold (M, g), we denote by Φ_e the restriction of Φ to the edge e, and define the *energy* of Φ by

$$\mathcal{E}(\Phi) = \mathcal{E}(\Phi, g) = \frac{1}{2} \sum_{e \in E_0} \int_0^1 \left\| \frac{d\Phi_e}{dt} \right\|^2 dt. \qquad (7.19)$$

To characterize critical maps of the functional $\mathcal{E}(\Phi)$, we shall compute the first variation formula. Let $\Phi(u, \cdot)$ ($|u| < \varepsilon$) be a smooth variation of a map $\Phi = \Phi(0, \cdot)$ and put

$$\frac{d\Phi_e}{dt} = \dot{\Phi}_e, \quad \frac{\partial \Phi_e}{\partial u}(u, t) = W_e(t), \quad \frac{\partial \Phi_e}{\partial u}(u, 0) = W(x),$$

where we abuse notation and write $\Phi_e = \Phi_e(u, t)$.

Theorem 7.11 (The First Variation Formula). *At $u = 0$, we have*

$$\frac{d\mathcal{E}(\Phi)}{du} = -2 \sum_{x \in V_0} \Big\langle W(x), \sum_{e \in E_{0x}} \dot{\Phi}_e(0) \Big\rangle - \sum_{e \in E_0} \int_0^1 \Big\langle W_e(t), \frac{D}{dt} \dot{\Phi}_e \Big\rangle dt,$$

where $\dfrac{D}{dt}$ stands for the covariant differentiation along a curve.

7.7 Notes

Proof. By routine computation, we have

$$\frac{d\mathcal{E}(\Phi)}{du} = \sum_{e \in E_0}\left(\left\langle\frac{\partial \Phi_e}{\partial u}, \frac{\partial \Phi_e}{\partial t}\right\rangle\bigg|_{t=0}^{t=1} - \int_0^1 \left\langle\frac{\partial \Phi_e}{\partial u}, \frac{D}{\partial t}\frac{\partial \Phi_e}{\partial t}\right\rangle dt\right).$$

Since

$$\frac{\partial \Phi_{\bar{e}}}{\partial u}(0,1) = \frac{\partial \Phi_e}{\partial u}(0,0) = W(o(e)), \quad \frac{\partial \Phi_{\bar{e}}}{\partial t}(0,1) = -\frac{\partial \Phi_e}{\partial t}(0,0) = -\dot{\Phi}_e(0),$$

$$\sum_{e \in E_0}\left\langle\frac{\partial \Phi_{\bar{e}}}{\partial u}, \frac{\partial \Phi_{\bar{e}}}{\partial t}\right\rangle(0,1) = \sum_{e \in E_0}\left\langle\frac{\partial \Phi_{\bar{e}}}{\partial u}, \frac{\partial \Phi_e}{\partial t}\right\rangle(0,1) = -\sum_{e \in E_0}\left\langle\frac{\partial \Phi_e}{\partial u}, \frac{\partial \Phi_e}{\partial t}\right\rangle(0,0),$$

we find that

$$\frac{d\mathcal{E}(\Phi)}{du}(0) = -\sum_{e \in E_0}\left(2\left\langle W(o(e)), \dot{\Phi}_e(0)\right\rangle + \int_0^1 \left\langle W_e, \frac{D}{dt}\dot{\Phi}_e\right\rangle dt\right),$$

from which the first variation formula easily follows. □

The *second variation formula* established in a similar way is expressed as

$$\frac{d^2\mathcal{E}(\Phi)}{du^2} = -2\sum_{x \in V_0}\left\langle D_W W(x), \sum_{e \in E_{0x}}\dot{\Phi}_e(0)\right\rangle + \sum_{e \in E_0}\int_0^1 \left(-\left\langle D_{W_e}W_e, \frac{D}{dt}\dot{\Phi}_e\right\rangle\right.$$
$$\left. + \left\|\frac{DW_e}{dt}\right\|^2 - \left\langle W_e, R(W_e, \dot{\Phi}_e)\dot{\Phi}_e\right\rangle\right)dt,$$

where R denotes the curvature tensor.

The second variation formula is used when we want to know whether a harmonic map gives a local minimum.

The usual argument in the variational problem leads us to the following theorem in which $\dot{\Phi}_e(0)$ stands for the velocity vector (tangent vector) of the curve $\Phi_e(t)$ at $t = 0$.

Theorem 7.12. *A piecewise smooth map Φ is a critical map if and only if Φ is a piecewise geodesic map, i.e., Φ_e is a geodesic for every edge e, and at each $x \in V_0$*

$$\sum_{e \in E_{0x}} \dot{\Phi}_e(0) = 0. \tag{7.20}$$

This theorem somehow tells us that if we think of Φ as a net "made of an elastic material" and spread out in M, then Φ is critical if and only if the net is in equilibrium.

As for the existence of a harmonic map, we have:

Theorem 7.13. *Each homotopy class of maps of X_0 into a compact Riemannian manifold M contains at least one harmonic map Φ. If M is non-positively curved, then Φ is energy minimizing. Furthermore, if M is a flat torus, then Φ is unique up to translations.*

The first statement is a generalization of Poincaré's theorem asserting that each homotopy class of closed curves contains at least one closed geodesic. Indeed, a harmonic map from the 1-bouquet graph into M is a closed geodesic. The last statement is equivalent to the existence and uniqueness of a harmonic realization, because $\mathrm{Per}(X,\rho)$ introduced in Sect. 7.4 is identified with the set of maps of X_0 into the flat torus $\mathbb{R}^d/\rho(L)$ in a fixed homotopy class.[20]

We shall give a proof of the first statement in the theorem. Let C be a homotopy class of maps of X_0 into M. For each piecewise smooth map $\Phi \in C$, one can find a piecewise geodesic map $\Phi_1 \in C$ with $\mathcal{E}(\Phi_1) \leq \mathcal{E}(\Phi)$. Indeed, it is enough to replace each Φ_e by the shortest geodesic homotopic to Φ_e relative to the end points. We denote by C^* the set of piecewise geodesic maps in C on which we introduce the C^1-topology so that the functional $\mathcal{E}$ is continuous on C^*. It is straightforward to see that for every $a > 0$, the sublevel set $\{\Phi \in C^* | \, \mathcal{E}(\Phi) \leq a\}$ is compact. Therefore one can find a $\Phi_0 \in C^*$ such that $\mathcal{E}(\Phi_0) \leq \mathcal{E}(\Phi)$ for every $\Phi \in C^*$. Since Φ_0 is a critical map, it gives a desired harmonic map.

The concept of standard realization suggests considering the following general set-up. We take a family $\mathcal{G}$ of Riemannian metrics on M with a *fixed volume*. Fix a homotopy class C of maps of finite graph X_0 into M, and consider the functional $\mathcal{E} = \mathcal{E}(\Phi, g)$ on $C \times \mathcal{G}$. Then we ask *whether $\mathcal{E}$ attains its minimum at some pair* (Φ_0, g_0)? The existence of a standard realization ensures that this is true when M is a torus, $\mathcal{G}$ is the family of flat metrics on M, and maps in C induce a surjective homomorphism of $\pi_1(X_0)$ onto $\pi_1(M)$. Having this case in mind, it is interesting to investigate the case of a surface M with the family $\mathcal{G}$ of constant negatively curved metrics. It is plausible in this case that there exists a unique minimizer (Φ_0, g_0). If so, the lifting of Φ_0 to the Poincaré disc (the universal covering surface of M) may deserve to be called the "non-Euclidean standard realization".

The notion of harmonic map was introduced by J. Eells and J.H. Sampson in 1964 in their fundamental paper [39] as a generalization of geodesics, harmonic functions, and holomorphic maps between Kähler manifolds. They showed the existence of harmonic maps from a compact Riemannian manifold to a complete Riemannian manifold of non-positive sectional curvature. Since then, harmonic maps have been playing a ubiquitous role in differential geometry. Harmonic maps between *singular spaces* have been investigated by many researchers as well (see [40,53]). In contrast to the case for harmonic maps from graphs, it is a fairly difficult problem to establish the existence of harmonic maps from singular spaces of high dimension.

---

[20] See Nagano and Smith [69] for harmonic maps from a compact manifold into a flat torus.

Chapter 8
Explicit Construction

In the previous chapter we gave a characterization of the standard realization by means of a minimal principle. The existence was also confirmed. What remains to be done is to establish its explicit construction fitting in with the enumeration of topological crystals. The idea is quite simple. We make up a candidate for the building block $\{\mathbf{v}_0(e)\}_{e \in E_0}$ by the method suggested in Notes (III) in Chap. 7. To this end, we shall equip the vector space $C_1(X_0, \mathbb{R})$ of 1-chains with the canonical Euclidean structure. Thus, homology theory plays again a key role in our discussion.

Our construction embraces the goal of topological crystallography as far as its applications are concerned, for it furnishes us with a practical procedure by means of matrix computations to materialize the logically constructed object, thereby giving a powerful approach to the systematic design and enumeration of crystal structures.[1] To convince the reader how the concept of standard realization is natural, we exhibit many examples, including the diamond crystal, K_4 crystal, and Lonsdaleite.

The learned reader who skims through the first section may find an analogy with the classical theory of *harmonic integrals* in the discussion. Actually the idea we explain here is, by and large, a discrete version of the *method of orthogonal projections* developed by Weyl, Hodge, and Kodaira. A proto-idea in the graph setting is already seen in Weyl's work in the 1920s applied to the problem of electric circuits (see [95]).

8.1 General Construction

Let $X = (V, E)$ be a topological crystal of dimension d over a finite graph $X_0 = (V_0, E_0)$ whose abstract period lattice is L, and let $\mu : H_1(X_0, \mathbb{Z}) \longrightarrow L$ be the surjective homomorphism associated with the covering map $\omega : X \longrightarrow X_0$ (see Sect. 6.1 for the terminology).

---

[1] As for other ways in crystallography, see [4, 13, 29, 73] for instance.

First, we provide the homology group $H_1(X_0, \mathbb{R})$ with a natural inner product (which allows us to identify $H_1(X_0, \mathbb{R})$ with the Euclidean space $\mathbb{R}^b$, $b = b_1(X_0)$). To achieve this, we start with an inner product on $C_1(X_0, \mathbb{R})$, the group of 1-chains with real coefficients.

For $e, e' \in E_0$, we set

$$\langle e, e' \rangle = \begin{cases} 1 & (e' = e) \\ -1 & (e' = \overline{e}) \\ 0 & (\text{otherwise}), \end{cases}$$

and extend to an inner product on $C_1(X_0, \mathbb{R})$ in a natural manner:

$$\left\langle \sum_{e \in E_0} a_e e, \sum_{e' \in E_0} b_{e'} e' \right\rangle = \sum_{e,e' \in E_0} a_e b_{e'} \langle e, e' \rangle.$$

An orientation E_0^o ($\subset E_0$) gives an orthonormal basis of $C_1(X_0, \mathbb{R})$, and

$$\left\| \sum_{e \in E_0^o} a_e e \right\|^2 = \sum_{e \in E_0^o} |a_e|^2.$$

Restricting this inner product to the subspace

$$H_1(X_0, \mathbb{R}) \ (= \mathrm{Ker}\, \partial \subset C_1(X_0, \mathbb{R})),$$

we get a Euclidean structure on $H_1(X_0, \mathbb{R})$, which will be the space where the maximal topological crystal X_0^{ab} is to be realized in the standard way.

Let $P_{\mathrm{ab}} : C_1(X_0, \mathbb{R}) \longrightarrow H_1(X_0, \mathbb{R})$ be the orthogonal projection, and put $\mathbf{v}_{\mathrm{ab}}(e) = P_{\mathrm{ab}}(e)$, regarding each edge as a 1-chain. Obviously $\mathbf{v}_{\mathrm{ab}}(\overline{e}) = -\mathbf{v}_{\mathrm{ab}}(e)$. The homomorphic extension

$$\widehat{\mathbf{v}}_{\mathrm{ab}} : C_1(X_0, \mathbb{Z}) \longrightarrow H_1(X_0, \mathbb{R})$$

of $\mathbf{v}_{\mathrm{ab}}$ coincides with $P_{\mathrm{ab}}\big|C_1(X_0, \mathbb{Z})$, and hence if $\alpha \in H_1(X_0, \mathbb{Z})$, then $\widehat{\mathbf{v}}_{\mathrm{ab}}(\alpha) = \alpha$.

We will see that $\{\mathbf{v}_{\mathrm{ab}}(e)\}_{e \in E_0}$ is a building block of the standard realization of the maximal topological crystal X_0^{ab}.

1. Fix a reference point $x_0 \in V_0^{\mathrm{ab}}$. Since Image $\widehat{\mathbf{v}}_{\mathrm{ab}}|H_1(X_0, \mathbb{Z})$ coincides with the lattice group $H_1(X_0, \mathbb{Z})$ of $H_1(X_0, \mathbb{R})$, as mentioned in Sect. 7.1, we have a well-defined piecewise linear map

$$\Phi^{\mathrm{ab}} : X_0^{\mathrm{ab}} \longrightarrow H_1(X_0, \mathbb{R})$$

by setting

$$\Phi^{\mathrm{ab}}(x) = \mathbf{v}_{\mathrm{ab}}\big(\omega^{\mathrm{ab}}(e_1)\big) + \cdots + \mathbf{v}_{\mathrm{ab}}\big(\omega^{\mathrm{ab}}(e_n)\big), \tag{8.1}$$

8.1 General Construction

where $c = (e_1, \ldots, e_n)$ is a path in X_0^{ab} with $o(c) = x_0, t(c) = x$. We easily observe that $\Phi^{ab}(\sigma x) = \Phi^{ab}(x) + \sigma$ for $\sigma \in H_1(X_0, \mathbb{Z})$. That is, the period homomorphism for Φ^{ab} is the identity homomorphism of $H_1(X_0, \mathbb{Z})$.

2. We shall prove that Φ^{ab} is a harmonic realization, namely,

$$\sum_{e \in E_{0x}} \mathbf{v}_{ab}(e) = \mathbf{0} \qquad (x \in V_0).$$

For this, it is enough to observe that

$$\sum_{e \in E_{0x}} \langle e, \langle c \rangle \rangle = 0 \tag{8.2}$$

for an arbitrary circuit $c = (e_1, \ldots, e_n)$ in X_0. Indeed, if this is true, then

$$\sum_{e \in E_{0x}} e \in H_1(X_0, \mathbb{R})^\perp = \mathrm{Ker} P_{ab}$$

since the set of circuits generates $H_1(X_0, \mathbb{R})$, and hence

$$\sum_{e \in E_{0x}} \mathbf{v}_{ab}(e) = P_{ab}\Big(\sum_{e \in E_{0x}} e\Big) = \mathbf{0}.$$

We shall check (8.2). If c does not contain any edge in $E_{0x} \cup \overline{E_{0x}}$, then (8.2) evidently holds. Suppose c contains an edge in $E_{0x} \cup \overline{E_{0x}}$. Let (e', e'') be a part of c such that $t(e') = x$ and $o(e'') = x$. Then

$$\sum_{e \in E_{0x}} \langle e, e' + e'' \rangle = 1 + (-1) = 0.$$

This leads us to (8.2).[2]

3. Finally, we show that Φ^{ab} is the normalized standard realization. That is, we check that

$$\sum_{e \in E_0} \langle \mathbf{x}, \mathbf{v}_{ab}(e) \rangle^2 = 2\|\mathbf{x}\|^2 \qquad (\mathbf{x} \in H_1(X_0, \mathbb{R})). \tag{8.3}$$

If we choose an orientation E_0^o, this is written as

$$\sum_{e \in E_0^o} \langle \mathbf{x}, \mathbf{v}_{ab}(e) \rangle^2 = \|\mathbf{x}\|^2 \qquad (\mathbf{x} \in H_1(X_0, \mathbb{R})).$$

Using the fact that E_0^o is an orthonormal basis of $C_1(X_0, \mathbb{R})$, we find that

$$\sum_{e \in E_0^o} \langle \mathbf{x}, \mathbf{v}_{ab}(e) \rangle^2 = \sum_{e \in E_0^o} \langle \mathbf{x}, P_{ab}(e) \rangle^2 = \sum_{e \in E_0^o} \langle \mathbf{x}, e \rangle^2 = \|\mathbf{x}\|^2$$

---

[2] One can prove this claim by using the adjoint operator $\partial^* : C_0(X_0, \mathbb{Z}) \longrightarrow C_1(X_0, \mathbb{R})$ of the boundary operator ∂ which will be introduced in Sect. 10.2.

[cf. Theorem 7.9 in Notes (III) in Chap. 7]. Thus Φ^{ab} is the normalized standard realization associated with the building block $\{\mathbf{v}_{ab}(e)\}_{e \in E_0}$, as desired.

We now go to the general case. Let $H_{\mathbb{R}}$ be the subspace of $H_1(X_0, \mathbb{R})$ spanned by the vanishing subgroup $H = \operatorname{Ker}\mu$, and $H_{\mathbb{R}}^{\perp}$ the orthogonal complement of $H_{\mathbb{R}}$ in $H_1(X_0, \mathbb{R})$:

$$H_1(X_0, \mathbb{R}) = H_{\mathbb{R}} \oplus H_{\mathbb{R}}^{\perp}.$$

Then $\dim H_{\mathbb{R}}^{\perp} = \operatorname{rank} L = d$. By choosing an orthonormal basis of $H_{\mathbb{R}}^{\perp}$, we identify $H_{\mathbb{R}}^{\perp}$ with the Euclidean space $\mathbb{R}^d$.

Let $P: H_1(X_0, \mathbb{R}) \longrightarrow H_{\mathbb{R}}^{\perp} = \mathbb{R}^d$ be the orthogonal projection. Then P is a rational projection; that is, $P(H_1(X_0, \mathbb{Z}))$ is a lattice group of $H_{\mathbb{R}}^{\perp}$. Indeed, taking a $\mathbb{Z}$-basis $\{\alpha_1, \ldots, \alpha_b\}$ of $H_1(X_0, \mathbb{Z})$ such that $\{\alpha_{d+1}, \ldots, \alpha_b\}$ is a $\mathbb{Z}$-basis of H (thus also a basis of the vector space $H_{\mathbb{R}}$), we observe that $\{P(\alpha_1), \ldots, P(\alpha_d)\}$ is a $\mathbb{Z}$-basis of $P(H_1(X_0, \mathbb{Z}))$. Since $\{P(\alpha_1), \ldots, P(\alpha_d)\}$ forms a basis of the vector space $H_{\mathbb{R}}^{\perp} = \mathbb{R}^d$, we conclude that $P(H_1(X_0, \mathbb{Z}))$ is a lattice group of $H_{\mathbb{R}}^{\perp}$.

Noting now that

$$\operatorname{Ker}\big(P | H_1(X_0, \mathbb{Z})\big) = H_{\mathbb{R}} \cap H_1(X_0, \mathbb{Z}) = H,$$

we see straight away that the rational projection Φ_0 of Φ^{ab} by P is the normalized standard realization of $X = X_0^{ab}/H$. The period lattice is given by $P(H_1(X_0, \mathbb{Z}))$, and the period homomorphism coincides with the homomorphism $\rho : L = H_1(X_0, \mathbb{Z})/H \longrightarrow H_{\mathbb{R}}^{\perp}$ induced from $P: H_1(X_0, \mathbb{Z}) \longrightarrow H_{\mathbb{R}}^{\perp}$. Putting $\mathbf{v}_0(e) = P(\mathbf{v}_{ab}(e))$, we obtain the building block $\{\mathbf{v}_0(e)\}_{e \in E_0}$ of Φ_0.

To sum up, we have the commutative diagram for the normalized standard realizations:

$$\begin{array}{ccc} X_0^{ab} & \xrightarrow{\Phi^{ab}} & \mathbb{R}^b \\ \omega_1 \downarrow & & \downarrow P \\ X & \xrightarrow{\Phi_0} & \mathbb{R}^d, \end{array}$$

where $\omega_1 : X_0^{ab} \longrightarrow X$ is the subcovering map of $\omega^{ab} : X_0^{ab} \longrightarrow X_0$. To coin a phrase, the standard realization of a topological crystal is obtained by projecting down the standard realization of the maximal topological crystal onto a suitable subspace.

At this point we should say a few words about edges e with $\mathbf{v}_{ab}(e) = \mathbf{0}$. One can prove that $\mathbf{v}_{ab}(e) = \mathbf{0}$ if and only if e is a separating edge (this is not surprising in view of the fact that separating edges do not play any role in the homology group; see Sect. 4.3). For, if e is a separating edge, then any circuit c does not pass through e so that $\langle e, \langle c \rangle \rangle = 0$. Recall that the homology group $H_1(X_0, \mathbb{R})$ has a basis consisting of homology classes of circuits. We thus conclude that $e \in H_1(X_0, \mathbb{R})^{\perp}$, and hence $\mathbf{v}_{ab}(e) = \mathbf{0}$. If e is not a separating edge, then there is a circuit c containing e so that $\langle \mathbf{v}_{ab}(e), \langle c \rangle \rangle = \langle e, \langle c \rangle \rangle = \pm 1$ and $\mathbf{v}_{ab}(e) \neq \mathbf{0}$.

Remember that gathering all separating edges of X_0, we obtain a forest Y_0 (Sect. 3.4). As described in Sect. 7.2, reducing each component of Y to a point, we

8.2 Computations

obtain a reduction of X_0. The inverse image $Y = (\omega^{ab})^{-1}(Y_0)$ is also a forest, and gives rise to a reduction of X_0^{ab}. The reduction of X_0^{ab} is the maximal topological crystal over the reduction of X_0.

Loop edges in X_0 play a special role in the building block $\{\mathbf{v}_{ab}(e)\}_{e \in E_0}$. Let $e_1, \overline{e_1}, \ldots, e_k, \overline{e_k}$ be the totality of loop edges in X_0, and let $X_0' = (V_0, E_0')$ be the graph obtained by removing these loop edges. Then $\mathbf{v}_{ab}(e_i) = e_i$ since $e_i \in H_1(X_0, \mathbb{R})$, and $\{e_1, \ldots, e_k\} \subset H_1(X_0', \mathbb{R})^\perp$ (see Sect. 4.3).

8.2 Computations

Let us turn our attention to an algorithm to construct the normalized standard realization. The key point is to select a $\mathbb{Z}$-basis $\{\alpha_1, \ldots, \alpha_b\}$ ($b = b_1(X_0)$) of the homology group $H_1(X_0, \mathbb{Z})$ such that $\{\alpha_{d+1}, \ldots, \alpha_b\}$ is a $\mathbb{Z}$-basis of the vanishing subgroup H. Once this basis is given, one may reduce the computation to that of matrices.

In what follows the set of matrices with m rows and n columns will be denoted by $M(m,n)$. An element A in $M(m,n)$ is also called a matrix of size (m,n).

1. We first handle the case of the maximal topological crystal over X_0, namely the case $H = \{0\}$. We write

$$\mathbf{v}_{ab}(e) = \sum_{i=1}^{b} a_i(e) \alpha_i.$$

Since $\mathbf{v}_{ab}(e) - e \in H_1(X_0, \mathbb{R})^\perp$, we obtain $\langle \mathbf{v}_{ab}(e) - e, \alpha_j \rangle = 0$, and so

$$\sum_{i=1}^{b} a_i(e) \langle \alpha_i, \alpha_j \rangle = \langle e, \alpha_j \rangle.$$

Put $A = (\langle \alpha_i, \alpha_j \rangle) \in M(b,b)$, and

$$\mathbf{a}(e) = {}^t(a_1(e), \ldots, a_b(e)), \quad \mathbf{b}(e) = {}^t(\langle e, \alpha_1 \rangle, \ldots, \langle e, \alpha_b \rangle).$$

Then $\mathbf{a}(e) = A^{-1} \mathbf{b}(e)$. The matrix A, which describes how the 1-cycles α_i's intersect each other, is referred to as the *intersection matrix* associated with the $\mathbb{Z}$-basis $\{\alpha_1, \ldots, \alpha_b\}$ since it describes how the 1-cycles α_i's intersect each other (see Sect. 10.4).

Notice that the entries of A and $\mathbf{b}(e)$ are integers, and hence the entries of $\mathbf{a}(e)$ are rational

2. Next we look at the general case. Let $A = (\langle \alpha_i, \alpha_j \rangle) \in M(b,b)$ as before.

If we put $\gamma_i = P(\alpha_i)$ ($i = 1, \ldots, d$), then $\{\gamma_1, \ldots, \gamma_d\}$ is a $\mathbb{Z}$-basis of the period lattice. Using $P(\alpha_i) - \alpha_i \in H_\mathbb{R}$, we may write

$$\gamma_i - \alpha_i = P(\alpha_i) - \alpha_i = \sum_{j=d+1}^{b} d_{ij} \alpha_j.$$

Substituting $P(\alpha_i) = \alpha_i + \sum_{j=d+1}^{b} d_{ij}\alpha_j$ into $\langle P(\alpha_i), \alpha_k \rangle = 0$ ($k = d+1, \ldots, b$), we have

$$-\langle \alpha_i, \alpha_k \rangle = \sum_{j=d+1}^{b} d_{ij}\langle \alpha_j, \alpha_k \rangle.$$

Now writing

$$A = \begin{pmatrix} A_{11} & A_{12} \\ A_{21} & A_{22} \end{pmatrix},$$

with $A_{11} \in M(d,d)$, $A_{12} = {}^t A_{21} \in M(d, b-d)$, $A_{22} \in M(b-d, b-d)$, and using these matrices of small size, we may compute the matrix $D = (d_{ij}) \in M(d, b-d)$ as $D = -A_{12}A_{22}^{-1}$. Furthermore, using the property $P^2 = P = {}^t P$ of the orthogonal projection P, we have

$$\langle \gamma_i, \gamma_j \rangle = \langle P(\alpha_i), \alpha_j \rangle = \left\langle \alpha_i + \sum_{k=d+1}^{b} d_{ik}\alpha_k, \alpha_j \right\rangle$$

$$= \langle \alpha_i, \alpha_j \rangle + \sum_{k=d+1}^{b} d_{ik}\langle \alpha_k, \alpha_j \rangle.$$

Therefore the matrix $\Gamma = (\langle \gamma_i, \gamma_j \rangle) \in M(d,d)$, which completely characterizes the period lattice (up to orthogonal transformations), is computed as

$$\Gamma = A_{11} + DA_{21} = A_{11} - A_{12}A_{22}^{-1}A_{21}. \tag{8.4}$$

On the other hand, writing $\mathbf{v}_{ab}(e) = \sum_{i=1}^{b} a_i(e)\alpha_i$, we have

$$\mathbf{v}_0(e) = P(\mathbf{v}_{ab}(e)) = \sum_{i=1}^{d} a_i(e)\gamma_i.$$

We thus obtain an algorithm to construct the normalized standard realization of the topological crystal $X = X_0^{ab}/H$.

We remark that $\mathbf{v}_0(e)$ is a linear combination of γ_i's with rational coefficients, which implies that $\{\mathbf{v}_0(e)\}_{e \in E_0}$ generates a lattice group in $\mathbb{R}^d$ (this is true for the building blocks of any harmonic realization because a harmonic realization is an affine image of the standard realization).

Remark. In the introduction, we referred to the program *SYSTRE* constructed by Delgado-Friedrichs. An input data for this program is a finite graph X_0 together with a system of vectors $\{\mathbf{v}(e)\}_{e \in E_0}$ such that $\mathbf{v}(e) \in \mathbb{Z}^d$ and $\mathbf{v}(\bar{e}) = -\mathbf{v}(e)$. As a matter of fact, giving data (X_0, H) in our algorithm, where H is a vanishing subgroup, is equivalent to giving data such as $(X_0, \{\mathbf{v}(e)\}_{e \in E_0})$. For, if $(X_0, \{\mathbf{v}(e)\}_{e \in E_0})$ is given, then

$$H = \mathrm{Ker}\,(\widehat{\mathbf{v}} : H_1(X_0, \mathbb{Z}) \to \mathbb{Z}^d \subset \mathbb{R}^d)$$

is a vanishing group as seen in Sect. 7.1. Conversely, given a vanishing subgroup H of $H_1(X_0, \mathbb{Z})$, we may choose a homomorphism f of $H_1(X_0, \mathbb{Z})$ onto $\mathbb{Z}^d$ whose kernel is H. Using Theorem 4.1, we extend f to a homomorphism F of $C_1(X_0, \mathbb{Z})$ onto $\mathbb{Z}^d$. This extension yields a system of vectors $\{F(e)\}_{e \in E_0}$ satisfying the condition for SYSTRE input data. □

8.3 Examples

We are now ready to produce several examples of standard realizations based on the general recipe explained in the previous section. Our method gives rise to a sort of "veritable cottage industry" of crystal models.[3] The reader with programming skills may have fun by creating his/her own computer program to produce the CG images of two and three-dimensional crystals.

It is perhaps worthwhile to start with two-dimensional classical lattices to convince the reader that the standard realization is the most canonical placement in view of symmetry. Figure 8.1 illustrates the *square lattice* (**sq1**),[4] the *honeycomb lattice* (**hcb**), the *regular triangular lattice* (**hx1**), and the *regular kagome lattice* (**kgm**), respectively. Taking a look at their building blocks, we can directly check that these realizations are actually standard.

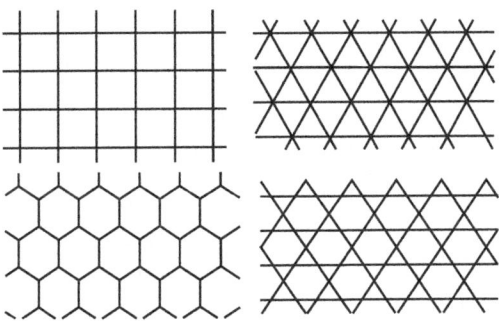

Fig. 8.1 Classical two-dimensional lattices

---

[3]The only problem we have when we want to describe standard realizations is that to say nothing of the higher dimensional case, it is not easy to imagine the complete geometric shape of a 3D crystal net when we display it on the plane (a computer screen). The best way to understand the shape is to use a crystalline compound kit (but such a kit is not always available for the crytal model we want to get).

[4]The three-letter names for crystal structures are proposed by O'Keeffe et al. [77].

Fig. 8.2 Cairo Pentagon

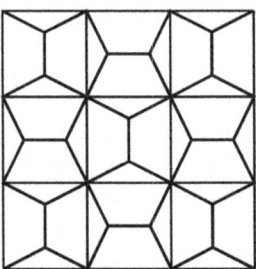

Fig. 8.3 Merging two square lattices

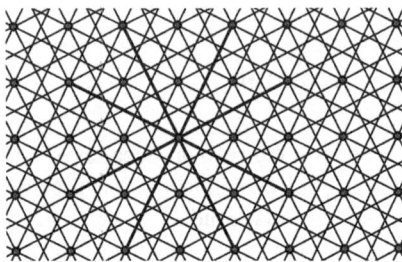

Figure 8.2 is a tiling of pentagons with picturesque properties that has become known as the *Cairo pentagon*.[5] Its 1-skeleton is the standard realization of a topological crystal over the finite graph depicted on the right.

As already mentioned, the dice lattice (**kgd**) in Fig. 3.9 (and Fig. 7.2) is also a standard realization.

All examples above are related to 2D tilings. Figure 8.3 which is a 2D standard realization obtained by merging two square lattices is a non-tiling example.[6]

We now turn to higher dimensional examples. Some of them are constructed as a methodical generalization of 2D examples.

(**I**) *Hypercubic lattice*. Let B_d be the d-bouquet graph with an orientation $\{e_1, \ldots, e_d\}$ (Example 4.2 in Sect. 4.3). If we put $\alpha_i = e_i$, then $\{\alpha_1, \ldots, \alpha_d\}$ is a $\mathbb{Z}$-basis of $H_1(B_d, \mathbb{Z})$ with $\langle \alpha_i, \alpha_j \rangle = \delta_{ij}$. Thus $\mathbf{v}_{ab}(e_i) = \alpha_i$. What we have as the standard realization of the maximal topological crystal over B_d is the *hypercubic lattice* (the net associated with the d-dimensional standard lattice group). It is known that NaCl forms a crystal whose net is the (primitive) cubic lattice (**pcu**).

(**II**) *Diamond of general dimension* (cf. [76]). Let D_d be the graph with two vertices joined by $d + 1$ parallel edges (see Fig. 8.4). The d-dimensional diamond is the

---

[5]It is given its name because several streets in Cairo are paved in this design (strictly speaking, it is a bit distorted). This is also called Macmahon's net, **mcm**, and a fourfold pentille (Conway).

[6]This figure is placed in my book [93] published in 2006, and is called a *Pythagorian lattice* because it is related to rational solutions of the equation $x^2 + y^2 = 1$. This is also an example of coincidence-site lattices [see Notes (I) in the present chapter].

8.3 Examples

Fig. 8.4 A base graph of the d-dimensional topological diamond

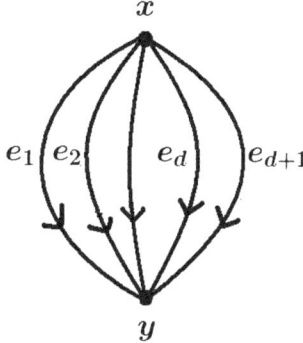

standard realization of the maximal topological crystal over the graph D_d. We shall compute its building block.

If we put $\alpha_i = e_{d+1} - e_i$ $(i = 1, \ldots, d)$, then $\{\alpha_1, \ldots, \alpha_d\}$ forms a $\mathbb{Z}$-basis of $H_1(D_d, \mathbb{Z})$, and

$$\langle \alpha_i, \alpha_j \rangle = \begin{cases} 2 & (i = j) \\ 1 & (i \neq j), \end{cases}$$

whence,

$$A = (\langle \alpha_i, \alpha_j \rangle) = \begin{pmatrix} 2 & 1 & 1 & \cdots & 1 & 1 \\ 1 & 2 & 1 & \cdots & 1 & 1 \\ \vdots & & & & & \vdots \\ 1 & 1 & 1 & \cdots & 2 & 1 \\ 1 & 1 & 1 & \cdots & 1 & 2 \end{pmatrix}. \tag{8.5}$$

Its inverse matrix is given by

$$B = A^{-1} = \frac{1}{d+1} \begin{pmatrix} d & -1 & -1 & \cdots & -1 & -1 \\ -1 & d & -1 & \cdots & -1 & -1 \\ \vdots & & & & & \vdots \\ -1 & -1 & -1 & \cdots & d & -1 \\ -1 & -1 & -1 & \cdots & -1 & d \end{pmatrix}.$$

We also have

$$\langle e_i, \alpha_k \rangle = \begin{cases} -\delta_{ik} & (1 \leq i \leq d) \\ 1 & (i = d+1), \end{cases}$$

and

$$a_{il} = \sum_{k=1}^{d} b_{kl} \langle e_i, \alpha_k \rangle = \begin{cases} -b_{il} & (1 \leq i \leq d) \\ \displaystyle\sum_{k=1}^{d} b_{kl} & (i = d+1), \end{cases}$$

Fig. 8.5 The building block of the diamond crystal

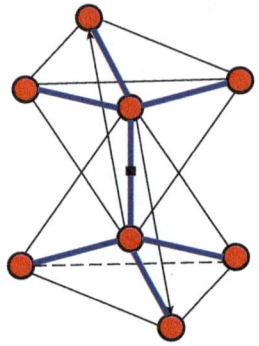

where $B = (b_{ij})$. We thus get

$$a_{il} = \begin{cases} -\dfrac{d}{d+1} & (1 \leq i \leq d,\ i = l) \\ \dfrac{1}{d+1} & (1 \leq i \leq d,\ i \neq l) \\ \dfrac{1}{d+1} & (i = d+1), \end{cases}$$

and

$$\mathbf{v}_{ab}(e_i) = \begin{cases} \dfrac{1}{d+1}(\alpha_1 + \cdots + \alpha_{i-1} - d\alpha_i + \alpha_{i+1} + \cdots + \alpha_d) & (1 \leq i \leq d) \\ \dfrac{1}{d+1}(\alpha_1 + \cdots + \alpha_d) & (i = d+1). \end{cases}$$

To describe $\mathbf{v}_{ab}(e_i)$ in an easy-to-see way, we put

$$\mathbf{p} = \dfrac{1}{d+1}(\alpha_1 + \cdots + \alpha_d).$$

Then the building block of the d-dimensional diamond is given by

$$\{\mathbf{v}_{ab}(e)\}_{e \in E_x} = \{\mathbf{v}_{ab}(e_i)\}_{i=1}^d = \{\mathbf{p}, \mathbf{p} - \alpha_1, \ldots, \mathbf{p} - \alpha_d\},$$
$$\{\mathbf{v}_{ab}(e)\}_{e \in E_y} = \{\mathbf{v}_{ab}(\overline{e_i})\}_{i=1}^d = \{-\mathbf{p}, -\mathbf{p} + \alpha_1, \ldots, -\mathbf{p} + \alpha_d\}.$$

Note that $\mathbf{0}, \alpha_1, \ldots, \alpha_d$ form the vertices of an equilateral simplex (a higher dimensional generalization of an equilateral triangle and an equilateral tetrahedron) whose barycenter is $\mathbf{p}$. Therefore, when $d = 2$ and $d = 3$, we have the honeycomb lattice and the net of the diamond crystal (**dia**) in the usual sense, respectively. Figure 8.5 is the building block for the 3D diamond crystal.

8.3 Examples

(III) *Triangular lattice of general dimension.* Example 6.2 in Sect. 6.2 suggests the following definition of the triangular lattice of general dimension. Let B_{d+1} be the $(d+1)$-bouquet graph, and let $\{e_1,\ldots,e_d,e_{d+1}\}$ be an orientation of B_{d+1}. We put

$$\alpha_i = \begin{cases} e_i & (1 \leq i \leq d) \\ e_1 + \cdots + e_d + e_{d+1} & (i = d+1), \end{cases}$$

which comprise a $\mathbb{Z}$-basis of $H_1(B_{d+1},\mathbb{Z})$. As a vanishing subgroup H of $H_1(B_{d+1},\mathbb{Z})$, we take the one generated by α_{d+1}. The matrix $A = (\langle \alpha_i, \alpha_j \rangle) \in M(d+1, d+1)$ has the form

$$A = \begin{pmatrix} & & & 1 \\ & I_d & & \vdots \\ & & & 1 \\ 1 & \cdots & 1 & d+1 \end{pmatrix}.$$

One may check that the following matrix gives the inverse of A.

$$A^{-1} = \begin{pmatrix} 2 & 1 & \cdots & 1 & -1 \\ 1 & 2 & \cdots & 1 & -1 \\ \cdots & \cdots & & & \\ 1 & 1 & \cdots & 2 & -1 \\ -1 & -1 & \cdots & -1 & 1 \end{pmatrix},$$

from which it follows that

$$\Gamma = \frac{d}{d+1} \begin{pmatrix} 1 & -\frac{1}{d} & -\frac{1}{d} & \cdots & -\frac{1}{d} & -\frac{1}{d} \\ -\frac{1}{d} & 1 & -\frac{1}{d} & \cdots & -\frac{1}{d} & -\frac{1}{d} \\ & \cdots & & & \cdots & \\ -\frac{1}{d} & -\frac{1}{d} & -\frac{1}{d} & \cdots & 1 & -\frac{1}{d} \\ -\frac{1}{d} & -\frac{1}{d} & -\frac{1}{d} & \cdots & -\frac{1}{d} & 1 \end{pmatrix}.$$

Easy computation leads us to

$$\mathbf{v}_0(e_i) = \gamma_i \quad (i = 1,\ldots,d),$$
$$\mathbf{v}_0(e_{d+1}) = -\gamma_1 - \cdots - \gamma_d.$$

In particular, for $d = 2$, the standard realization is nothing but the regular triangular lattice. For $d = 3$, the standard realization is related to the body-centered cubic lattice (**bcu**), as shown in Fig. 8.6. As in the case of the regular triangular lattice, this is formed by straight lines. It is known that CsCl forms a crystal whose net is the one associated with the body-centered cubic lattice.

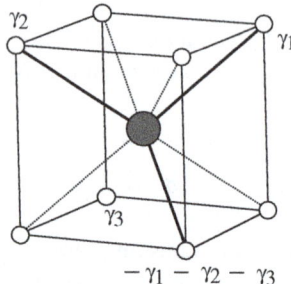
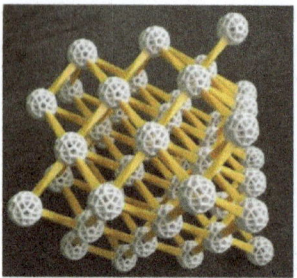

Fig. 8.6 The body-centered cubic lattice

Fig. 8.7 A_3-crystal net

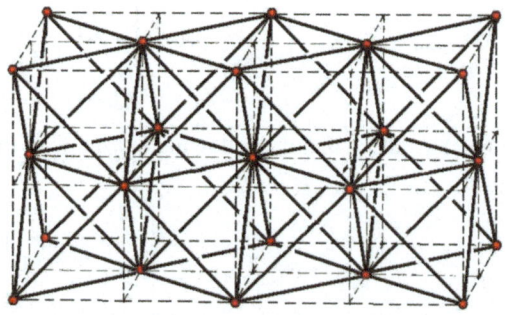

Another series of crystal nets that are regarded as a higher dimensional analogue of the regular triangular lattice is introduced by using what we call the *root lattice*[7] A_d. Precisely speaking, A_d is the lattice group in the subspace

$$W = \{(x_1,\ldots,x_{d+1}) \in \mathbb{R}^{d+1} \mid x_1 + \cdots + x_{d+1} = 0\}$$

of $\mathbb{R}^{d+1}$ defined by

$$A_d = \{(x_1,\ldots,x_{d+1}) \in \mathbb{Z}^{d+1} \mid x_1 + \cdots + x_{d+1} = 0\}.$$

The A_d-*crystal* is the net in $W = \mathbb{R}^d$ such that the set of vertices is A_d, and $\mathbf{x}, \mathbf{y} \in A_d$ are adjacent if and only if $\|\mathbf{x} - \mathbf{y}\|^2 = 2$. One can check that the A_d-crystal is the standard realization, and that the A_2-crystal is the regular triangular lattice. Figure 8.7 illustrates the A_3-crystal (thick lines represent edges), which is related to the densest sphere packing in $\mathbb{R}^3$ [see Notes (II) in this chapter].

We should point out that the lattice group A_d is the period lattice for the d-dimensional diamond. In crystallography, A_3 is called the *face-centered cubic lattice*, and the A_3-crystal is called **fcu**. The copper crystal is a typical one having the A_3-crystal structure.

---

[7] See Notes in this chapter and Ebeling [38] for general root lattices.

8.3 Examples 137

Fig. 8.8 3D kagome lattice of type I and its base graph

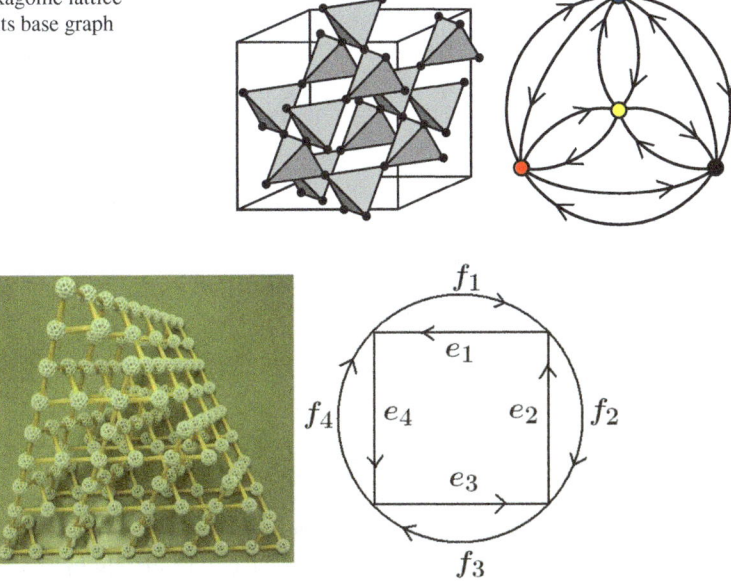

Fig. 8.9 3D kagome lattice of type II

Practically, the A_3-crystal structure has been employed to span large areas with few interior supports in architecture which is referred to as an *isotropic vector matrix* in Fuller's words[8] or an *octet truss* (see Fig. 1.5 in the introduction). Alexander Graham Bell is the first who used this structure to make rigid frames for man-lifting-kites.

(IV) *Kagome lattice*. Figure 8.8 (more precisely, the 1-skeleton of the solid figure in it) is the crystal structure of the 3D *kagome lattice* (**crs**). This is a 3D version of the ordinary kagome lattice in the sense that it is comprised of corner-sharing regular tetrahedra (note that the 2D kagome lattice is a network of corner-sharing triangles). We shall call this the 3D kagome lattice of type I. The base graph for its maximal period lattice is given on the right. The 3D kagome lattice of type I is used for a sandwich panel in building and ship construction.

There is another 3D analogue of the kagome lattice (**lvt**) which we shall call the 3D kagome lattice of type II. This is the standard realization of the topological crystal over the graph depicted on the right in Fig. 8.9 with the vanishing subgroup

---

[8]Richard Buckminster Fuller (1895–1983) is an American engineer, architectural designer, and inventor, and is best known for his geodesic dome. The carbon molecule "Fullerene" (C_{60}) was named after him for its resemblance to the geodesic dome.

Fig. 8.10 A base graph for the kagome lattice of type II

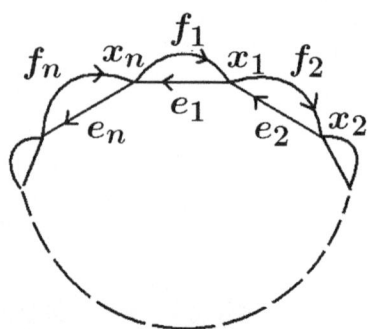

$$H = \mathbb{Z}(e_1 + e_2 + e_3 + e_4) + \mathbb{Z}(f_1 + f_2 + f_3 + f_4)$$

(compare with the description of the kagome lattice given in Example 6.5).

To give a higher dimensional analogue of the kagome lattice of type II, consider the graph P_n obtained by adding a parallel edge to each edge of the n-polygon (Fig. 8.10), having in mind Example 6.5 in Sect. 6.2. Evidently $b_1(P_n) = n + 1$.

We put $d = n - 1$, and define $\{\alpha_1, \ldots, \alpha_d, \alpha_{d+1}, \alpha_{d+2}\}$ by

$$\alpha_i = e_i + f_i \quad (i = 1, \ldots, d),$$
$$\alpha_{d+1} = e_1 + \cdots + e_{d+1}, \quad \alpha_{d+2} = f_1 + \cdots + f_{d+1},$$

which comprise a $\mathbb{Z}$-basis of $H_1(P_n, \mathbb{Z})$.

The *d-dimensional kagome lattice of type II* is defined to be the topological crystal over P_{d+1} associated with the vanishing subgroup H of $H_1(P_{d+1}, \mathbb{Z})$ generated by α_{d+1} and α_{d+2}. The matrix associated with the basis

$$\{\alpha_1, \ldots, \alpha_d, \alpha_{d+1}, \alpha_{d+2}\}$$

is given by

$$A = \begin{pmatrix} & & & 1 & 1 \\ & 2I_d & & \vdots & \vdots \\ & & & 1 & 1 \\ 1 & \cdots & 1 & d+1 & 0 \\ 1 & \cdots & 1 & 0 & d+1 \end{pmatrix}.$$

8.3 Examples

By easy (but tedious) computation, we find that

$$A^{-1} = \begin{pmatrix} 1 & \frac{1}{2} & \cdots & \frac{1}{2} & \frac{1}{2} & -\frac{1}{2} & -\frac{1}{2} \\ \frac{1}{2} & 1 & \cdots & \frac{1}{2} & \frac{1}{2} & -\frac{1}{2} & -\frac{1}{2} \\ & & \cdots \cdots & & & \vdots & \vdots \\ \frac{1}{2} & \frac{1}{2} & \cdots & 1 & \frac{1}{2} & -\frac{1}{2} & -\frac{1}{2} \\ \frac{1}{2} & \frac{1}{2} & \cdots & \frac{1}{2} & 1 & -\frac{1}{2} & -\frac{1}{2} \\ -\frac{1}{2} & -\frac{1}{2} & \cdots & -\frac{1}{2} & -\frac{1}{2} & \frac{d+2}{2(d+1)} & \frac{d}{2(d+1)} \\ -\frac{1}{2} & -\frac{1}{2} & \cdots & -\frac{1}{2} & -\frac{1}{2} & \frac{d}{2(d+1)} & \frac{d+2}{2(d+1)} \end{pmatrix}.$$

Using $\Gamma = A_{11} - A_{12} A_{22}^{-1} A_{21}$, we also have

$$\Gamma = \begin{pmatrix} 2 - \frac{2}{d+1} & -\frac{2}{d+1} & \cdots & -\frac{2}{d+1} \\ -\frac{2}{d+1} & 2 - \frac{2}{d+1} & \cdots & -\frac{2}{d+1} \\ & \cdots & \cdots & \\ -\frac{2}{d+1} & -\frac{2}{d+1} & \cdots & 2 - \frac{2}{d+1} \end{pmatrix}.$$

To describe the building block, we note that

$$E_{0x_i} = \{e_i, \overline{e_{i+1}}, \overline{f_i}, f_{i+1}\} \quad (i = 1, \ldots, n-1),$$
$$E_{0x_n} = \{e_n, \overline{e_1}, \overline{f_n}, f_1\}.$$

Applying the general recipe, we get

$$\begin{cases} \mathbf{v}_0(e_i) = \frac{1}{2}\gamma_i \\ \mathbf{v}_0(\overline{e_{i+1}}) = -\frac{1}{2}\gamma_{i+1} \\ \mathbf{v}_0(\overline{f_i}) = -\frac{1}{2}\gamma_i \\ \mathbf{v}_0(f_{i+1}) = \frac{1}{2}\gamma_{i+1} \end{cases} (i \le n-2), \quad \begin{cases} \mathbf{v}_0(e_{n-1}) = \frac{1}{2}\gamma_{n-1} \\ \mathbf{v}_0(\overline{e_n}) = \frac{1}{2}(\gamma_1 + \cdots + \gamma_{n-1}) \\ \mathbf{v}_0(\overline{f_{n-1}}) = -\frac{1}{2}\gamma_i \\ \mathbf{v}_0(f_n) = -\frac{1}{2}(\gamma_1 + \cdots + \gamma_{n-1}) \end{cases}$$

$$\begin{cases} \mathbf{v}_0(e_n) = -\frac{1}{2}(\gamma_1 + \cdots + \gamma_{n-1}) \\ \mathbf{v}_0(\overline{e_1}) = -\frac{1}{2}\gamma_1 \\ \mathbf{v}_0(\overline{f_n}) = \frac{1}{2}(\gamma_1 + \cdots + \gamma_{n-1}) \\ \mathbf{v}_0(f_1) = \frac{1}{2}\gamma_1. \end{cases}$$

When $d = n - 1 = 2$, the standard realization is the regular kagome lattice. Figure 8.11 shows the building block in the case $d = n - 1 = 3$. This is also related to the body-centered cubic lattice.

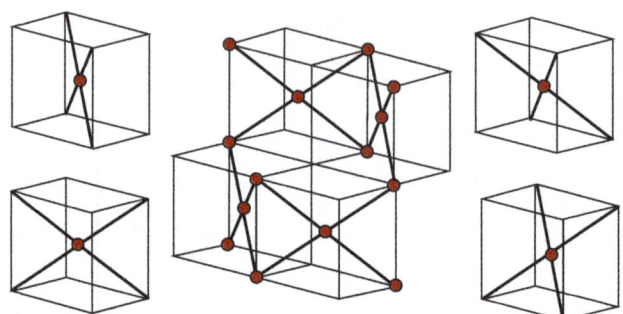

Fig. 8.11 The building block for the 3D kagome lattice of type II

Fig. 8.12 A base graph for the K_4 crystal

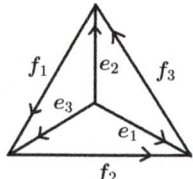

(V) K_4 *crystal.* Let X be the maximal topological crystal over the complete graph K_4. See Fig. 8.12 for the labeling of edges.

Take three closed paths $c_1 = (e_2, f_1, \overline{e_3})$, $c_2 = (e_3, f_2, \overline{e_1})$, $c_3 = (e_1, f_3, \overline{e_2})$ in K_4. Then $\alpha_1 = \langle c_1 \rangle$, $\alpha_2 = \langle c_2 \rangle$, $\alpha_3 = \langle c_3 \rangle$ constitute a $\mathbb{Z}$-basis of $H_1(K_4, \mathbb{Z})$ satisfying $\|\alpha_1\|^2 = \|\alpha_2\|^2 = \|\alpha_3\|^2 = 3$, $\alpha_i \cdot \alpha_j = -1$ $(i \neq j)$. An easy computation gives us

$$\mathbf{v}_0(e_1) = -\frac{1}{4}\alpha_2 + \frac{1}{4}\alpha_3, \quad \mathbf{v}_0(e_2) = \frac{1}{4}\alpha_1 - \frac{1}{4}\alpha_3, \quad \mathbf{v}_0(e_3) = -\frac{1}{4}\alpha_1 + \frac{1}{4}\alpha_2,$$

$$\mathbf{v}_0(f_1) = \frac{1}{2}\alpha_1 + \frac{1}{4}\alpha_2 + \frac{1}{4}\alpha_3, \quad \mathbf{v}_0(f_2) = \frac{1}{4}\alpha_1 + \frac{1}{2}\alpha_2 + \frac{1}{4}\alpha_3,$$

$$\mathbf{v}_0(f_3) = \frac{1}{4}\alpha_1 + \frac{1}{4}\alpha_2 + \frac{1}{2}\alpha_3.$$

The standard realization of the maximal topological crystal over K_4 is what we call the K_4 *crystal* (Fig. 1.4 in the introduction). See Notes (IV) in this chapter for many interesting features about the K_4 crystal.

(VI) In the sequel, we shall look at the 3D maximal topological crystal mentioned in Example 4.5 (excluding the diamond, cubic and K_4 crystals). We omit detailed computation. The CG images in Figs. 8.13–8.19 were created by Hisashi Naito.

(a) In the case of the finite graph X_0 depicted in Fig. 8.13, as a $\mathbb{Z}$-basis, we choose $\alpha_1 = e_1$, $\alpha_2 = e_2$, $\alpha_3 = f_1 + f_2$. Then for X_0^{ab}, we have

8.3 Examples

Fig. 8.13 Example (a)

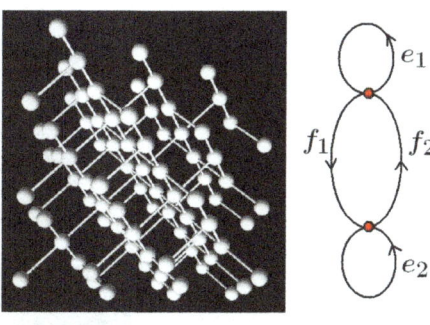

Fig. 8.14 Example (b)

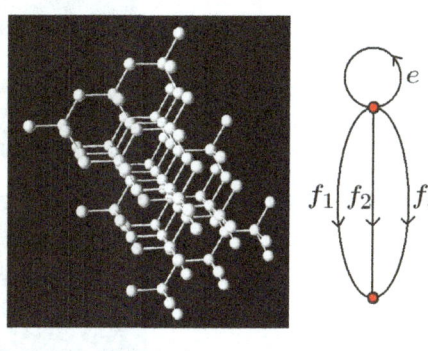

Fig. 8.15 Example (c)

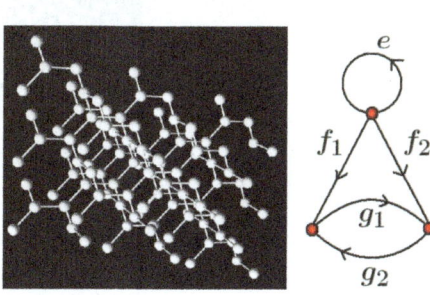

Fig. 8.16 Example (d)

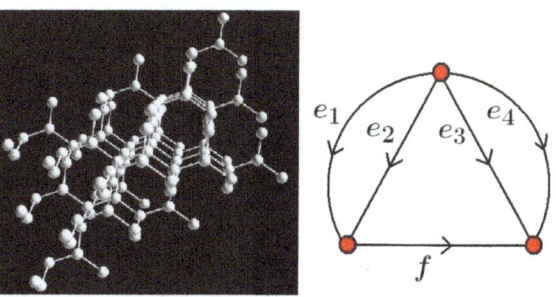

Fig. 8.17 Example (e)

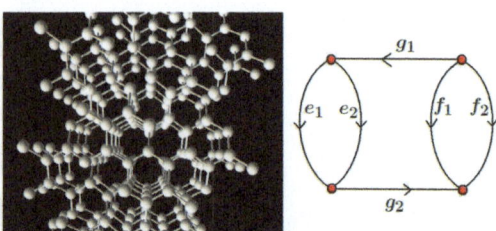

Fig. 8.18 3D example

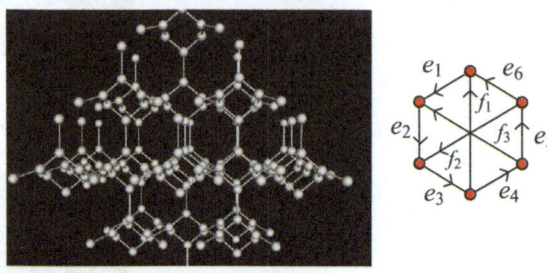

Fig. 8.19 3D example

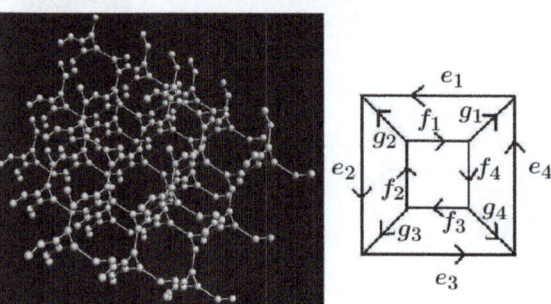

$$A = \begin{pmatrix} 1 & 0 & 0 \\ 0 & 1 & 0 \\ 0 & 0 & 2 \end{pmatrix}, \quad A^{-1} = \begin{pmatrix} 1 & 0 & 0 \\ 0 & 1 & 0 \\ 0 & 0 & 1/2 \end{pmatrix},$$

and

$$\mathbf{v}_0(e_1) = \alpha_1, \quad \mathbf{v}_0(e_2) = \alpha_2, \quad \mathbf{v}_0(f_1) = \frac{1}{2}\alpha_3, \quad \mathbf{v}_0(f_2) = \frac{1}{2}\alpha_3.$$

(b) In the case of the finite graph X_0 depicted in Fig. 8.14, as a $\mathbb{Z}$-basis, we choose $\alpha_1 = e$, $\alpha_2 = f_1 + \overline{f_2}$, $\alpha_3 = f_2 + \overline{f_3}$. Then for X_0^{ab}, we have

$$A = \begin{pmatrix} 1 & 0 & 0 \\ 0 & 2 & -1 \\ 0 & -1 & 2 \end{pmatrix}, \quad A^{-1} = \begin{pmatrix} 1 & 0 & 0 \\ 0 & 2/3 & 1/3 \\ 0 & 1/3 & 2/3 \end{pmatrix},$$

8.3 Examples

and

$$\mathbf{v}_0(e) = \alpha_1, \quad \mathbf{v}_0(f_1) = \frac{2}{3}\alpha_2 + \frac{1}{3}\alpha_3,$$

$$\mathbf{v}_0(f_2) = -\frac{1}{3}\alpha_2 + \frac{1}{3}\alpha_3, \quad \mathbf{v}_0(f_3) = -\frac{1}{3}\alpha_2 - \frac{2}{3}\alpha_3.$$

This is a graphite-like realization. Each layer has the honeycomb lattice whose building block is obtained from $\mathbf{v}_0(f_1)$, $\mathbf{v}_0(f_2)$, $\mathbf{v}_0(f_3)$.

(c) In the case of the finite graph X_0 depicted in Fig. 8.15, as a $\mathbb{Z}$-basis, we choose $\alpha_1 = e$, $\alpha_2 = f_1 + g_1 + \overline{f_2}$, $\alpha_3 = g_1 + g_2$. Then for X_0^{ab}, we have

$$A = \begin{pmatrix} 1 & 0 & 0 \\ 0 & 3 & 1 \\ 0 & 1 & 2 \end{pmatrix}, \quad A^{-1} = \begin{pmatrix} 1 & 0 & 0 \\ 0 & 2/5 & -1/5 \\ 0 & -1/5 & 3/5 \end{pmatrix},$$

and

$$\mathbf{v}_0(e) = \alpha_1, \quad \mathbf{v}_0(f_1) = \frac{2}{5}\alpha_2 - \frac{1}{5}\alpha_3, \quad \mathbf{v}_0(f_2) = -\frac{2}{5}\alpha_2 + \frac{1}{5}\alpha_3,$$

$$\mathbf{v}_0(g_1) = \frac{1}{5}\alpha_2 + \frac{2}{5}\alpha_3, \quad \mathbf{v}_0(g_2) = -\frac{1}{5}\alpha_2 + \frac{3}{5}\alpha_3.$$

This is also a graphite-like realization. In each layer, a distorted honeycomb lattice appears.

(d) In the case of the finite graph X_0 depicted in Fig. 8.16, as a $\mathbb{Z}$-basis, we choose $\alpha_1 = e_1 + \overline{e_2}$, $\alpha_2 = e_3 + \overline{e_4}$, $\alpha_3 = e_2 + f + \overline{e_3}$. Then for X_0^{ab}, we have

$$A = \begin{pmatrix} 2 & 0 & -1 \\ 0 & 2 & -1 \\ -1 & -1 & 3 \end{pmatrix}, \quad A^{-1} = \begin{pmatrix} 5/8 & 1/8 & 1/4 \\ 1/8 & 5/8 & 1/4 \\ 1/4 & 1/4 & 1/2 \end{pmatrix},$$

and

$$\mathbf{v}_0(e_1) = \frac{5}{8}\alpha_1 + \frac{1}{8}\alpha_2 + \frac{1}{4}\alpha_3, \quad \mathbf{v}_0(e_2) = -\frac{3}{8}\alpha_1 + \frac{1}{8}\alpha_2 + \frac{1}{4}\alpha_3,$$

$$\mathbf{v}_0(e_3) = -\frac{1}{8}\alpha_1 + \frac{3}{8}\alpha_2 - \frac{1}{4}\alpha_3, \quad \mathbf{v}_0(e_4) = -\frac{1}{8}\alpha_1 - \frac{5}{8}\alpha_2 - \frac{1}{4}\alpha_3,$$

$$\mathbf{v}_0(f) = \frac{1}{4}\alpha_1 + \frac{1}{4}\alpha_2 + \frac{1}{2}\alpha_3.$$

See Fig. 8.16 for the standard realization (called **tfa**).

(e) In the case of the finite graph X_0 depicted in Fig. 8.17, as a $\mathbb{Z}$-basis, we choose $\alpha_1 = e_1 + \overline{e_2}$, $\alpha_2 = f_1 + \overline{f_2}$, $\alpha_3 = e_2 + g_2 + \overline{f_1} + g_1$. Then for X_0^{ab}, we have

Fig. 8.20 A base graph for Lonsdaleite

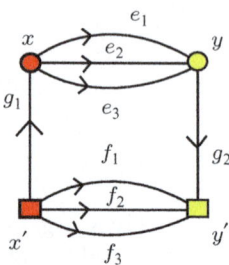

$$A = \begin{pmatrix} 2 & 0 & -1 \\ 0 & 2 & -1 \\ -1 & -1 & 4 \end{pmatrix}, \quad A^{-1} = \begin{pmatrix} 7/12 & 1/12 & 1/6 \\ 1/12 & 7/12 & 1/6 \\ 1/6 & 1/6 & 1/3 \end{pmatrix},$$

and

$$v_0(e_1) = \frac{7}{12}\alpha_1 + \frac{1}{12}\alpha_2 + \frac{1}{6}\alpha_3, \quad v_0(e_2) = -\frac{5}{12}\alpha_1 + \frac{1}{12}\alpha_2 + \frac{1}{6}\alpha_3,$$

$$v_0(f_1) = -\frac{1}{12}\alpha_1 + \frac{5}{12}\alpha_2 - \frac{1}{6}\alpha_3, \quad v_0(f_2) = -\frac{1}{12}\alpha_1 - \frac{7}{12}\alpha_2 - \frac{1}{6}\alpha_3,$$

$$v_0(g_1) = \frac{1}{6}\alpha_1 + \frac{1}{6}\alpha_2 + \frac{1}{3}\alpha_3, \quad v_0(g_2) = \frac{1}{6}\alpha_1 + \frac{1}{6}\alpha_2 + \frac{1}{3}\alpha_3.$$

The maximal topological crystal is what crystallographers call the ThSi$_2$ structure (**ths**), which is realized in a compound of Thorium and Silicide, and found in a number of other materials.

(**VII**) Figure 8.18 is the standard realization of the topological crystal over the bipartite complete graph K_{33} (see Fig. 8.23 for another representation of K_{33}). The vanishing subgroup is

$$H = \mathbb{Z}(e_1 + e_2 + e_3 + e_4 + e_5 + e_6).$$

(**VIII**) This is the case that X_0 is the graph depicted on the right in Fig. 8.19 with the vanishing subgroup

$$H = \mathbb{Z}(e_1 + e_2 + e_3 + e_4) + \mathbb{Z}(f_1 + f_2 + f_3 + f_4).$$

(**IX**) There are several relatives of Diamond. A typical one is Lonsdaleite.[9] The crystal structure of Lonsdaleite is given by the following building block (see Fig. 8.20 for the labeling of vertices and edges in the base graph).

---

[9] However Lonsdaleite is not isotropic.

8.3 Examples

$$\{v_0(e)\}_{e \in E_x} = \{v_0(e_1), v_0(e_2), v_0(e_3), v_0(\overline{g_1})\}$$
$$= \{\mathbf{a}_1, \mathbf{a}_2, -\mathbf{a}_1 - \mathbf{a}_2 - \mathbf{a}_3, \mathbf{a}_3\},$$
$$\{v_0(e)\}_{e \in E_y} = \{v_0(\overline{e_1}), v_0(\overline{e_2}), v_0(\overline{e_3}), v_0(g_2)\}$$
$$= \{-\mathbf{a}_1, -\mathbf{a}_2, \mathbf{a}_1 + \mathbf{a}_2 + \mathbf{a}_3, -\mathbf{a}_3\},$$
$$\{v_0(e)\}_{e \in E_{x'}} = \{v_0(f_1), v_0(f_2), v_0(f_3), v_0(g_1)\}$$
$$= \left\{\mathbf{a}_1 + \frac{2}{3}\mathbf{a}_3,\ \mathbf{a}_2 + \frac{2}{3}\mathbf{a}_3,\ -\mathbf{a}_1 - \mathbf{a}_2 - \frac{1}{3}\mathbf{a}_3,\ -\mathbf{a}_3\right\},$$
$$\{v_0(e)\}_{e \in E_{y'}} = \{v_0(\overline{f_1}), v_0(\overline{f_2}), v_0(\overline{f_3}), v_0(\overline{g_2})\}$$
$$= \left\{-\mathbf{a}_1 - \frac{2}{3}\mathbf{a}_3,\ -\mathbf{a}_2 - \frac{2}{3}\mathbf{a}_3,\ \mathbf{a}_1 + \mathbf{a}_2 + \frac{1}{3}\mathbf{a}_3,\ \mathbf{a}_3\right\},$$

where $\mathbf{a}_1, \mathbf{a}_2, \mathbf{a}_3$ are vectors satisfying

$$\langle \mathbf{a}_i, \mathbf{a}_i \rangle = 1, \quad \langle \mathbf{a}_i, \mathbf{a}_j \rangle = -1/3 \ (i \neq j).$$

We easily observe that each of

$$\{v_0(e)\}_{e \in E_x},\ \{v_0(e)\}_{e \in E_y},\ \{v_0(e)\}_{e \in E_x'},\ \{v_0(e)\}_{e \in E_y'}$$

forms an equilateral tetrahedron. Thus the net associated with Lonsdaleite is the standard realization.

To see the structure more explicitly, consider the *reflection* (mirror operation) $T_\mathbf{a}$ with respect to the hyperplane through the origin, orthogonal to a vector $\mathbf{a}$. It is given by

$$T_\mathbf{a}(\mathbf{x}) = \mathbf{x} - 2\frac{\langle \mathbf{x}, \mathbf{a} \rangle}{\|\mathbf{a}\|^2}\mathbf{a}.$$

Then we observe

$$T_{\mathbf{a}_3}(v_0(e_i)) = v_0(f_i) \ (i = 1, 2, 3), \quad T_{\mathbf{a}_3}(v_0(\overline{g_i})) = v_0(g_i) \ (i = 1, 2).$$

In the diamond crystal, only centrosymmetry $((x,y,z) \mapsto (-x,-y,-z))$ plays a role when we describe the building block in terms of the equilateral tetrahedron, whereas, in Lonsdaleite, we need the mirror operation besides centrosymmetry.

One can easily verify that this building block yields the standard realization. Its period lattice is generated by vectors $\mathbf{a}_1 - \mathbf{a}_2,\ \mathbf{a}_1 + 2\mathbf{a}_2 + \mathbf{a}_3,\ -\frac{2}{3}\mathbf{a}_3$, which correspond to homology classses $e_1 + \overline{e_2},\ e_2 + \overline{e_3},\ e_3 + g_2 + \overline{f_3} + g_1$, respectively. The vanishing subgroup H of $H_1(X_0, \mathbb{Z})$ corresponding to Lonsdaleite is given by

$$H = \mathbb{Z}(e_1 + \overline{e_3} + \overline{f_1} + f_3) + \mathbb{Z}(e_2 + \overline{e_3} + \overline{f_2} + f_3).$$

> **Summary.** A topological crystal X over $X_0 = (V_0, E_0)$ and its normalized standard realization Φ are constructed in the following way.
> 1. Take a $\mathbb{Z}$-basis $\{\alpha_1, \ldots, \alpha_b\}$ ($b = b_1(X_0)$) of the homology group $H_1(X_0, \mathbb{Z})$. Pick up $\alpha_{d+1}, \ldots, \alpha_b$ ($d \leq b$), and form a subgroup H of $H_1(X_0, \mathbb{Z})$ generated by these $(b-d)$ elements. We then have the d-dimensional topological crystal $X = X_0^{\mathrm{ab}}/H$.
> 2. Compute the square matrix of size b:
>
> $$A = (\langle \alpha_i, \alpha_j \rangle) = \begin{pmatrix} A_{11} & A_{12} \\ A_{21} & A_{22} \end{pmatrix} \quad (A_{11} \in M(d,d),\ A_{22} \in M(b-d, b-d)).$$
>
> Put $\Gamma = A_{11} - A_{12} A_{22}^{-1} A_{21} \in M(d,d)$, and take vectors $\gamma_1, \ldots, \gamma_d \in \mathbb{R}^d$ such that $\Gamma = (\langle \gamma_i, \gamma_j \rangle)$, which is to be a $\mathbb{Z}$-basis of the period lattice of the crystal $\Phi(X)$.
> 3. Compute $\mathbf{b}(e) = {}^t(\langle e, \alpha_1 \rangle, \ldots, \langle e, \alpha_b \rangle)$ ($e \in E_0$), and $\mathbf{a}(e) = A^{-1} \mathbf{b}(e) = {}^t(a_1(e), \ldots, a_b(e))$. Then putting $\mathbf{v}_0(e) = \sum_{i=1}^{d} a_i(e) \gamma_i$, we obtain the building block $\{\mathbf{v}_0(e)\}_{e \in E_0}$ of the standard realization $\Phi: X \longrightarrow \mathbb{R}^d$.

8.4 Notes

(I) One of the special features of standard realizations is that their period lattices are rational. Here a lattice L in $\mathbb{R}^d$ is *rational* if $\langle L, L \rangle \subset \lambda \mathbb{Q}$ for some positive number λ. Indeed, Eq. (8.4) tells us that the period lattices of the normalized standard realizations (and hence any standard realization) are rational.

The rationality of lattices is closely related to the theory of *coincidence-site lattices* (CSL), an important area in crystallography, initiated by material scientists in connection with crystalline interfaces and grain boundaries in polycrystalline materials (see [14, 81] for earlier work).

Mathematically, a coincidence site-lattice is nothing but the intersection of two *commensurable* lattices in space, where two lattices L_1, L_2 are said to be commensurable if $L_1 \cap L_2$ is a lattice. Symbolically we write $L_1 \sim L_2$. The relation $\sim$ is an equivalence relation.

In the theory of grain boundaries, different domains of a crystal across a boundary are related by having a sublattice (of full rank) in common. In this set-up, the CSL can be viewed as the intersection of a lattice with a rotated copy of itself.[10] This view leads us to the problem of how many rotated copies can be commensurable to the original lattice. Namely, we want to know the size of the *coincidence*

---

[10] A grain boundary is represented as a two-dimensional section of the CSL.

8.4 Notes

symmetry group $G(L) = \{g \in SO(d)|\ gL \sim L\}$ for a lattice L (note that $K(L) = \{g \in SO(d)|\ gL = L\}$, the symmetry group preserving L, is a finite group). For instance,

$$G(\mathbb{Z}^2) = \left\{ \begin{pmatrix} p & q \\ -q & p \end{pmatrix} \middle|\ p^2 + q^2 = 1,\ p, q \in \mathbb{Q} \right\}.$$

As is known, primitive Pythagorean triples[11] yield pairs $(p,q) \in \mathbb{Q} \times \mathbb{Q}$ with $p^2 + q^2 = 1$. Using this we conclude that $G(\mathbb{Z}^2)$ is dense in $SO(2)$. This fact is generalized in the following way.

Theorem 8.1. *$G(L)$ is dense in $SO(d)$ if and only if L is rational.*

The "if" part is verified by using Cayley's parametrization of rotations. The "only if" part is a consequence of transitivity of the $SO(d)$-action on the unit sphere $S^{d-1} = \{\mathbf{x} \in \mathbb{R}^d|\ \|\mathbf{x}\| = 1\}$. See Sunada [97] for details.

(II) In the previous section, we gave the definition of the A_3-crystal. We shall introduce its generalization.

In general, a *root lattice* is a lattice group L in $\mathbb{R}^d$ with the properties:

1. $\langle \mathbf{x}, \mathbf{x} \rangle \in 2\mathbb{Z}$ for every $\mathbf{x} \in L$.
2. The set $R = \{\mathbf{x} \in L|\ \langle \mathbf{x}, \mathbf{x} \rangle = 2\}$ generates L.

Elements in R are called *roots*, and R is called the *root system* of the root lattice L. We easily see that $\langle \mathbf{x}, \mathbf{y} \rangle \in \mathbb{Z}$ for every $\mathbf{x}, \mathbf{y} \in L$.

For a root lattice L, the *L-crystal* is the net in $\mathbb{R}^d$ defined by setting

$$L = \text{the set of vertices,}$$

$$\mathbf{x}, \mathbf{y} \in L \text{ are adjacent if and only if } \mathbf{x} - \mathbf{y} \in R.$$

Because L itself is the period lattice of the L-crystal, the base graph of the L-crystal is the bouquet graph. The root system R is a building block of the L-crystal. The A_d-crystal is the crystal net constructed in this way.

For each root $\mathbf{x}$, the reflection $T_\mathbf{x}$ leaves R invariant since $T_\mathbf{x}(\mathbf{y}) = \mathbf{y} - \langle \mathbf{x}, \mathbf{y} \rangle \mathbf{x}$. Let $W(R)$ be the subgroup of $O(d)$ generated by the reflections $T_\mathbf{x}$ ($\mathbf{x} \in R$). This group is called the *Weyl group* associated with R.

A root lattice is said to be *irreducible* if it cannot be decomposed as an orthogonal direct sum of two non-trivial root lattices. The following two facts are known (cf. [38]): if L is an irreducible root lattice, then

1. The action of $W(R)$ on $\mathbb{R}^d$ is irreducible.
2. $W(R)$ acts transitively on R.

---

[11] A triple of positive integers (x, y, z) is called *primitive Pythagorean* if they are pair-wise coprime, and $x^2 + y^2 = z^2$.

Fig. 8.21 Johannes Kepler

Therefore, in view of Theorem 7.10, the realization of the L-crystal is standard. Property (2) tells us that the L-crystal is *isotropic* (see (IV) in Notes).

Irreducible root lattices are classified into two series A_d, ($d = 2, 3, \ldots$), D_d ($d = 4, 5, \ldots$)[12] and three exceptional ones E_6, E_7, E_8 (see the aforementioned reference).

(III) At the beginning of Chap. 6, we mentioned that Greek mathematics started from their curiosity regarding the shapes of crystals. Since the Hellenistic period, geometry has taken its own path, and the study of crystals has not been the central theme in mathematics. An exception is the work of Johannes Kepler (1571–1630) published in the short pamphlet[13] entitled *New-Year's gift concerning six-cornered snow* ("Strena Seu de Nive Sexangula" in Latin) in 1611 (Fig. 8.21).

This pamphlet contains a problem closely related to the lattice A_3, and is considered the first work on the problem of crystal structures, although he did not refer at all to the atomistic viewpoint[14] that dates back to Ancient Greece. Indeed, he speculated, in the pamphlet that the densest packing of equally sized spheres is attained by the *hexagonal arrangement*, which later became known as the Kepler conjecture.[15] The hexagonal arrangement is obtained from the lattice A_3 as illustrated in Fig. 8.22, in which the nodes (vertices) are located at centers of equal spheres, and the line segments (edges) indicate that two spheres with centers located at their end points touch each other.[16]

---

[12] Note that the symbol D_d used here does not mean the graph introduced in Sect. 8.3 (II).

[13] This pamphlet was dedicated on the occasion of the New Year of 1611 to his friend and patron, the scholar and imperial privy councilor Johannes Matthäus Wackher von Wackenfels.

[14] The atomism was advocated by Leucippus and Democritus in the fifth century BC.

[15] In 1998 Thomas Hales, following an approach suggested by Fejes Tóth (1953), announced that he proved the Kepler conjecture.

[16] In connection with the hexagonal arrangement, Kepler noticed that the honeycomb structure maximizes the number of wax walls each bee shares with his neighbor, thereby allowing bees to collaborate in constructing the shared walls of a cell. The hexagon also turns out to be most efficient

8.4 Notes

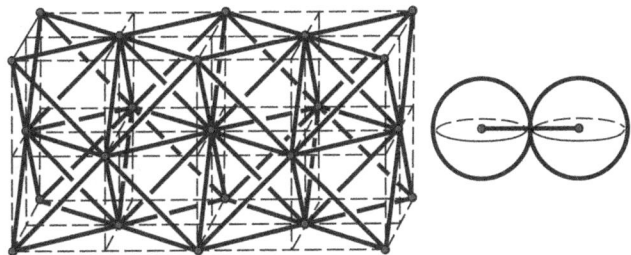

Fig. 8.22 The hexagonal sphere packing

In hindsight, Kepler's observation also includes indirect pointers to the law of constant angles for a six-sided snow crystal; thus he is regarded as a forerunner of N. Steno (1669), M.W. Lomonosov (1749) and Romé de l'Isle (1783), the discoverers of the law.[17]

It is noteworthy that in *Harmonice Mundi* (1619) including his discovery of the third law of planetary motion, Kepler accomplished a complete classification of *semi-regular polyhedra*[18] [25].

See [56] for a relation between sphere packings and crystallography.

(IV) The K_4 crystal is a purely mathematical object at present, not having been known as yet to exist in nature as a pure crystal like diamond,[19] but possessing remarkable mathematical properties.

1. **Homogeniety**. *For any two vertices, there is a motion preserving the crystal net which brings one vertex to another.*
2. **Strong isotropic property**. *For any two directed edges with the same origin, there is a motion preserving the crystal net and exchanging these two edges, while other directed edges (with the same origin) are fixed.*

---

in terms of exploiting the maximum space using the minimal amount of wax. Pappus (290 AD–350 AD) of Alexandria had already noticed that bees have foresight at their disposal in allowing them to understand that a hexagonal partition is more effective than a triangular or square partition.

[17] Needless to say, Kepler is renowned as an astronomer who discovered the three laws of planetary motion. His essay demonstrates that he had been thinking of not only the law of the vast universe but also the smallest aspects of nature. It should be emphasized, however, that he was an astrologer and theologian as well, and always sought an interpretation of the harmony of forms as God's choice [106]. That is, he asks the snowflakes the same question as he did the planets: which form follows God's order? Kepler's achievement tells us, though his study of geometric figures cannot be separated from his theological dogmatism, that he was one of the most outstanding mathematicians of his day.

[18] Pappus testifies in his *Collection* that Archimedes (287 BC–212 BC) proposed, in a now-lost work, the notion of *semi-regular polyhedron* as a generalization of regular polyhedron, and found 13 such solids. Here a polyhedron is said to be semi-regular if it has regular faces and a symmetry group that is transitive on its vertices. Precisely speaking, there are two infinite series of convex prisms and convex antiprisms satisfying this condition.

[19] For the evaluation of its various physical properties by first principles calculations, see [52] (for an sp^2 carbon crystal) and [27] (for a Boron crystal).

Fig. 8.23 K_{33}

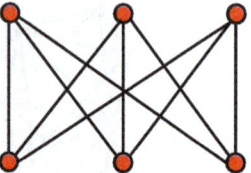

Ordinarily, the *isotropic*[20] property would describe that there is no distinction in any direction in the sense that there is a motion preserving the crystal net and exchanging these two directed edges (but not necessarily fixing other directed edges). The strong isotropic property is the strongest among all possible meanings of isotropy.

3. The K_4 crystal is a web of the same decagonal rings, and the number of decagonal rings meeting at each vertex is 15 (for the diamond crystal, the number of hexagonal rings passing through each vertex is 12).
4. The K_4 crystal has chirality; that is, its mirror image cannot be superimposed on the original one by a parallel translation or rotation.

The nets of the K_4 and diamond crystals are the only two 3D crystal net having *maximal symmetry*[21] together with properties (1) and (2). In this sense, the K_4 crystal deserves to be called the *diamond twin* [93, 94].

It is perhaps worthwhile to stress that the (primitive) cubic lattice, a crystal net with big symmetry, has property (1) and the isotropic property, but does not have the strong isotropic property.

The crystallographer who discovered for the first time this crystal structure as a hypothetical crystal is believed to be Fritz H. Laves (1933). Coxeter [22] called it "Laves' graph of girth ten". I rediscovered this structure by chance while I was studying random walks on topological crystals. In Conway et al. [21], this is called the *triamond net*.

The K_4 crystal is also referred to as $(\mathbf{10}, \mathbf{3})$-**a** [109] or **srs** net [31] by crystallographers. The name **srs** comes from the fact that it occurs in the compound $SrSi_2$ as the Si substructure. The K_4 crystal has a close relationship with the *gyroid*, an infinitely connected triply periodic minimal surface discovered by Alan Schoen in 1970. See [50] for this interesting story about the K_4 crystal and also [28, 33].

We are tempted to list all homogeneous and strongly isotropic crystal nets in general dimension. In 2D the honeycomb lattice is the only example. A typical example in $\mathbb{R}^d$ is the d-dimensional diamond. The maximal topological crystal over the complete graph K_n also yields an example, whose dimension is $(n-1)(n-2)/2$. Besides the four-dimensional diamond, the maximal topological crystal over the graph in Fig. 8.23 (K_{33} graph) gives an example of dimension 4.

---

[20]The term "isotropic" is used in a different context in crystallography. That is, an isotropic crystal is a crystal which has the same optical properties in all directions.

[21]See Sect. 9.6.

8.4 Notes

Compared with the strong isotropy, the ordinary isotropy is a much weaker property so that there are many examples of homogeneous and isotropic crystal nets. In the two-dimensional case, such crystal nets are the classical lattices mentioned at the beginning of this section; that is, the square lattice, honeycomb lattice, regular triangular lattice, and regular kagome lattice, together with three series of pathological crystal nets.[22] Here are a few (non-pathological) three-dimensional examples[23]:

(a) K_4 crystal,   degree 3
(b) Diamond crystal,   degree 4
(c) 3D kagome lattice of type II   degree 4
(d) (Primitive) cubic lattice,   degree 6
(e) 3D kagome lattice of type I,   degree 6
(f) Body-centered cubic lattice,   degree 8
(g) A_3-crystal,   degree 12

(V) In the introduction, we mentioned that the difference in nomenclature makes the communication between mathematicians and crystallographers difficult. For the convenience of the reader, we shall give a "dictionary" of terminology used in crystallography [9, 42] and mathematics.

(a) Quotient graph (of a periodic graph) = base graph of a topological crystal.
(b) Labelled (quotient) graph = a finite graph with a building block.
(c) Cycle space = homology group $H_1(X_0, \mathbb{R})$.
(d) Cocycle space = the orthogonal complement of $H_1(X_0, \mathbb{R})$ in the 1-chain group $C_1(X_0, \mathbb{R})$ with respect to the canonical inner product.
(e) Cyclomatic number = the first Betti number.
(f) Minimal net = maximal topological crystal.
(g) Edge space = 1-chain group $C_1(X_0, \mathbb{Z})$.
(h) Equilibrium placement (barycentric drawing) = harmonic realization.
(i) Archetypical representation = standard realization (canonical placement).

(VI) We shall give a strange but innocuous view of standard realizations.

As mentioned at the beginning of Chap. 5, the theory of covering maps is parallely developed as Galois theory of field extensions that originally stemmed from the problem of solvability of algebraic equations.

Galois theory is the base for *class field theory*, synonymous with the study of abelian extensions of number fields, whose origin lies in the quadratic reciprocity law proved by Gauss. After the refinement of Gauss's work by Dirichlet and Kummer, and the legitimate formulation by Hilbert in a special case, Teiji Takagi (1875–1960) gave a general formulation of class field theory and established a

---

[22] The two-dimensional crystal net in Fig. 8.3 is such an example.

[23] See [31, 34, 35] for a related question in which a slightly weaker version (*edge-transitivity*) of isotropy is discussed.

crucial theorem which asserts that any abelian extension is perfectly described in terms of *ideal class groups*. A remaining important issue, historically entitled *Kronecker's Dream in youth* ("meinen liebsten Jugendtraum") and not yet resolved in full generality,[24] is to construct all abelian extensions *in an explicit way* (for instance, by using special values of transcendental functions[25] or by certain structures in algebraic geometry).

The gist of my view is in the fact that one can make a "miniature" of algebraic number theory in the graph setting. As I mentioned in Notes (II), Chap. 6, one may even formulate class field theory in the discrete setting by considering prime cycles as a discrete analogue of prime ideals (cf. [88]). The homology group $H_1(X_0, \mathbb{Z})$ is considered an analogue of the ideal class group. Thus the issue of an explicit construction of an abelian covering graph (an analogue of an abelian extension), especially a topological crystal, is regarded as a counterpart of Kronecker's Dream in the framework of graphs. In this connection, it is interesting to point out that, as will be explained in Chap. 10, standard realizations are closely related to a discrete analogue of Abel–Jacobi maps in classical algebraic geometry.

As Barry Mazur says in his essay (2008) "Visions, Dreams, and Mathematics", we may think of Kronecker's Dream as the very broad class of visions of the following kind:

> One mathematical field can be a source of explanation by providing explicit solutions to problems posed in another mathematical field.

Therefore, to stretch a point, Kronecker's Dream in the framework of graphs is accomplished by the concept of standard realization.

---

[24]Leopold Kronecker (1823–1891). His dream was described in a letter to Dedekind in 1880.

[25]This problem was officially posed by Hilbert in his famous 12th problem. The exponential function can be used to describe abelian extensions of the rational number field $\mathbb{Q}$ (the Kronecker–Weber theorem). There are partial results in the cases of *CM*-fields or real quadratic fields, but the problem is largely still open.

Part III
Advanced Topics

Chapter 9
Random Walks on Topological Crystals

The most exciting moment we encounter while studying mathematics is when we observe that a seemingly isolated subject turns out to be connected with other fields in an unexpected way. In the present and next chapters, we shall give two such examples in connection with standard realizations. The first is asymptotic analysis of random walks on topological crystals which motivated the author to introduce the concept of standard realization [58, 60], and is the theme of this chapter. The second, to be explained in the next chapter, is a discrete (combinatorial) analogue of classical algebraic geometry, a field of more recent vintage, in which the standard realizations in a processed form show up as an analogue of the Abel–Jacobi map.

Thus our focus moves away from applications to crystallography (except for Sect. 9.6 where symmetry of topological crystals together with the uniqueness of standard realizations is dealt with). The discussion includes much background material in pure mathematics. The reader is expected to have some prior familiarity with advanced mathematics.

9.1 Simple Random Walks

By way of introduction and motivation, let us start with the general issue of random walks.

A *random walk*, a synonym in a loose sense for a time homogeneous Markov chain on a countable state space, is an object that probabilists are handling as a typical stochastic process. In terms of a graph $X = (V, E)$, a random walk is a stochastic process associated with a positive-valued function p on E satisfying $\sum_{e \in E_x} p(e) = 1$. We think of $p(e)$ as the *transition probability* that a random walker at $o(e)$ moves to $t(e)$ along the edge e. The *transition operator* $P : C(V) \longrightarrow C(V)$, which plays an important role in our discussion, is defined by

$$Pf(x) = \sum_{e \in E_x} p(e) f(t(e)),$$

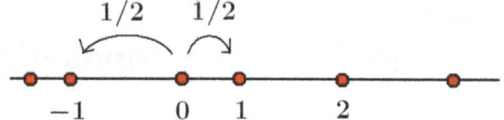

Fig. 9.1 Random walk on the one-dimensional lattice

where $C(V)$ is the space of functions on V. The *n-step transition probability* $p(n,x,y)$ is the probability that, after the n-step movement, a random walker at the initial site x is found at y, which is computed as

$$p(n,x,y) = \sum_{\substack{c=(e_1,\ldots,e_n) \\ o(c)=x, t(c)=y}} p(e_1)\cdots p(e_n),$$

and turns out to be the kernel function of P^n, namely,

$$(P^n f)(x) = \sum_{y \in V} p(n,x,y) f(y).$$

A central theme in the theory of random walks is the properties of $p(n,x,y)$ as n goes to infinity.[1] A typical sample is the classical *Laplace–de Moivre theorem* asserting that for the random walk on the one-dimensional lattice with the probability to move in the direction of left and right being $1/2$ each, we have

$$p(n,x,y) = \frac{\sqrt{2}}{\sqrt{\pi n}} \exp\left(-\frac{|x-y|^2}{2n}\right)(1+r_n(x-y)) \quad (x,y \in \mathbb{Z},\ n \equiv x-y \bmod 2)$$

where, for each positive constant A, $\lim_{n\to\infty} r_n(x) = 0$ uniformly for x with $|x| \leq A\sqrt{n}$. Here we should emphasize that the integers x and y are used only for labeling vertices (sites) on the one-dimensional lattice, thereby initially $p(n,x,y)$ has nothing to do with the magnitude $|x-y|$. Nevertheless we can read the magnitude $|x-y|$ from the asymptotic behavior of $p(n,x,y)$. In Fig. 9.1, in order to stress this phenomenon, we exhibit non-uniform spaces between two adjacent vertices. What should be noted is that the inclusion map of $\mathbb{Z}$ into $\mathbb{R}$ is the standard realization of the one-dimensional lattice, which we denote by Φ following our convention. Then it is more natural to write $|\Phi(x) - \Phi(y)|$ instead of $|x-y|$ to convey what we mean.

The walk we consider hereafter is the *simple random walk*, a generalization of the walk on $\mathbb{Z}$ we looked at above. This is the random walk on $X = (V,E)$ with the transition probability $p(e) = 1/\deg o(e)$; that is, the walk such that the probabilities moving along out-going edges from a vertex are the same (Fig. 9.2). This being the case, the transition operator P acting on functions with finite support is symmetric with respect to the inner product defined by

---

[1] See Woess [112].

9.1 Simple Random Walks

Fig. 9.2 Simple random walk on the hexagonal lattice

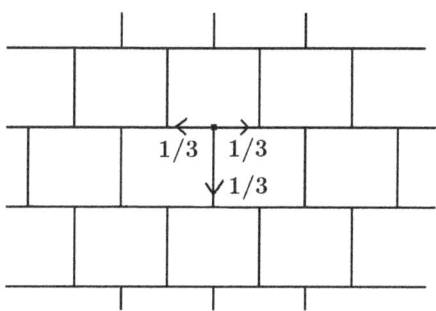

$$\langle f_1, f_2 \rangle = \sum_{x \in V} f_1(x) f_2(x) (\deg x),$$

and hence $p(n,x,y)(\deg y)^{-1} = p(n,y,x)(\deg x)^{-1}$.

The observation above is generalized to the case of the simple random walk on a d-dimensional topological crystal X [60, 61]. Indeed, we can establish the *central limit theorem*, a generalization of the Laplace–de Moivre theorem.[2] This theorem involves, as expected, the standard realization Φ of X in the form $\exp\left(-\dfrac{C}{n} \|\Phi(x) - \Phi(y)\|^2\right)$ [see Notes (III) in this chapter].

Furthermore we may prove the following asymptotic formula wherein the standard realization shows up in a bit tricky way.[3]

$$p(n,x,y)(\deg y)^{-1} \sim n^{-d/2} C_0 \left[1 + c_1(x,y) n^{-1} + c_2(x,y) n^{-2} + \cdots \right]. \qquad (9.1)$$

Here the symbol $\sim$ means that for any integer $k \geq 0$, there is a positive constant c such that

$$\left| p(n,x,y)(\deg y)^{-1} - n^{-d/2} C_0 \left(1 + c_1(x,y) n^{-1} + \cdots + c_k(x,y) n^{-k}\right) \right|$$
$$\leq c n^{-k-1}.$$

The coefficients $c_k(x,y)$, which is uniquely determined by $p(n,x,y)$, satisfy $c_k(x,y) = c_k(y,x)$ since $p(n,x,y)(\deg y)^{-1} = p(n,y,x)(\deg x)^{-1}$. As predicted, the shape of $c_k(x,y)$, which can be computed in principle at least, tends to be awfully complicated as k increases. Fortunately, $c_1(x,y)$ is tractable to give an explicit shape. In Sects. 9.3

---

[2] It is not a difficult task to establish the central limit theorem if we do not care about the explicit and geometric shape of the Gaussian function involved in the formula.

[3] Strictly speaking, this asymptotic expansion holds in the non-bipartite case since, in the bipartite case, we have $p(n,x,y) = 0$ if x,y have different colors and n is even. Thus a slight modification is required in the bipartite case. See Sect. 9.5.

and 9.4, we shall show that $c_1(x,y)$ is expressed in terms of the normalized standard realization Φ as

$$c_1(x,y) = -\frac{|E_0|}{4}\|\Phi(x) - \Phi(y)\|^2 + g(x) + g(y) + C$$

with a certain computable function $g(x)$ and a constant C. In this formula E_0 is the set of directed edges of a base graph for X. Using this, we may establish the following direct relationship between the standard realization and the transition probability (see Sect. 9.6).

$$|E_0|\|\Phi(x) - \Phi(y)\|^2 = \lim_{n\to\infty} 2n\left\{\frac{p(n,x,x)}{p(n,y,x)} + \frac{p(n,y,y)}{p(n,x,y)} - 2\right\}. \tag{9.2}$$

What is significant here is that because of the nature of simple random walk, the right-hand side depends only on the graph structure of X. In clear language, equality (9.2) is rephrased as "a random walker can detect, after walking around for a long time, the most natural way for his topological crystal to sit in space". As will be seen in Sect. 9.6, this observation is a great benefit when we prove that the standard realization Φ, which is originally associated with an abstract period lattice L (or equivalently a base graph), is uniquely determined by the graph structure of X up to similar transformations. One can also use (9.2) for studying the symmetry of topological crystals including pathological ones.

The reader might still be puzzled by the abrupt appearance of the standard realization. A clue is in the integral expression of $p(n,x,y)$, to which we apply the Laplace method in an appropriate way. Our technique is, as will be seen in the next section, very much in the spirit of the classical theory of Fourier series. We only have to modify the argument in a suitable geometric form.

It should be noticed that the operator $P - I$ is the discrete Laplacian associated with the weight functions $m_V(x) = \deg x$, $m_E \equiv 1$ [see Notes (II) in Chap. 7]. In fact

$$((P-I)f)(x) = \frac{1}{\deg x}\sum_{e \in E_x}[f(t(e)) - f(o(e))].$$

This observation will be implicitly incorporated into the subsequent discussion.

9.2 Integral Formula

From now on we consider the simple random walk on a topological crystal $X = (V,E)$ over a finite graph $X_0 = (V_0, E_0)$. Let L be the abstract period lattice. Then P is L-equivariant, and is related to the transition operator P_0 associated with the simple random walk on X_0 as

9.2 Integral Formula

$$P(f \circ \omega) = (P_0(f)) \circ \omega,$$

where f is an arbitrary function on V_0, and $\omega : X \longrightarrow X_0$ is the covering map.

Let Φ be a general periodic realization of X, and let $\rho : L \longrightarrow \mathbb{R}^d$ be its period homomorphism (thus $\Phi(\sigma x) = \Phi(x) + \rho(\sigma)$ for $\sigma \in L$). We put $\Gamma = \rho(L)$ (the period lattice for Φ), and let $\Gamma^\#$ be the *dual* (reciprocal) *lattice* of Γ; that is, $\Gamma^\# = \{\alpha \in \mathbb{R}^d | \ \langle \alpha, \beta \rangle \in \mathbb{Z} \text{ for all } \beta \in \Gamma\}$. For $\mathbf{x} \in \mathbb{R}^d$, define the unitary character[4] $\chi_\mathbf{x}$ of L by

$$\chi_\mathbf{x}(\sigma) = \exp 2\pi \sqrt{-1} \langle \mathbf{x}, \rho(\sigma) \rangle.$$

Note $\chi_{\mathbf{x}+\alpha} = \chi_\mathbf{x}$ for every $\alpha \in \Gamma^\#$, and the correspondence $\mathbf{x} \mapsto \chi_\mathbf{x}$ yields an isomorphism of the torus group $\mathbb{R}^d/\Gamma^\#$ onto the group of unitary characters for L.

We introduce the finite-dimensional Hilbert space[5]

$$\ell_\mathbf{x}^2(V) = \{\mathbf{s} : V \to \mathbb{C} | \ \mathbf{s}(\sigma x) = \chi_\mathbf{x}(\sigma) \mathbf{s}(x) \quad \text{for } \sigma \in L\}$$

with the norm

$$\|\mathbf{s}\| = \left(\sum_{x \in \mathcal{F}} |\mathbf{s}(x)|^2 (\deg x) \right)^{1/2},$$

where $\mathcal{F}(\subset V)$ is a fundamental set for the action of L; we should note that $|\mathbf{s}|^2$ is regarded as a function on V_0 because $|\mathbf{s}|^2$ is invariant under the action of L, and hence the above definition does not depend on the choice of $\mathcal{F}$. Evidently the dimension of $\ell_\mathbf{x}^2(V)$ is equal to $|V_0|(= |\mathcal{F}|)$. For $\mathbf{x} = \mathbf{0}$, $\ell_\mathbf{x}^2(V)$ is the space of L-invariant (periodic) functions so that it is identified with the space $C(V_0)$ consisting of complex-valued functions on V_0 with the norm

$$\|f\| = \left(\sum_{x \in V_0} |f(x)|^2 (\deg x) \right)^{1/2}.$$

Now take a look at the transition operator P on X. As easily checked, $P(\ell_\mathbf{x}^2(V)) \subset \ell_\mathbf{x}^2(V)$, and hence by restricting P to $\ell_\mathbf{x}^2(V)$, we obtain the hermitian operator[6] $P_\mathbf{x} : \ell_\mathbf{x}^2(V) \longrightarrow \ell_\mathbf{x}^2(V)$. Obviously $P_\mathbf{0}$ is identified with the transition operator P_0 on X_0.

---

[4] A *unitary character* of L is a one-dimensional unitary representation of L, i.e., a homomorphism of L into $U(1) = \{z \in \mathbb{C} | \ |z| = 1\}$. The set of all unitary characters has a natural group structure.

[5] A finite-dimensional vector space with an inner product.

[6] $P_\mathbf{x} - I$ is a discrete analogue of the *twisted Laplacian* [91].

We enumerate the eigenvalues $\mu_i(\mathbf{x}) \in [-1, 1]$ $(i = 1, \ldots, N = |V_0|)$ of the operator $P_\mathbf{x}$ with repetition according to the multiplicity:

$$\mu_1(\mathbf{x}) \geq \mu_2(\mathbf{x}) \geq \cdots \geq \mu_N(\mathbf{x}).$$

Each $\mu_i(\mathbf{x})$ is regarded as a function on $\mathbb{R}^d/\Gamma^\#$ and is continuous. Let $\{\mathbf{s}_{\mathbf{x},1}, \ldots, \mathbf{s}_{\mathbf{x},N}\}$ be an orthonormal basis of $\ell_\mathbf{x}^2(V)$ with $P_\mathbf{x} \mathbf{s}_{\mathbf{x},i} = \mu_i(\mathbf{x})\mathbf{s}_{\mathbf{x},i}$ $(i = 1, \ldots, N)$. Note that $\mu_1(\mathbf{0}) = 1$ is a simple eigenvalue of $P_\mathbf{0} = P_0$ whose eigenfunctions are constant. Hence, in view of perturbation theory, one can take $\mathbf{s}_{\mathbf{x},1}$ satisfying the following conditions:

1. $\mathbf{s}_{\mathbf{0},1} \equiv |E_0|^{-1/2}$ (remember the formula $\sum_{x \in V_0} \deg x = |E_0|$; (3.1) in Sect. 3.1, from which $\|\mathbf{s}_{\mathbf{0},1}\| = 1$ follows).
2. $\mathbf{s}_{\mathbf{x},1}$ depends smoothly on $\mathbf{x}$ around $\mathbf{0}$.

Furthermore other $\mathbf{s}_{\mathbf{x},i}$'s can be assumed to be integrable in $\mathbf{x}$. A significant remark which we will use later is that one can change $\mathbf{s}_{\mathbf{x},1}$ by multiplying any smooth complex-valued function $f(\mathbf{x})$ with $|f(\mathbf{x})| \equiv 1$.

The following theorem is key to establish the mysterious relationship between random walks and standard realizations.

Theorem 9.1.

$$p(n,x,y)(\deg y)^{-1} = \mathrm{vol}(\mathbb{R}^d/\Gamma^\#)^{-1} \int_{\mathbb{R}^d/\Gamma^\#} \sum_{i=1}^N \mu_i(\mathbf{x})^n \mathbf{s}_{\mathbf{x},i}(x)\overline{\mathbf{s}_{\mathbf{x},i}(y)}d\mathbf{x}, \qquad (9.3)$$

In this theorem, $\overline{\mathbf{s}_{\mathbf{x},i}(y)}$ is the complex conjugate of $\mathbf{s}_{\mathbf{x},i}(y)$, and $\mathrm{vol}(\mathbb{R}^d/\Gamma^\#)$ denotes the volume of the flat torus $\mathbb{R}^d/\Gamma^\#$, which is equal to the volume of a fundamental parallelotope $D_{\Gamma^\#}$. It should be noted that $\mathrm{vol}(\mathbb{R}^d/\Gamma^\#) = \bigl(\mathrm{vol}(\mathbb{R}^d/\Gamma)\bigr)^{-1}$. Indeed, for a $\mathbb{Z}$-basis $\{\mathbf{a}_1, \ldots, \mathbf{a}_d\}$ of Γ, the set of vectors $\{\mathbf{b}_1, \ldots, \mathbf{b}_d\}$ characterized by $\langle \mathbf{a}_i, \mathbf{b}_j \rangle = \delta_{ij}$ comprises a $\mathbb{Z}$-basis of $\Gamma^\#$ (called the *dual basis*). If we put $A = (\mathbf{a}_1, \ldots, \mathbf{a}_d)$, $B = (\mathbf{b}_1, \ldots, \mathbf{b}_d)$, then ${}^t AB = I_d$ so that

$$\mathrm{vol}(\mathbb{R}^d/\Gamma^\#) = |\det B| = |\det A|^{-1} = \mathrm{vol}(\mathbb{R}^d/\Gamma)^{-1}. \qquad (9.4)$$

We shall give a proof of the above theorem. Let $x, y \in V$. We can take a fundamental set $\mathcal{F}$ containing both x and y. For each $z \in \mathcal{F}$, define the modified delta function $f_z \in \ell_\mathbf{x}^2(V)$ by

$$f_z(w) = \begin{cases} \chi_\mathbf{x}(\sigma) & \text{if } w = \sigma z \\ 0 & \text{otherwise,} \end{cases}$$

with which we have

9.3 How to Get the Asymptotic Expansion

$$\langle P_{\mathbf{x}}^n f_y, f_x \rangle = \sum_{z \in \mathcal{F}} \sum_{w \in V} p(n, z, w) f_y(w) \overline{f_x(z)} (\deg z)$$

$$= \sum_{w \in V} p(n, x, w) f_y(w) (\deg x)$$

$$= \sum_{w \in \mathcal{F}} \sum_{\sigma \in \Gamma} p(n, x, \sigma w) f_y(\sigma w) (\deg x)$$

$$= \sum_{\sigma \in L} \chi_{\mathbf{x}}(\sigma) p(n, x, \sigma y) (\deg x).$$

Using the *orthogonal relation of unitary characters*,

$$\frac{1}{\mathrm{vol}(\mathbb{R}^d/\Gamma^{\#})} \int_{\mathbb{R}^d/\Gamma^{\#}} \chi_{\mathbf{x}}(\sigma) d\mathbf{x} = \begin{cases} 1 & \sigma = 1, \\ 0 & \sigma \neq 1, \end{cases}$$

we obtain

$$p(n, x, y)(\deg x) = \frac{1}{\mathrm{vol}(\mathbb{R}^d/\Gamma^{\#})} \int_{\mathbb{R}^d/\Gamma^{\#}} \langle P_{\mathbf{x}}^n f_y, f_x \rangle d\mathbf{x}. \tag{9.5}$$

Since $\langle f_x, \mathbf{s}_{\mathbf{x},i} \rangle = \overline{\mathbf{s}_{\mathbf{x},i}(x)} (\deg x)$, we have

$$f_x = \sum_{i=1}^{N} \mathbf{s}_{\mathbf{x},i} \langle f_x, \mathbf{s}_{\mathbf{x},i} \rangle = \sum_{i=1}^{N} \mathbf{s}_{\mathbf{x},i} \overline{\mathbf{s}_{\mathbf{x},i}(x)} (\deg x),$$

$$P_{\mathbf{x}}^n f_y = \sum_{i=1}^{N} \mu_i(\mathbf{x})^n \mathbf{s}_{\mathbf{x},i} \overline{\mathbf{s}_{\mathbf{x},i}(y)} (\deg y).$$

Therefore we get

$$\langle P_{\mathbf{x}}^n f_y, f_x \rangle = \sum_{i=1}^{N} \mu_i(\mathbf{x})^n \mathbf{s}_{\mathbf{x},i}(x) \overline{\mathbf{s}_{\mathbf{x},i}(y)} (\deg x)(\deg y),$$

which, combined with (9.5), leads to (9.3). □

9.3 How to Get the Asymptotic Expansion

We shall apply the *Laplace method* to the integral in (9.3) in order to establish the asymptotic expansion of $p(n,x,y)$. Here the Laplace method is a technique used to approximate integrals of the form

$$\int_{\mathbb{R}^d} e^{-nf(\mathbf{x})} g(\mathbf{x}) d\mathbf{x} \qquad (n \gg 1). \tag{9.6}$$

More explicitly, the following statement holds[7]:

If $f'(\mathbf{x}_0) = 0$, and $\det f''(\mathbf{x}_0) > 0, f' \neq 0$ in $(\mathrm{supp}\ g)\backslash\{\mathbf{x}_0\}$, then, as n goes to infinity,

$$\int_{\mathbb{R}^d} e^{-nf(\mathbf{x})} g(\mathbf{x}) d\mathbf{x} \sim e^{-nf(\mathbf{x}_0)} \left(\frac{n}{2\pi}\right)^{-d/2} \left(\det(f''(\mathbf{x}_0))\right)^{-1/2} g(\mathbf{x}_0)$$
$$\times \left(1 + a_1(\mathbf{x}_0)n^{-1} + a_2(\mathbf{x}_0)n^{-2} + \cdots\right),$$

where $f''(\mathbf{x}_0)$ is the Hessian of f at $\mathbf{x}_0$, and $a_i(\mathbf{x}_0)$ can be calculated in terms of derivatives of f and g.

Before working in our situation, we shall explain a heuristic technique to calculate the coefficient $a_1(\mathbf{x}_0)$ in a special case, which is justified by carefully handling the remainder terms, and which will be used in the discussion just after Theorem 9.2. Suppose that $\mathbf{x}_0 = \mathbf{0}$, and $f(\mathbf{x})$, $g(\mathbf{x})$ have the following forms:

$$f(\mathbf{x}) = a\|\mathbf{x}\|^2 + R(\mathbf{x}) + O(\|\mathbf{x}\|^5), \tag{9.7}$$
$$g(\mathbf{x}) = b + c_1 \langle \mathbf{a}, \mathbf{x}\rangle + c_2 \langle \mathbf{a}, \mathbf{x}\rangle^2 + Q(\mathbf{x}) + O(\|\mathbf{x}\|^3), \tag{9.8}$$

where $R(\mathbf{x})$ and $Q(\mathbf{x})$ are certain homogeneous polynomials of degree 4 and degree 2, respectively. Then

$$\int_{\mathbb{R}^d} e^{-nf(\mathbf{x})} g(\mathbf{x}) d\mathbf{x} = n^{-d/2} \int_{\mathbb{R}^d} e^{-nf(\mathbf{y}/\sqrt{n})} g(\mathbf{y}/\sqrt{n}) d\mathbf{x}$$
$$= n^{-d/2} \int_{\mathbb{R}^d} e^{-a\|\mathbf{y}\|^2} e^{n^{-1} R(\mathbf{y}) + \cdots} \left[b + n^{-1/2} c_1 \langle \mathbf{a}, \mathbf{y}\rangle \right.$$
$$\left. + n^{-1} c_2 \langle \mathbf{a}, \mathbf{y}\rangle^2 + n^{-1} Q(\mathbf{y}) + \cdots\right] d\mathbf{y}$$
$$= n^{-d/2} \int_{\mathbb{R}^d} e^{-a\|\mathbf{y}\|^2} \left[b + n^{-1/2} c_1 \langle \mathbf{a}, \mathbf{y}\rangle + n^{-1} c_2 \langle \mathbf{a}, \mathbf{y}\rangle^2\right.$$
$$\left. + n^{-1} Q(\mathbf{y}) + b n^{-1} R(\mathbf{y}) + \cdots\right) d\mathbf{y}.$$

Here we use the well-known formulae

$$\int_{\mathbb{R}^d} e^{-a\|\mathbf{x}\|^2} d\mathbf{x} = \left(\frac{\pi}{a}\right)^{d/2}, \quad \int_{\mathbb{R}^d} x_i e^{-a\|\mathbf{x}\|^2} d\mathbf{x} = 0,$$
$$\int_{\mathbb{R}^d} x_i x_j e^{-a\|\mathbf{x}\|^2} d\mathbf{x} = \frac{1}{2a}\left(\frac{\pi}{a}\right)^{d/2} \delta_{ij}$$

to deduce

---

[7] See Theorem 7.7.5 (p. 220) in Hörmander [49].

9.3 How to Get the Asymptotic Expansion

$$\int_{\mathbb{R}^d} e^{-nf(\mathbf{x})} g(\mathbf{x}) d\mathbf{x} = n^{-d/2} \left[b\left(\frac{\pi}{a}\right)^{d/2} + n^{-1} c_2 \frac{1}{2a} \left(\frac{\pi}{a}\right)^{d/2} \|\mathbf{a}\|^2 \right.$$
$$\left. + n^{-1} \int_{\mathbb{R}^d} e^{-a\|\mathbf{y}\|^2} \left(Q(\mathbf{y}) + bR(\mathbf{y}) \right) d\mathbf{y} + \cdots \right],$$

which gives us

$$a_1(0) = \frac{c_2}{2ab} \|\mathbf{a}\|^2 + b^{-1} \left(\frac{\pi}{a}\right)^{-d/2} \int_{\mathbb{R}^d} e^{-a\|\mathbf{y}\|^2} \left(Q(\mathbf{y}) + bR(\mathbf{y}) \right) d\mathbf{y}. \quad (9.9)$$

Now we go back to our situation. To apply the Laplace method to the integral in (9.3), we need to take care of points $\mathbf{x} \in \mathbb{R}^d / \Gamma^\#$ with $\mu_1(\mathbf{x}) = 1$ or $\mu_N(\mathbf{x}) = -1$. As will be described in Sect. 9.5,

1. $\mu_1(\mathbf{x}) = 1 \iff \mathbf{x} = \mathbf{0}$,
2. $\mu_N(\mathbf{x}) = -1$ for some $\mathbf{x}$ $\implies$ X is bipartite[8].

For a while we confine ourselves to the non-bipartite case. This being the case, the term in (9.3) contributing to the asymptotic expansion of $p(n,x,y)$ is

$$\int_{U(\mathbf{0})} \mu_1(\mathbf{x})^n \mathbf{s}_{\mathbf{x},1}(x) \overline{\mathbf{s}_{\mathbf{x},1}(y)} d\mathbf{x} \quad (9.10)$$

because

$$\int_{(\mathbb{R}^d / \Gamma^\#) \setminus U(\mathbf{0})} \mu_1(\mathbf{x})^n \mathbf{s}_{\mathbf{x},1}(x) \overline{\mathbf{s}_{\mathbf{x},1}(y)} d\mathbf{x}$$

and

$$\int_{\mathbb{R}^d / \Gamma} \mu_i(\mathbf{x})^n \mathbf{s}_{\mathbf{x},i}(x) \overline{\mathbf{s}_{\mathbf{x},i}(y)} d\mathbf{x} \quad (i=2,\ldots,N)$$

have exponential decay. To look more closely at the integrand in (9.10), define the function $\mathbf{s}_{\mathbf{x}}$ by

$$\mathbf{s}_{\mathbf{x}}(x) = \exp(2\pi\sqrt{-1}\langle \mathbf{x}, \Phi(x)\rangle).$$

We easily find that $\mathbf{s}_{\mathbf{x}} \in \ell_{\mathbf{x}}^2(V)$. Define a unitary map $S: \ell_0^2(V) = C(V_0) \to \ell_{\mathbf{x}}^2(V)$ by $S(f) = f\mathbf{s}_{\mathbf{x}}$, and put $Q_{\mathbf{x}} = S^{-1} P_{\mathbf{x}} S : \ell_0^2(V) \to \ell_{\mathbf{x}}^2(V)$. It is straightforward to check that

$$(Q_{\mathbf{x}} f)(x) = (\deg x)^{-1} \sum_{e \in E_{0x}} \exp(2\pi\sqrt{-1}\langle \mathbf{x}, \mathbf{v}(e)\rangle) f(t(e)).$$

Furthermore, if we put $\phi_{\mathbf{x}} = S^{-1} \mathbf{s}_{\mathbf{x},1}$, i.e., $\mathbf{s}_{\mathbf{x},1} = \phi_{\mathbf{x}} \mathbf{s}_{\mathbf{x}}$, then integral (9.10) is expressed as

$$\int_{U(\mathbf{0})} \mu_1(\mathbf{x})^n \phi_{\mathbf{x}}(x) \overline{\phi_{\mathbf{x}}(y)} \exp\left[2\pi\sqrt{-1}\langle \mathbf{x}, (\Phi(x) - \Phi(y))\rangle\right] d\mathbf{x},$$

---

[8] The converse is also true. See Notes (II) in this chapter.

where, strictly speaking, $\phi_{\mathbf{x}}$ is identified with a periodic function on V by using the covering map. This integral has the form of (9.6) with the functions

$$f(\mathbf{x}) = -\log\left(1 - (1 - \mu_1(\mathbf{x}))\right),$$
$$g(\mathbf{x}) = \phi_{\mathbf{x}}(x)\overline{\phi_{\mathbf{x}}(y)} \exp\left[2\pi\sqrt{-1}\langle \mathbf{x}, (\Phi(x) - \Phi(y))\rangle\right]$$

(if we want to apply the above argument to this case, we have to multiply g by a smooth cut-off function which is identically 1 around $\mathbf{0}$, but the omission of this procedue does not affect the result). Therefore our business boils down to the computation of the derivatives at $\mathbf{x} = \mathbf{0}$ of $\mu_1(\mathbf{x})$ and $\phi_{\mathbf{x}}(x)\overline{\phi_{\mathbf{x}}(y)}$ (in this computation, we are not going to use the assumption that X is non-bipartite).

Up to now we have not assumed that the realization Φ is standard. From now on Φ is assumed to be the normalized standard realization; thus

$$\sum_{e \in E_{0x}} \langle \mathbf{x}, \mathbf{v}(e)\rangle = 0, \quad \sum_{e \in E_0} \langle \mathbf{x}, \mathbf{e}(e)\rangle^2 = 2\|\mathbf{x}\|^2.$$

This assumption allows us to comupute explicitly the derivatives at $\mathbf{x} = \mathbf{0}$ of $\mu_1(\mathbf{x})$ and $\phi_{\mathbf{x}}(x)\overline{\phi_{\mathbf{x}}(y)}$ up to the required order, using the following notations.

We write $\mathbf{x}(e)$ for $\langle \mathbf{x}, \mathbf{v}(e)\rangle$, and denote by $|\mathbf{x}|^2$ the function on V_0 (and hence the periodic function on V) defined by

$$|\mathbf{x}|^2(x) = \frac{1}{2} \frac{1}{\deg x} \sum_{e \in E_{0x}} \mathbf{x}(e)^2$$

(note $\sum_{x \in V_0} |\mathbf{x}|^2(x)(\deg x) = \|\mathbf{x}\|^2$).

We also denote by $G_0 : C(V_0) \longrightarrow C(V_0)$ the *discrete Green operator* for the discrete Laplacian $P_0 - I$. Namely $f = G_0(g)$ is a (unique) solution of the Poisson equation

$$(P_0 - I)f = g - \frac{1}{|E_0|} \sum_{x \in V_0} g(x)(\deg x), \quad \sum_{x \in V_0} f(x)(\deg x) = 0$$

(cf. Theorem 7.8). Note $G_0(1) = 0$.

Theorem 9.2. *1.*

$$\mu_1(\mathbf{x}) = 1 - \frac{4\pi^2}{|E_0|}\|\mathbf{x}\|^2 + \left(\frac{2}{3}\frac{\pi^4}{|E_0|} \sum_{e \in E_0} \mathbf{x}(e)^4 - \frac{16\pi^4}{|E_0|}\langle G_0|\mathbf{x}|^2, |\mathbf{x}|^2\rangle\right)$$
$$+ O(\|\mathbf{x}\|^6) \qquad (9.11)$$

(remember that $\langle \cdot, \cdot \rangle$ is the inner product on $\ell_0^2(V) = C(V_0)$).

9.3 How to Get the Asymptotic Expansion

2. By a suitable choice of $s_{x,1}$, we have

$$\phi_x(x)\overline{\phi_x(y)} = |E_0|^{-1} + 4\pi^2|E_0|^{-1}[(G_0|x|^2)(x) + (G_0|x|^2)(y)]$$
$$+ O(\|x\|^3). \quad (9.12)$$

Before proceeding to the proof of Theorem 9.2, we shall compute the coefficient $c_1(x,y)$ in (9.1) using (9.11) and (9.12).

We first have

$$f(x) = -\log\left(1 - (1 - \mu_1(x))\right) = 1 - \mu_1(x) + \frac{1}{2}(1 - \mu_1(x))^2 + \cdots$$

$$= \frac{4\pi^2}{|E_0|}\|x\|^2 + \left(\frac{16\pi^4}{|E_0|}\langle G_0|x|^2, |x|^2\rangle - \frac{2}{3}\frac{\pi^4}{|E_0|}\sum_{e \in E_0} x(e)^4 + \frac{8\pi^4}{|E_0|^2}\|x\|^4\right) + O(\|x\|^6),$$

$$g(x) = \phi_x(x)\overline{\phi_x(y)}\exp\left[2\pi\sqrt{-1}\langle x, (\Phi(x) - \Phi(y))\rangle\right]$$
$$= |E_0|^{-1} + 2\pi\sqrt{-1}|E_0|^{-1}\langle x, (\Phi(x) - \Phi(y))\rangle$$
$$+ 4\pi^2|E_0|^{-1}[(G_0|x|^2)(x) + (G_0|x|^2)(y)]$$
$$- 2\pi^2|E_0|^{-1}\langle x, (\Phi(x) - \Phi(y))\rangle^2 + O(\|x\|^3).$$

Therefore f and g are of the forms (9.7) and (9.8), respectively, where

$$a = \frac{4\pi^2}{|E_0|},\ b = |E_0|^{-1},\ c_1 = 2\pi\sqrt{-1}|E_0|^{-1},\ c_2 = -2\pi^2|E_0|^{-1},$$

$$\mathbf{a} = \Phi(x) - \Phi(y),$$

$$R(x) = \frac{16\pi^4}{|E_0|}\langle G_0|x|^2, |x|^2\rangle - \frac{2}{3}\frac{\pi^4}{|E_0|}\sum_{e \in E_0} x(e)^4 + \frac{8\pi^4}{|E_0|^2}\|x\|^4,$$

$$Q(x) = 4\pi^2|E_0|^{-1}[(G_0|x|^2)(x) + (G_0|x|^2)(y)],$$

Here the terms involving x, y are $\mathbf{a}$ and $Q(x)$; other terms give constants after performing the integration (9.9). Thus, discarding the detailed shapes of constant terms in $a_1(\mathbf{0})$, and putting

$$g(x) = 4\pi^2|E_0|^{-1}b^{-1}\left(\frac{\pi}{a}\right)^{-d/2}\int_{\mathbb{R}^d} e^{-a\|x\|^2}(G_0|x|^2)(x)dx,$$

we find that

$$c_1(x,y) = a_1(\mathbf{0}) = C_1\|\Phi(x) - \Phi(y)\|^2 + g(x) + g(y) + C_2. \quad (9.13)$$

One may elaborate the computation to get

$$C_0 = \frac{|E_0|^{d/2-1}}{(4\pi)^{d/2}}\mathrm{vol}(\mathbb{R}^d/\Gamma), \quad C_1 = -\frac{|E_0|}{4}, \quad g(x) = \frac{|E_0|}{2}\sum_{i=1}^{d}\left(G_0|v_i|^2\right)(x),$$

where Γ is the period lattice of the normalized standard realization, $\mathbf{v}(e) = {}^t\!\left(v_1(e),\ldots,v_d(e)\right)$, and

$$|v_i|^2(x) = \frac{1}{2}(\deg x)^{-1}\sum_{e\in E_{0x}} v_i(e)^2.$$

9.4 Proof of Theorem 9.2

To prove Theorem 9.2, we make use of a perturbation technique. Fixing $\mathbf{x}$, we put $Q_t = Q_{t\mathbf{x}}$, $\phi_t = \phi_{t\mathbf{x}}$ and $\mu_t = \mu_1(t\mathbf{x})$. Since $\overline{Q_{\mathbf{x}}f} = Q_{-\mathbf{x}}\overline{f}$, we have $\mu_1(-\mathbf{x}) = \mu_1(\mathbf{x})$, and hence μ_t is an even function of t (thus the derivative $\mu_0^{(2k-1)}$ of odd order vanishes). Moreover, $\phi_0 \equiv |E_0|^{-1/2}$ and

$$\|\phi_t\|^2 = \sum_{x\in V_0} \phi_t(x)\overline{\phi_t(x)}(\deg x) = 1 \tag{9.14}$$

since $\|\mathbf{s}_{\mathbf{x},1}\|^2 = 1$. By differentiating both sides of

$$(\deg x)^{-1}\sum_{e\in E_{0x}} e^{2\pi\sqrt{-1}t\mathbf{x}(e)}\phi_t(t(e)) = \mu_t\phi_t(x), \tag{9.15}$$

we obtain

$$(\deg x)^{-1}\sum_{e\in E_{0x}} \left[2\pi\mathbf{x}(e)e^{2\pi\sqrt{-1}t\mathbf{x}(e)}\phi_t(t(e)) + e^{2\pi\sqrt{-1}t\mathbf{x}(e)}\dot{\phi}_t(t(e))\right]$$
$$= \dot{\mu}_t\phi_t(x) + \mu_t\dot{\phi}_t(x). \tag{9.16}$$

Therefore

$$(P\dot{\phi}_0)(x) = (\deg x)^{-1}\sum_{e\in E_{0x}} \dot{\phi}_0(t(e)) = \dot{\phi}_0(x)$$

because ϕ_0 is constant, $\dot{\mu}_0 = 0$, and $\sum_{e\in E_{0x}} \mathbf{x}(e) = \sum_{e\in E_{0x}} \langle \mathbf{x}, \mathbf{v}(e)\rangle = 0$. This implies that $\dot{\phi}_0$ is a constant function.

Next we compute $\ddot{\mu}_0$. Differentiating (9.16), we have

9.4 Proof of Theorem 9.2

$$(\deg x)^{-1} \sum_{e \in E_{0x}} e^{2\pi\sqrt{-1}tx(e)} \left[-4\pi^2 \mathbf{x}(e)^2 \phi_t(t(e)) + 4\pi\sqrt{-1}\mathbf{x}(e)\dot{\phi}_t(t(e)) + \ddot{\phi}_t(t(e)) \right]$$
$$= \ddot{\mu}_t \phi_t(x) + 2\dot{\mu}_t \dot{\phi}_t(x) + \mu_t \ddot{\phi}_t(x). \tag{9.17}$$

We thus obtain

$$(\deg x)^{-1} \sum_{e \in E_{0x}} \left[-4\pi^2 |E_0|^{-1/2} \mathbf{x}(e)^2 + \ddot{\phi}_0(t(e)) \right] = |E_0|^{-1/2} \ddot{\mu}_0 + \ddot{\phi}_0(x) \tag{9.18}$$

because $\dot{\phi}_0$ is constant (we again use $\sum_{e \in E_{0x}} \mathbf{x}(e) = 0$). Multiplying $\deg x$ to both sides of (9.18) and summing up over V_0, we have

$$-4\pi^2 \sum_{x \in V_0} \sum_{e \in E_{0x}} \mathbf{x}(e)^2 |E_0|^{-1/2} + \langle P_0 \ddot{\phi}_0, 1 \rangle = \ddot{\mu}_0 |E_0|^{-1/2} |E_0| + \langle \ddot{\phi}_0, 1 \rangle.$$

Using here $\sum_{e \in E_0} \mathbf{x}(e)^2 = \sum_{e \in E_0} \langle \mathbf{x}, \mathbf{v}(e) \rangle^2 = 2\|\mathbf{x}\|^2$ and $\langle P_0 \ddot{\phi}_0, 1 \rangle = \langle \ddot{\phi}_0, 1 \rangle$, we obtain

$$\ddot{\mu}_0 = -\frac{8\pi^2}{|E_0|} \|\mathbf{x}\|^2,$$

from which we get the second order term in the Taylor expansion of $\mu_1(\mathbf{x})$. Furthermore, substituting this for $\ddot{\mu}_0$ in (9.18), we obtain

$$(P - I)\ddot{\phi}_0 = 8\pi^2 |E_0|^{-1/2} \left(|\mathbf{x}|^2 - \frac{1}{|E_0|} \|\mathbf{x}\|^2 \right). \tag{9.19}$$

We now calculate the 4th derivative $\mu_0^{(4)}$. To do this, we need to change the function $\phi_\mathbf{x}$ (and hence $s_{\mathbf{x},1}$) around $\mathbf{0}$ in such a way that ϕ_t satisfies

1. $\dot{\phi}_0 \equiv 0$,
2. $\ddot{\phi}_0$ is real-valued,
3. $\langle \ddot{\phi}_0, 1 \rangle = 0$.

Indeed, this is possible. As seen above, $\dot{\phi}_0$ is a constant function. On the other hand, by differentiating both sides of (9.14) with respect to t at $t = 0$, we see straight away that $\dot{\phi}_0$ takes a value in purely imaginary numbers, say $\sqrt{-1}a(\mathbf{x})$ for $a(\mathbf{x}) \in \mathbb{R}$. Taking a look at (9.19), we see that the imaginary part $\mathrm{Im}\,\ddot{\phi}_0$ of $\ddot{\phi}_0$ satisfies $(P - I)(\mathrm{Im}\,\ddot{\phi}_0) = 0$, and hence $\mathrm{Im}\,\ddot{\phi}_0$ is a constant, which we denote by $b(\mathbf{x})$. It should be noted that $a(t\mathbf{x}) = ta(\mathbf{x})$ and $b(t\mathbf{x}) = t^2 b(\mathbf{x})$. Replacing $\phi_\mathbf{x}$ with $\phi_\mathbf{x} \exp\bigl(-\sqrt{-1}(a(\mathbf{x}) + \frac{1}{2}b(\mathbf{x}))|E_0|^{1/2}\bigr)$, we can assume the first two properties above for $\dot{\phi}_0$ and $\ddot{\phi}_0$. The third property is a consequence of $\langle \ddot{\phi}_0, \phi_0 \rangle + + \langle \phi_0, \ddot{\phi}_0 \rangle = 0$ which is derived from (9.14).

From now on we assume the above properties for $\phi_{\mathbf{x}}$. In view of (9.19) and the definition of G_0, this assumption leads us to $\ddot{\phi}_0 = 8\pi^2 |E_0|^{-1/2} G_0 |\mathbf{x}|^2$, and hence

$$\phi_{\mathbf{x}}(x) = |E_0|^{-1/2} + 4\pi^2 |E_0|^{-1/2} (G_0|\mathbf{x}|^2)(x) + O(\|\mathbf{x}\|^3),$$

from which (9.12) follows.

The computation of $\mu_0^{(4)}$ is now ready. We differentiate (9.17) twice at $t = 0$ and use $\dot{\phi}_0 \equiv 0$ to get

$$(\deg x)^{-1} \sum_{e \in E_{0x}} \left[16\pi^4 \mathbf{x}(e)^4 \phi_0(t(e)) - 24\pi^2 \mathbf{x}(e)^2 \ddot{\phi}_0(t(e)) \right.$$

$$\left. + 8\pi\sqrt{-1}\mathbf{x}(e)\phi_0^{(3)}(t(e)) + \phi_0^{(4)}(t(e)) \right]$$

$$= \mu_0^{(4)} \phi_0(x) + 6\ddot{\mu}_0 \ddot{\phi}_0(x) + \mu_0 \phi_0^{(4)}(x).$$

Multiplying $\deg x$ and summing up over V_0 again, we find that

$$\frac{16\pi^2}{|E_0|^{1/2}} \sum_{e \in E_0} \mathbf{x}(e)^4 - 24\pi^2 \sum_{e \in E_0} \mathbf{x}(e)^2 \ddot{\phi}_0(t(e)) + 8\pi\sqrt{-1} \sum_{e \in E_0} \mathbf{x}(e) \phi_0^{(4)}(t(e))$$

$$= |E_0|^{1/2} \mu_0^{(4)} + 6\ddot{\mu}_0 \sum_{x \in V_0} \ddot{\phi}_0(x)(\deg x).$$

In this computation we have used $\langle P\phi_0^{(4)}, 1\rangle = \langle \phi_0^{(4)}, 1\rangle$. Furthermore noting that

$$\sum_{e \in E_0} \mathbf{x}(e)^2 \ddot{\phi}_0(t(e)) = \sum_{e \in E_0} \mathbf{x}(e)^2 \ddot{\phi}_0(o(e)) = \sum_{x \in V_0} \ddot{\phi}_0(x) \sum_{e \in E_{0x}} \mathbf{x}(e)^2$$

$$= 2\langle |\mathbf{x}|^2, \ddot{\phi}_0\rangle = \frac{16\pi^2}{|E_0|^{1/2}} \langle G_0|\mathbf{x}|^2, |\mathbf{x}|^2\rangle,$$

$$\sum_{e \in E_0} \mathbf{x}(e)\phi_0^{(4)}(t(e)) = -\sum_{e \in E_0} \mathbf{x}(e)\phi_0^{(4)}(o(e)) = \sum_{x \in V_0} \phi_0^{(4)}(x) \sum_{e \in E_{0x}} \mathbf{x}(e) = 0,$$

and

$$\sum_{x \in V_0} \ddot{\phi}_0(x)(\deg x) = \frac{8\pi^2}{|E_0|^{1/2}} \sum_{x \in V_0} (G_0|\mathbf{x}|^2)(x)(\deg x) = 0,$$

we obtain

$$\mu_0^{(4)} = \frac{16\pi^4}{|E_0|} \sum_{e \in E_0} \mathbf{x}(e)^4 - \frac{24 \cdot 16\pi^4}{|E_0|} \langle G_0|\mathbf{x}|^2, |\mathbf{x}|^2\rangle.$$

This completes the proof of Theorem 9.2.

9.5 Bipartiteness and the Eigenvalue μ_N

We return to the issue of the eigenvalue $\mu_N(\mathbf{x})$. We first establish:

Lemma 9.5.1. *Let* $\mathbf{s} \neq 0 \in \ell^2_\mathbf{x}(V)$.

1. *If* $P_\mathbf{x}\mathbf{s} = \pm \mathbf{s}$, *then* $|\mathbf{s}| \equiv const$,
2. *If* $P_\mathbf{x}\mathbf{s} = \mathbf{s}$, *then* $\mathbf{x} = \mathbf{0} (\in \mathbb{R}^d/\Gamma^\#)$,
3. *If* $P_\mathbf{x}\mathbf{s} = -\mathbf{s}$, *then* $\chi_\mathbf{x}(\sigma) = \pm 1$ *for every* $\sigma \in L$.

Thus we have

(i) $\mu_1(\mathbf{x}) = 1$ *if and only if* $\mathbf{x} = \mathbf{0}$,
(ii) *If* $\mu_N(\mathbf{x}) = -1$, *then* $\chi_\mathbf{x}(\sigma) = \pm 1$ *for every* $\sigma \in L$.

Proof. The proof is divided into 3 steps.

1. From the equation $P_\mathbf{x}\mathbf{s} = \pm \mathbf{s}$, we obtain $(P|\mathbf{s}|)(x) \geq |\mathbf{s}|(x)$. Since $|\mathbf{s}(\sigma x)| = |\mathbf{s}(x)|$ for every $\sigma \in L$, the function $|\mathbf{s}|$ is regarded as a function of V_0. So $|\mathbf{s}|$ attains its maximum at a vertex, say, x_0, and

$$|\mathbf{s}(x_0)| \geq \sum_{e \in E_{x_0}} p(e)|\mathbf{s}(t(e))| \geq |\mathbf{s}(x_0)|,$$

where $p(e) = 1/(\deg o(e))$, from which we conclude that $|\mathbf{s}(t(e))| = |\mathbf{s}(x_0)|$ (cf. Theorem 7.7). Since X is connected, we have $|\mathbf{s}| \equiv |\mathbf{s}(x_0)|$ (Lemma 3.4.1).

2. Suppose $P_\mathbf{x}\mathbf{s} = \mathbf{s}$. We may assume without loss of generality that $|\mathbf{s}| \equiv 1$. The corresponding eigenfunction, also denoted by $\mathbf{s}$, of the unitarily equivalent operator $Q_\mathbf{x}$ satisfies $|\mathbf{s}| \equiv 1$ and

$$\sum_{e \in E_x} p(e) \exp(2\pi\sqrt{-1}\langle \mathbf{x}, \mathbf{v}(e)\rangle) \mathbf{s}(t(e)) = \mathbf{s}(x).$$

Note that if complex numbers z_i ($i = 1, \ldots, n$) with $|z_i| = 1$ and positive numbers p_i ($i = 1, \ldots, n$) with $\sum_{i=1}^n p_i = 1$ satisfy $\left|\sum_{i=1}^n p_i z_i\right| = 1$, then $z_1 = \cdots = z_n$. Applying this elementary fact to the above equality, we find that $\exp(2\pi\sqrt{-1}\langle \mathbf{x}, \mathbf{v}(e)\rangle)\mathbf{s}(t(e)) = \mathbf{s}(o(e))$ and $\exp(2\pi\sqrt{-1}\langle \mathbf{x}, \mathbf{v}(c)\rangle) = 1$ for every closed path c. Hence $\mathbf{x} = \mathbf{0} (\in \mathbb{R}^d/\Gamma^\#)$.

3. Using the same argument as above, we get

$$\exp(2\pi\sqrt{-1}\langle \mathbf{x}, \mathbf{v}(e)\rangle)\mathbf{s}(t(e)) = -\mathbf{s}(o(e)),$$

and

$$\chi_\mathbf{x}(c) = \begin{cases} 1 & \text{if } |c| \text{ is even} \\ -1 & \text{if } |c| \text{ is odd.} \end{cases}$$

□

Lemma 9.5.2. *If there are an element* $\mathbf{x} \in \mathbb{R}^d/\Gamma^{\#}$ *with* $\mu_N(\mathbf{x}) = -1$, *then* X *is bipartite (the converse is true; see Notes (II)).*

Proof. By the above lemma, $\chi_{\mathbf{x}}(\sigma) = \pm 1$. If $\mathbf{x} = \mathbf{0}$, then identifying $P_{\mathbf{0}}$ with the transition operator P_0 on X_0, and letting f be the real part or the imaginary part of $\mathbf{s}$, we have $P_0 f = -f$ and $|f| = $ constant $\neq 0$. Then putting $A = \{x \in V_0 | f(x) > 0\}$ and $B = \{x \in V_0 | f(x) < 0\}$, we have a bipartition $V_0 = A \cup B$ of X_0, so that X_0 is bipartite. Next suppose $\mathbf{x} \neq \mathbf{0}$. Let $L_1 = \operatorname{Ker} \chi_{\mathbf{x}}$. Then $|L/L_1| = 2$. Put $X_1 = (V_1, E_1) = X/L_1$. The function $\mathbf{s}$ is regarded as a function on V_1 since $\mathbf{s}(\sigma x) = \mathbf{s}(x)$ for every $\sigma \in L_1$. The function f defined in the same way as above satisfies $Pf = -f$. Therefore X_1 is bipartite. In view of Theorem 5.6, we conclude that X is bipartite. □

We finally handle the asymptotic expansion in the bipartite case. Let X be a bipartite topological crystal over X_0. In view of Theorem 5.6, we may assume that X_0 is bipartite. This being the case, if $V = A \cup B$ is the bipartition of X, then both A and B are L-invariant. Thus defining the unitary transformation $U : \ell^2_{\mathbf{x}}(V) \longrightarrow \ell^2_{\mathbf{x}}(V)$ by

$$(Uf)(x) = \begin{cases} f(x) & x \in A, \\ -f(x) & x \in B, \end{cases}$$

we have $U P_{\mathbf{x}} U^{-1} = -P_{\mathbf{x}}$. Therefore $\mu_N(\mathbf{x}) = -\mu_1(\mathbf{x})$ (hence $\mu_N(\mathbf{x}) = -1$ if and only if $\mathbf{x} = \mathbf{0}$), and one may assume $s_{\mathbf{x},N} = U s_{\mathbf{x},1}$ so that

$$s_{\mathbf{x},N}(x)\overline{s_{\mathbf{x},N}(y)} = \begin{cases} s_{\mathbf{x},1}(x)\overline{s_{\mathbf{x},1}(y)} & x,y \in A \text{ or } x,y \in B, \\ -s_{\mathbf{x},1}(x)\overline{s_{\mathbf{x},1}(y)} & x \in A, y \in B \text{ or } x \in B, y \in A. \end{cases}$$

We have three cases for n,x,y when we look at $p(n,x,y)$.

1. $x,y \in A$ or $x,y \in B$, and $n = 2m$ (even).
2. $x \in A, y \in B$ or $x \in B, y \in A$, and $n = 2m - 1$ (odd).
3. other cases.

In the first and second cases, the terms in the right-hand side of (9.3) to which we must pay attention is

$$\int_{U(\mathbf{0})} \left[\mu_1(\mathbf{x})^n s_{\mathbf{x},1}(x)\overline{s_{\mathbf{x},1}(y)} d\mathbf{x} + \mu_N(\mathbf{x})^n s_{\mathbf{x},N}(x)\overline{s_{\mathbf{x},N}(y)} \right] d\mathbf{x}$$

$$= 2 \int_{U(\mathbf{0})} \mu_1(\mathbf{x})^n s_{\mathbf{x},1}(x)\overline{s_{\mathbf{x},1}(y)} d\mathbf{x}. \tag{9.20}$$

Therefore the asymptotic expansion of $p(n,x,y)$ for the first two cases differs only by a factor of 2 from the one in the non-bipartite case.

In the third case the integral (9.20) vanishes. This fits in with the fact that $p(n,x,y) = 0$, which is deduced from the fact that there is no path of length n joining x and y in this case.

9.6 Symmetry

In Sect. 7.6, we took a look at $\text{Aut}(X/X_0)$, the group of symmetries *preserving the abstract period lattice*, which is found to be *nearly* a crystallographic group. To be exact, a factor group of $\text{Aut}(X/X_0)$ modulo a finite normal subgroup is a crystallographic group. This section is devoted to a generalization of this fact to the full automorphism group $\text{Aut}(X)$ by applying the results we derived in the previous sections. We also revisit, as promised, the issue of uniqueness of standard realizations.

In general, when a graph X is infinite, it is natural to provide $\text{Aut}(X)$ with the compact-open topology defined by the system of neighborhoods $\{U(K)\}_K$ of the identity element, where K runs over all finite subsets of V, and $U(K) = \{g \in \text{Aut}(X) | \ g|_K = I_K\}$. With this topology, $\text{Aut}(X)$ is a topological group.[9]

In the following theorem, the motion group $\mathbf{M}(d)$ is, as a set, identified with the product $O(d) \times \mathbb{R}^d$.

Theorem 9.3 ([61]). *Let $\Phi : X \longrightarrow \mathbb{R}^d$ be the standard realization of a d-dimensional topological crystal X associated with an abstract period lattice L. Then there exists a homomorphism $M : \text{Aut}(X) \longrightarrow \mathbf{M}(d)$ such that*

1. *When we write $M(g) = \bigl(A(g), \mathbf{b}(g)\bigr) \in O(d) \times \mathbb{R}^d$,*

$$\Phi(gx) = A(g)\Phi(x) + \mathbf{b}(g) \qquad (x \in V),$$

2. *The image $M\bigl(\text{Aut}(X)\bigr)$ is a crystallographic group,*
3. *$\text{Ker}\, M = \{g \in \text{Aut}(X) | \ \Phi(gx) = \Phi(x) \text{ for every } x \in V\}$ is a compact subgroup of $\text{Aut}(X)$. The subgroup $\text{Ker}\, M$ is characterized as the (unique) maximal compact normal subgroup of $\text{Aut}(X)$.*

Since, if Φ is standard, then so is $\Phi \circ g$, the reader might think that claim (1) is obvious in view of Theorem 7.5 in which the uniqueness of standard realization is stated. However this uniqueness holds only when we fix an abstract period lattice. The gist of the matter is in the fact that the abstract period lattice $\Phi \circ g$ is $g^{-1}Lg$, which may be different from the original period lattice L, and may be even non-commensurable with L in general so that one cannot appeal to the uniqueness we have proved in Sect. 7.5 to relate $\Phi \circ g$ to Φ in a direct way. This is the difficulty we must overcome. For this, we shall exploit (9.13).

We first consider the non-bipartite case. One may assume that Φ is the normalized standard realization. Let $X_0 = (V_0, E_0)$ be the base graph. Since the n-step transition probability $p(n, x, y)$ for the simple random walk depends only on the graph structure of a topological crystal, we have

$$p(n, gx, gy)(\deg gy)^{-1} = p(n, x, y)(\deg y)^{-1}$$

---

[9] A topological group is a topological space with a group structure such that group operations (product and inverse) are continuous. As a topological space, $\text{Aut}(X)$ is *totally disconnected*.

for $g \in \text{Aut}(X)$, from which it follows that $c_1(gx, gy) = c_1(x, y)$. On the other hand, using (9.13), we find the neat formula:

$$2c_1(x,y) - c_1(x,x) - c_1(y,y) = -\frac{|E_0|}{2}\|\Phi(x) - \Phi(y)\|^2. \qquad (9.21)$$

Therefore

$$\|\Phi(gx) - \Phi(gy)\| = \|\Phi(x) - \Phi(y)\|. \qquad (9.22)$$

Suppose $\Phi(x_0) = \mathbf{0}$, and put $\Phi_g(x) = \Phi(gx) - \Phi(gx_0)$. Then

$$\|\Phi_g(x) - \Phi_g(y)\| = \|\Phi(x) - \Phi(y)\|,$$

and $\|\Phi_g(x)\| = \|\Phi(x)\|$ $(x, y \in V)$. Hence

$$\langle \Phi_g(x), \Phi_g(y) \rangle = \langle \Phi(x), \Phi(y) \rangle \qquad (9.23)$$

for $x, y \in V$. Take vertices $x_1, \ldots, x_d \in V$ such that $\Phi(x_1), \ldots, \Phi(x_d)$ form a basis of $\mathbb{R}^d$. In view of (9.23), $\Phi_g(x_1), \ldots, \Phi_g(x_d)$ form a basis of $\mathbb{R}^d$ as well, and there exists an orthogonal matrix $A(g)$ with $A(g)\Phi(x_i) = \Phi_g(x_i)$ for $i = 1, \ldots, d$. Again by (9.23), $A(g)\Phi(x) = \Phi_g(x)$ for all $x \in V$. Thus, putting $\mathbf{b}(g) = \Phi(gx_0)$, we obtain $\Phi(gx) = A(g)\Phi(x) + \mathbf{b}(g)$. The map $M : \text{Aut}(X) \longrightarrow \mathbf{M}(d)$ defined by $M(g) = (A(g), \mathbf{b}(g))$ is a homomorphism. Evidently $M(\sigma) = (I, \rho(\sigma))$ for $\sigma \in L$, where ρ is the period homomorphism for Φ.

In the bipartite case, (9.21), and hence (9.22) and (9.23), hold for x and y with the same color. Hence $\langle \Phi_g(x), \Phi_g(y) \rangle = \langle \Phi_g(x), \Phi_g(y) \rangle$ as far as x and y have the same color. Suppose x and y have different colors. Then the color of $t(e)$ for any edge e with $o(e) = x$ is the same as that of y. Thus $\langle \Phi_g(t(e)), \Phi_g(y) \rangle = \langle \Phi(t(e)), \Phi(y) \rangle$. Since Φ and Φ_g are harmonic,

$$\langle \Phi(x), \Phi(y) \rangle = (\deg x)^{-1} \sum_{e \in E_x} \langle \Phi(t(e)), \Phi(y) \rangle$$

$$= (\deg x)^{-1} \sum_{e \in E_x} \langle \Phi_g(t(e)), \Phi_g(y) \rangle = \langle \Phi_g(x), \Phi_g(y) \rangle.$$

This implies that (9.23) holds for every $x, y \in V$. The rest of the proof of (1) is carried out in the same way as in the non-bipartite case.

The claim that Image M is discrete and co-compact in $\mathbf{M}(d)$ (and hence is a crystallographic group) boils down to the following easy observations:

1. The group Ω generated by $\Phi(V)$ is a lattice group of $\mathbb{R}^d$. In fact, Ω is contained in the lattice group generated by the building block $\{\mathbf{v}(e)\}_{e \in E_0}$ of Φ (see Sect. 8.2).
2. $\rho(L) \subset \text{Image } \mathbf{b} \subset \Phi(V) \ (\subset \Omega)$. For, $\Phi(gx_0) = A(g)\Phi(x_0) + \mathbf{b}(g) = \mathbf{b}(g)$.
3. The subgroup Image A of $O(d)$ leaves the lattice group Ω invariant. Indeed, $A(g)\Phi(x) = \Phi(gx) - \mathbf{b}(g) \in \Omega$.

9.6 Symmetry

We shall prove the claim (3). The compactness of Ker M is obvious. For simplicity, we put $d_0(x,y) = \|\Phi(x) - \Phi(y)\|$. Let $\mathcal{F}(\subset V)$ be a fundamental set for the L-action on V. If H is a compact normal subgroup of $\mathrm{Aut}(X)$, then there exists a positive constant C such that $d_0(gx,x) \leq C$ for every $g \in H$ and $x \in X$. This is so because, writing $x = \sigma y$ with $y \in \mathcal{F}$, we get

$$d_0(g\sigma y, \sigma y) = d_0(\sigma^{-1} g \sigma y, y) \leq \sup_{y \in \mathcal{F}, g \in H} d_0(gy, y) < \infty.$$

Now for $g \in H$ and $\sigma \in L$,

$$d_0(g\sigma^n x_0, \sigma^n x_0) = \|n(A(g) - I)\rho(\sigma) + \mathbf{b}(g)\|,$$

which is bounded with respect to n, and so $A(g)\rho(\sigma) = \rho(\sigma)$ for every $\sigma \in L$. Thus $A(g) = I$. We also have $\Phi(g^n x) = \Phi(x) + n\mathbf{b}(g)$, and hence $d_0(g^n x, x) = n\|\mathbf{b}(g)\|$. Because $d_0(g^n x, x)$ is bounded, $\mathbf{b}(g) = 0$ and $H \subset \mathrm{Ker}\, M$.

The above idea is available for the proof of the following theorem which tells us that the similarity class of standard realizations of a topological crystal X does not depend on the choice of an abstract period lattice (or of a base graph).

Theorem 9.4. *Let $X_0 = (V_0, E_0)$ and $X_0' = (V_0', E_0')$ be two base graphs[10] of a topological crystal X, and let Φ and Φ' be the normalized standard realizations associated with X_0 and X_0', respectively. If $\Phi(x_0) = \Phi'(x_0) = \mathbf{0}$, then there exists an orthogonal matrix $U \in O(d)$ such that*

$$\Phi' = \left(\frac{|E_0|}{|E_0'|}\right)^{1/2} U \Phi.$$

This follows from the equality

$$|E_0|\|\Phi(x) - \Phi(y)\|^2 = |E_0'|\|\Phi'(x) - \Phi'(y)\|^2,$$

which is derived from (9.21).

In connection with Theorem 9.3, we shall say something about symmetry of crystals.

The network associated with a crystal has two distinct notions of symmetry; one is extrinsic symmetry, the same as the classical notion bound up directly with the beauty of the spatial object, which thus depends on realizations and is described in terms of the motion group; another is intrinsic symmetry, the notion irrelevant to realizations, which is solely explained in terms of $\mathrm{Aut}(X)$, and hence somehow denotes beauty enshrined inward. In general, intrinsic symmetry is "larger than" extrinsic symmetry since motions leaving the crystal invariant induce

---

[10] When X is bipartite, we take bipartite graphs X_0 and X_0' as base graphs.

Fig. 9.3 A one-dimensional topological crystal

automorphisms, but not vice versa. The theorem above tells us that the standard realization has the property

"extrinsic symmetry" = "intrinsic symmetry",

and, in this sense, has *maximal symmetry* among all periodic realizations.[11]

At first sight, the fact that symmetry is connected with randomness might sound mysterious because of the big conceptual discrepancy between "chance" and "order". However once we perceive that "laws of randomness" are solidly present in the world, it is no wonder that symmetry favored by the world is naturally connected with randomness, just like the relation between symmetry and minimum principles.

If $\operatorname{Ker} M = \{1\}$ in Theorem 9.3, the automorphism group $\operatorname{Aut}(X)$ is isomorphic to a crystallographic group. However, $\operatorname{Ker} M$ is not trivial in general even if $\Phi|V$ is injective. For instance, for the one-dimensional topological crystal X depicted in Fig. 9.3, $\operatorname{Ker} M$ is isomorphic to the product group $\prod_{-\infty}^{\infty} \mathbb{Z}_2$. Notice that when we realize X in $\mathbb{R}$, parallel edges are mapped to a single edge.

More generally, we have the following theorem which is easily proved by noting that if $g \in \operatorname{Ker} M$ is not the identity element 1, then there exists an edge $e \in E$ with $ge \neq e$, and such e must be a loop edge or parallel edge.

Theorem 9.5. *Suppose that the standard realization Φ restricted to V is injective. Then* $\operatorname{Ker} M = \{1\}$ *if and only if X is a combinatorial graph.*

Before closing this section, we shall present a proof for (9.2) mentioned in Sect. 9.1 for the sake of completeness. Writing[12]

$$p(n,x,y)(\deg y)^{-1} = p(n,y,x)(\deg y)^{-1} = 1 + c_1(x,y)n^{-1} + O_0(n^{-2}),$$
$$p(n,x,x)(\deg x)^{-1} = 1 + c_1(x,x)n^{-1} + O_1(n^{-2}),$$
$$p(n,y,y)(\deg y)^{-1} = 1 + c_1(x,x)n^{-1} + O_2(n^{-2}),$$

we get

$$\frac{p(n,x,x)}{p(n,y,x)} = \frac{1 + c_1(x,x)n^{-1} + O_1(n^{-2})}{1 + c_1(x,y)n^{-1} + O_0(n^{-2})},$$
$$\frac{p(n,y,y)}{p(n,x,y)} = \frac{1 + c_1(y,y)n^{-1} + O_2(n^{-2})}{1 + c_1(x,y)n^{-1} + O_0(n^{-2})}.$$

---

[11] Roughly speaking, maximal symmetry means that no structural deformation of its periodic arrangement of atoms in a crystal will make the structure more symmetrical than it is.

[12] In general, $f(n) = O(n^{-k})$ means that $|f(n)| \leq An^{-k}$ for some constant A.

Putting $O(n^{-2}) = O_1(n^{-2}) + O_2(n^{-2}) - 2O_0(n^{-2})$, we have

$$\frac{p(n,x,x)}{p(n,y,x)} + \frac{p(n,y,y)}{p(n,x,y)} - 2 = \frac{[c_1(x,x) + c_1(y,y) - 2c_1(x,y)]n^{-1} + O(n^{-2})}{1 + c_1(x,y)n^{-1} + O_0(n^{-2})}.$$

Consequently

$$\lim_{n \to \infty} 2n \left(\frac{p(n,x,x)}{p(n,y,x)} + \frac{p(n,y,y)}{p(n,x,y)} - 2 \right) = 2[c_1(x,x) + c_1(y,y) - 2c_1(x,y)]$$
$$= |E_0| \|\Phi(x) - \Phi(y)\|^2.$$

9.7 Notes

(I) One can establish similar results about asymptotic behaviors for the *heat kernel* on an abelian covering manifold over a compact Riemannian manifold. The heat kernel is the fundamental solution of the heat equation $\frac{\partial f}{\partial t} - \Delta f = 0$, and is thought of as the transition probability for the *Brownian motion*, a typical stochastic process with the continuous-time parameter. The idea has much in common with the case of topological crystals [60].

(II) Our discussion in Sect. 9.2 is closely related to the spectral analysis of the transition operator.

Denote by $\ell^2(V)$ the Hilbert space of complex-valued functions f on V with $\sum_{x \in V} |f(x)|^2 \deg x < \infty$, and by $\int_{\mathbb{R}^d/\Gamma^{\#}}^{\oplus} \ell_{\mathbf{x}}^2(V) \, \widehat{d\mathbf{x}}$ as the Hilbert space of functions $\mathbf{s}$ on $(\mathbb{R}^d/\Gamma^{\#}) \times V$ such that $\mathbf{s}(\mathbf{x}, \cdot) \in \ell_{\mathbf{x}}^2(V)$ and

$$\|\mathbf{s}\|^2 := \int_{\mathbb{R}^d/\Gamma^{\#}} \sum_{x \in V_0} |\mathbf{s}(\mathbf{x},x)|^2 (\deg x) \widehat{d\mathbf{x}} < \infty,$$

where $\widehat{d\mathbf{x}} = \frac{1}{\text{vol}(\mathbb{R}^d/\Gamma^{\#})} d\mathbf{x}$, the normalized Lebesgue measure on $\mathbb{R}^d/\Gamma^{\#}$. For a function f on V with finite support, define $(Uf)(\mathbf{x},x)$ by setting

$$(Uf)(\mathbf{x},x) = \sum_{\sigma} \chi_{\mathbf{x}}(\sigma) f(\sigma^{-1}x).$$

Then U extends to a unitary map

$$U : \ell^2(V) \longrightarrow \int_{\mathbb{R}^d/\Gamma^{\#}}^{\oplus} \ell_{\mathbf{x}}^2(V) \, \widehat{d\mathbf{x}},$$

and satisfies $(UPf)(\mathbf{x},x) = ((P_\mathbf{x}(Uf(\mathbf{x},\cdot))))(x)$. This means that $(P,\ell^2(V))$ is decomposed into the *direct integral*[13] [37]

$$\int_{\mathbb{R}^d/\Gamma^\#}^\oplus (P_\mathbf{x}, \ell_\mathbf{x}^2(V))\widehat{d\mathbf{x}}.$$

The general theory of direct integrals gives the following relationship between the *spectrum*[14] of P and eigenvalues of $P_\mathbf{x}$:

$$\mathrm{Spec}(P) = \bigcup_{i=1}^N \{\mu_i(\mathbf{x}) |\ \mathbf{x} \in \mathbb{R}^d/\Gamma^\#\}.$$

In particular, the spectrum of P has *band structure*; that is, the spectrum is a union of finitely many disjoint, possibly degenerate, closed intervals.

This fact enables us to prove that if X is bipartite, then there exists $\mathbf{x}$ such that $\mu_N(\mathbf{x}) = -1$. Let $V = A \cup B$ be the bipartition, and define the unitary map $U : \ell^2(V) \to \ell^2(V)$ by

$$(Uf)(x) = \begin{cases} f(x) & (x \in A) \\ -f(x) & (x \in B). \end{cases} \quad (9.24)$$

Then we have $UP = -PU$, and hence P and $-P$ are unitarily equivalent. In particular, $-1 \in \mathrm{Spec}(P)$ because $1 \in \mathrm{Spec}(P)$, and therefore one can find $\mathbf{x}$ such that -1 is an eigenvalue of $P_\mathbf{x}$.

(III) A careful analysis of the integral formula (9.3) leads to the *local central limit theorem* [60].

Theorem 9.6. *Let X be a topological crystal over $X_0 = (V_0, E_0)$, and let (Φ, ρ) be the normalized standard realization of X. Then*

$$\lim_{n\to\infty} \Big[(4\pi n)^{d/2} p(n,x,y)(\deg y)^{-1}$$

$$- \delta |E_0|^{d/2-1} \mathrm{vol}(\mathbb{R}^d/\rho(L)) \exp\Big(-\frac{|E_0|}{4n} \|\Phi(x) - \Phi(y)\|^2\Big) \Big] = 0$$

---

[13]This is a disguised form of Bloch–Floquet theory applied to the operator P. As for Bloch–Floquet theory for differential operators, see [64].

[14]For a hermitian operator $T : H \longrightarrow H$ of a Hilbert space H, the spectrum of T, symbolically $\mathrm{Spect}(T)$, is the set of complex numbers λ such that $T - \lambda I$ is not invertible as a bounded operator. It is known that $\mathrm{Spect}(T)$ is a bounded closed subset of $\mathbb{R}$.

9.7 Notes

uniformly for all $x, y \in V$.[15] Here $\delta = 1$ if X is non-bipartite, and $\delta = 2$ if X is bipartite (this being the case, X_0 is supposed to be bipartite).

This theorem or the asymptotic expansion of $p(n,x,x)$ leads to the *local limit formula* (cf. [63]):

$$\lim_{n \to \infty} (4\pi n)^{d/2} p(n,x,x) (\deg x)^{-1} = \delta |E_0|^{d/2-1} \mathrm{vol}(\mathbb{R}^d / \rho(L)).$$

The constant $C(X) = \delta |E_0|^{d/2-1} \mathrm{vol}(\mathbb{R}^d / \rho(L))$, depending only on the graph structure of X, is surely an interesting quantity in its own right. The following table gives the explicit numerical values of $C(X)$ for a few 2D crystal lattices.

X	Hexagonal	Triangular	Quadrilateral	Kagome
$C(X)$	$2\sqrt{3}$	$\dfrac{\sqrt{3}}{3}$	2	$\dfrac{2\sqrt{3}}{3}$

Along with these limit formulae, we may establish another significant asymptotic property of the simple random walk on a topological crystal which is described in terms of *large deviations*. An interesting fact is that a *convex polytope*[16] shows up in its geometric description.

In general, the theory of large deviations concerns the remote tails of sequences of probability distributions. In our case the problem is to find an asymptotic behavior of the transition probability $p(n,x,y)$ while y is kept near the boundary $\partial B_n(x) = \{ y \in V \mid d(x,y) = n \}$ of the ball

$$B_n(x) = \{ y \in V \mid d(x,y) \leq n \},$$

where $d(x,y)$ is the *graph distance* between x and y, i.e., the minimal length of paths joining x and y. Note that $B_n(x)$ is the reachable region of the random walker starting from x in n steps.

At the outset we look at the shape of $B_n(x)$ when n goes to infinity. More precisely, we fix a periodic realization $\Phi : X \longrightarrow \mathbb{R}^d$ with $\Phi(x) = 0$, and consider the scaling-limit figure

$$D = \lim_{n \to \infty} \frac{1}{n} \Phi(B_n(x)).$$

For several two-dimensional examples, $\dfrac{1}{n} \Phi(B_n(x))$ looks like convex polygons (Fig. 9.4 is the case of the hexagonal lattice with a brick-type realization).

---

[15] As usual, in bipartite case, n runs over even numbers when x and y have the same color, or odd numbers when x, y have different colors.

[16] A higher dimensional analogue of convex polygons in plane and polyhedra in space.

Fig. 9.4 Limit figure

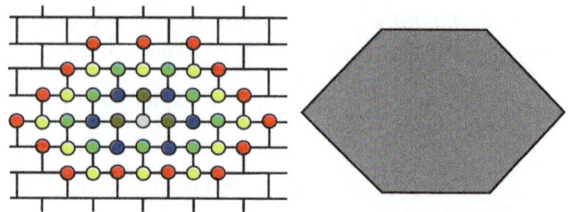

From this observation we anticipate that D is a convex polytope in general. Indeed one can prove that D coincides with the unit ball for the norm on $\mathbb{R}^d$ defined by

$$\|\xi\|_1 = \inf\{\|\mathbf{x}\|_1 \mid \mathbf{x} \in H_1(X_0, \mathbb{R}), \widehat{v}_{0\mathbb{R}}(\mathbf{x}) = \xi\},$$

where $\widehat{v}_{0\mathbb{R}} : H_1(X_0, \mathbb{R}) \longrightarrow \mathbb{R}^d$ is the linear extension of the homomorphism $\widehat{v}_0 : H_1(X_0, \mathbb{Z}) \longrightarrow \mathbb{R}^d$ introduced in Sect. 7.1, and $\|\mathbf{x}\|_1$ is the norm of $\mathbf{x}$ introduced in Sect. 6.2. In other words, D is the image of the unit ball $D^{ab} = \{\mathbf{x} \in H_1(X_0, \mathbb{R}) \mid \|\mathbf{x}\|_1 \leq 1\}$ by the linear operator $\widehat{v}_{0\mathbb{R}}$. Since D^{ab} is certainly a convex polytope, so is D.

Now what about the large deviation asymptotic of $p(n,x,y)$? The answer is the following (see [62] for details).

Theorem 9.7. *There exists a convex function $H : \mathbb{R}^d \longrightarrow \mathbb{R} \cup \{+\infty\}$ satisfying*

1. $D = \{\xi \mid H(\xi) < \infty\}$,
2. $\lim_{n \to \infty} \frac{1}{n} \log p(n, x, y_n) = -H(\xi)$ *for every ξ in the interior of D, and for every sequence $\{y_n\}_{n=1}^{\infty}$ in V such that $\{\Phi(y_n) - n\xi\}$ is bounded.*

The boundary ∂D is somehow a sort of "ideal boundary" in the sense that it characterizes the behavior of the random walk as time goes to infinity (it is, of course, irrelevant to the notion of *Martin boundary* used to interpret the "final" behavior of walks). The facets of D correspond to "generic" directions toward which a random walker tends to walk more surely. The most singular direction in this sense corresponds to extreme points of D.

The metric space $\mathbb{R}^d$ with the distance $d_1(\xi_1, \xi_2) = \|\xi_1 - \xi_2\|_1$ coincides with the *Gromov–Hausdorff limit* of $(X, n^{-1}d)$ as n goes to infinity [46, 62]. This implies that the "continuum limit" of a d-dimensional topological crystal is a d-dimensional normed space.[17]

(IV) The subjects over all in this book belong to "commutative mathematics", the general term expressing areas in which abelian groups play a significant role in various ways. Mathematically, it is natural to enter the "non-commutative world"; for instance, a *non-commutative crystal*, which means a regular covering graph over

---

[17]The "physical" continuum limit of the net corresponding to a real crystal is an elastic body (cf. [85]).

9.7 Notes

a finite graph with a non-abelian covering transformation group, is an object in the non-commutative world. Cayley–Serre graphs introduced in Notes (II) in Chap. 5 provide prime examples.

An immediate generalization is partially possible for the simple random walks on *nilpotent crystals*, covering graphs whose covering transformation groups are *nilpotent*. Tanaka [98, 99] generalized the large deviation property in the nilpotnet case.

A non-commutative nature appears even in analysis on topological crystals. As an example, we consider a *discrete Schrödinger operator with a periodic magnetic field* [92]. This is a generalization of the following operator on $\mathbb{Z}^2$ introduced by Harper [47]:

$$(Hf)(m,n) = \frac{1}{4}\big(e^{-\sqrt{-1}\alpha_1 n}f(m+1,n) + e^{\sqrt{-1}\alpha_1 n}f(m-1,n)$$
$$+ e^{-\sqrt{-1}\alpha_2 m}f(m,n+1) + e^{\sqrt{-1}\alpha_2 m}f(m,n-1)\big).$$

The operator $H - I$ is a discretization of the Schrödinger operator

$$\big(\frac{\partial}{\partial x} + \sqrt{-1}\alpha_1 y\big)^2 + \big(\frac{\partial}{\partial y} + \sqrt{-1}\alpha_2 x\big)^2$$

with the uniform magnetic field $B = (\alpha_2 - \alpha_1)dx \wedge dy$. The spectrum $\sigma(H)$ is known to be quite sensitive to the *magnetic flux* $\theta = \alpha_2 - \alpha_1$, in contrast to the continuous case. For instance, if $\theta/2\pi$ is rational, then $\sigma(H)$ has band structure, while if $\theta/2\pi$ is a Liouville number, then $\sigma(H)$ is a Cantor set.

To generalize the notion of discrete Schrödinger operators with uniform magnetic fields, we begin with a regular covering graph $X = (V, E)$ over a finite graph X_0 with covering transformation group Γ, and a 1-cochain $f \in C^1(X, U(1))$ on X whose cohomology class $[f] \in H^1(X, U(1))$ is Γ-invariant. Then we define, as a generalization of the Harper's operator, the bounded self-adjoint operator $H_f : \ell^2(V) \longrightarrow \ell^2(V)$ by

$$(H_f \varphi)(x) = \sum_{e \in E_x} \big(\deg o(e)\big)^{-1} f(e) \varphi(t(e)),$$

which we call a *magnetic transition operator*. From the assumption that $[f]$ is Γ-invariant, it follows that there exists a $U(1)$-valued function s_σ on V such that $f(\sigma e) = f(e)s_\sigma(t(e))s_\sigma(o(e))^{-1}$. Put

$$\Theta_f(\sigma, \gamma) = \frac{s_\sigma(\gamma x) s_\gamma(x)}{s_{\sigma\gamma}(x)}.$$

It is observed that the right-hand side does not depend on x and $\Theta_f(\sigma, \gamma)$ is a group 2-cocycle of Γ with coefficients in $U(1)$, that is,

$$\Theta_f(\sigma_1, \sigma_2\sigma_3)\Theta_f(\sigma_2, \sigma_3) = \Theta_f(\sigma_1, \sigma_2)\Theta_f(\sigma_1\sigma_2, \sigma_3)$$

(see [16] for group cohomology). Thus we obtain $[\Theta_f] \in H^2(\Gamma, U(1))$, which we call the *magnetic flux class*. With this notion, a partial generalization of what we mentioned for H is stated as:

If Γ is abelian and $[\Theta_f] \in H^2(\Gamma, \mathbb{Q}/\mathbb{Z})$, then the spectrum of H_f has band structure.

The magnetic flux class $[\Theta_f]$ depends only on $[f]$ so that we have a homomorphism $\Theta : H^1(X, U(1))^\Gamma \longrightarrow H^2(\Gamma, U(1))$. In this connection, we have the following exact sequence

$$1 \to H^1(\Gamma, U(1)) \to H^1(X_0, U(1)) \xrightarrow{\omega^*} H^1(X, U(1))^\Gamma \xrightarrow{\Theta} H^2(\Gamma, U(1)) \to 1,$$

where $H^1(\Gamma, U(1)) \to H^1(X_0, U(1))$ is the homomorphism induced from the surjective homomorphism $\mu : \pi_1(X_0) \longrightarrow \Gamma$ associated with the covering map $\omega : X \longrightarrow X_0$. If X is the maximal topological crystal over X_0, then $H^1(\Gamma, U(1)) \to H^1(X_0, U(1))$ is an isomorphism so that $H^1(X, U(1))^\Gamma \xrightarrow{\Theta} H^2(\Gamma, U(1))$ is also an isomorphism. Therefore the magnetic flux class $[\Theta_f]$ determines the unitary equivalence class of H_f. This is not the case in general (an analogue of the *Aharonov-Bohm effect*).

Another example is a *quantum walk* on a topological crystal. In general, a quantum walk is a quantum version of a classical random walk defined by replacing the transition probability $\{p(e)\}$ by a family of complex square matrices $\{P(e)\}$ satisfying a certain condition. The non-commutative nature of the matrix-valued transition probability reflects on the asymptotic behaviors of the n-th step transition probability which are much different from the classical case [100].

(IV) In this chapter we concentrated on the case of simple random walks. It is possible to generalize our discussion to the asymptotic behaviors of a *symmetric random walk*, which, for a general graph $X = (V, E)$, is defined to be a random walk with a transition probability $\{p(e)\}$ together with a positive-valued function m on V satisfying $p(e)m(o(e)) = p(\bar{e})m(t(e))$ (in this case, $P - I$ is the discrete Laplacian associated with the weight functions $m_V = m$ and $m_E(e) = p(e)m(o(e))$). What we need to assume for a symmetric random walk on a topological crystal is that p and m are invariant under the action of the abstract period lattice. The notion of standard realization can be introduced in this general set-up. See [60] for details.

Chapter 10
Discrete Abel–Jacobi Maps

The normalized standard realization Φ^{ab} of the maximal topological crystal X_0^{ab} has a special feature, which is properly interpreted by analogy with classical algebraic geometry.

The origin of algebraic geometry[1] is found in the work on elliptic integrals by Euler, Legendre, Gauss, Abel, and Jacobi. They were concerned with integration of irrational functions such as $\int \frac{1}{\sqrt{1-x^4}} dx$ which shows up when we want to compute the arc-length of the lemniscate and cannot be expressed in terms of elementary functions. A paradigmatic idea proposed in Gauss's unpublished work is to consider the inverse function of the indefinite integral and to treat it as a function of a complex variable. His epoch-making discovery is that it is a meromorphic function on the complex plane, periodic in two directions. After the rediscovery and generalization of his achievement by Abel and Jacobi, Gauss's genius disciple Georg Friedrich Bernhard Riemann (1826–1866) extended their work to more general integrals by initiating the theory of complex algebraic curves.

The aim of this chapter is to develop the idea essentially due to Riemann (and his successors) in the realm of discrete geometric analysis and to feed it back to crystallography. The reader may usefully consult the references [3,5,12,59,95] for more about "discrete algebraic geometry", an active field still in a state of flux.[2]

---

[1] An area of mathematics that combines techniques of abstract algebra with the problems of geometry.

[2] Algebraic geometry over a finite field looks similar to discrete algebraic geometry. But there is no direct connection between them.

10.1 Classical Algebraic Geometry

For the convenience of the reader, we shall start with a brief review of a relevant part in classical algebraic geometry.

Given an algebraic curve[3] S with a reference point p_0, we denote by $\Omega^1(S)$ the space of holomorphic 1-forms on S, and think of the first homology group $H_1(S,\mathbb{Z})$ as a subgroup of the dual space $(\Omega^1(S))^*$ by using the pairing map

$$([\alpha], \omega) = \int_\alpha \omega,$$

where $[\alpha]$ stands for the homology class of a 1-cycle α. Since $H_1(S,\mathbb{Z})$ is a lattice group of $(\Omega^1(S))^*$, the factor group

$$J(S) = (\Omega^1(S))^* / H_1(S,\mathbb{Z})$$

is a complex torus, which we call *Jacobian* (torus). The *Albanese map*

$$\Phi : S \longrightarrow J(S)$$

is the holomorphic map defined by the pairing

$$(\Phi(p), \omega) \equiv \int_{p_0}^p \omega \quad (\mathrm{mod}\ H_1(S,\mathbb{Z})). \tag{10.1}$$

The right-hand side of (10.1) is the line integral along a curve joining p_0 and p. The line integral is dependent on a curve from p_0 to p. However, any two such curves yield a closed loop in S, so the difference of integrations along these curves is written as $\langle [\alpha], \omega \rangle$ with a 1-cycle α. This is the meaning of the symbol "mod $H_1(S,\mathbb{Z})$" in the right-hand side.

Now let $\mathrm{Div}^0(S)$ be the group of divisors with degree 0, namely

$$\mathrm{Div}^0(S) = \left\{ \sum_{p \in S} a_p p \in C_0(S,\mathbb{Z}) \,\Big|\, \sum_{p \in S} a_p = 0 \right\},$$

and let $\mathrm{Prin}(S)$ be the *group of principal divisors*, the subgroup of $\mathrm{Div}^0(S)$ consisting of divisors associated with meromorphic functions f on S:

$$(f) = \sum_{p \in S} \mathrm{ord}_p(f) p. \tag{10.2}$$

---

[3]Precisely speaking, we are considering a non-singular complex projective algebraic curve, or equivalently a closed Riemann surface.

Here $\operatorname{ord}_p(f)$ is the order of a pole or zero of f. What we need to note is the classical result in complex analysis that tells us

$$\sum_{p \in S} \operatorname{ord}_p(f) = 0,$$

and hence the divisor (10.2) belongs to $\operatorname{Div}^0(S)$.

The *Picard group*[4] is the factor group $\operatorname{Pic}(S) = \operatorname{Div}^0(S)/\operatorname{Prin}(S)$. The Abel–Jacobi map

$$\Psi : S \longrightarrow \operatorname{Pic}(S)$$

is a holomorphic map defined by

$$\Psi(p) \equiv p - p_0 \quad (\operatorname{mod} \operatorname{Prin}(S)).$$

With these terminology, *Abel's theorem*, one of the culmination of classical algebraic geometry, is stated as:

The correspondence $p - p_0 \mapsto \Phi(p)$ yields an isomorphism of $\operatorname{Pic}(S)$ onto $J(S)$. Thus under the identification between $\operatorname{Pic}(S)$ and $J(S)$, the Abel–Jacobi map coincides with the Albanese map.

10.2 Discrete Set-up

Having the above review in mind, we now embark on a description of an analogue of Jacobian torus in discrete set-up. Let $X_0 = (V_0, E_0)$ be a finite graph with a reference point x_0. Define the *Jacobian torus $J(X_0)$* by setting

$$J(X_0) = H_1(X_0, \mathbb{R})/H_1(X_0, \mathbb{Z}),$$

which possesses the flat metric induced from the inner product on $H_1(X_0, \mathbb{R})$ introduced in Sect. 8.1.[5] The flat torus $J(X_0)$ has the structure of an abelian group at the same time.

Let $\Phi^{\mathrm{ab}} : X_0^{\mathrm{ab}} = (V^{\mathrm{ab}}, E^{\mathrm{ab}}) \longrightarrow H_1(X_0, \mathbb{R})$ be the normalized standard realization of the maximal topological crystal X_0^{ab} satisfying $\Phi^{\mathrm{ab}}(x_0) = \mathbf{0}$. Since

$$\Phi^{\mathrm{ab}}(\alpha x) = \Phi^{\mathrm{ab}}(x) + \alpha \quad (\alpha \in H_1(X_0, \mathbb{Z})),$$

---

[4] Both Jacobian and Picard groups are known to be abelian varieties. A complex torus $\mathbb{C}^n/L$ is called an abelian variety if it is holomorphically embedded in the complex projective space $P^N(\mathbb{C})$.

[5] This is also called the Albanese torus.

we obtain a piecewise linear map $\Phi_0^{ab} : X_0 \longrightarrow J(X_0)$. This is what we regard as an analogue of the Albanese map. Later we shall give another version of the Albanese map fitting in with the discrete set-up even better.

To see a stronger resemblance between a holomorphic Albanese map and its graph-theoretic analogue, we introduce the notion of *harmonic 1-forms*[6] as an analogue of holomorphic 1-forms. A harmonic 1-form on X_0 is a function $\omega : E_0 \longrightarrow \mathbb{R}$ satisfying

$$\omega(\bar{e}) = -\omega(e) \quad (e \in E_0), \tag{10.3}$$

$$\sum_{e \in E_{0x}} \omega(e) = 0 \quad (x \in V_0). \tag{10.4}$$

The legitimacy of the name "harmonic 1-form" will be explained below.

Now an analogue of (10.1) is stated as

$$(\Phi_0^{ab}(x), \omega) \equiv \omega(e_1) + \cdots + \omega(e_n) \pmod{H_1(X_0, \mathbb{Z})} \tag{10.5}$$

where ω is an arbitrary harmonic 1-form, and $(e_1, \ldots, e_n)$ is a path in X_0 joining x_0 and x. The pairing on the left-hand side is defined by

$$\left(\sum_{e \in E_0} a_e e, \omega \right) = \sum_{e \in E_0} a_e \omega(e).$$

The right hand side of (10.5) is regarded as an analogue of a line integral along a curve.

To prove (10.5), it suffices to check

$$\big(\mathbf{v}_{ab}(e), \omega\big) = \omega(e) \big(= (e, \omega) \big).$$

Since $\mathbf{v}_{ab}(e) - e \in H_1(X_0, \mathbb{R})^\perp$, what we have to show is that $(\alpha, \omega) = 0$ for any $\alpha \in H_1(X_0, \mathbb{R})^\perp$. To this end, we use

$$\partial^* : C_0(X_0, \mathbb{R}) \longrightarrow C_1(X_0, \mathbb{R}),$$

the *adjoint* of the boundary operator $\partial : C_1(X_0, \mathbb{R}) \longrightarrow C_0(X_0, \mathbb{R})$. Here the inner products on $C_0(X_0, \mathbb{R})$ is defined by

$$\langle x, y \rangle = \begin{cases} 1 & (x = y) \\ 0 & (x \neq y), \end{cases}$$

---

[6] The notion of harmonic forms was originally introduced in Riemannian geometry. It is related to the *cohomology theory* of manifolds.

10.2 Discrete Set-up

and the inner product on $C_1(X_0, \mathbb{R})$ is the one defined in Sect. 8.1. We then have $H_1(X_0, \mathbb{R})^\perp = (\mathrm{Ker}\, \partial)^\perp = \mathrm{Image}\, \partial^*$. Since V_0 generates $C_0(X_0, \mathbb{R})$, our claim is equivalent to that $\langle \partial^* x, \omega \rangle = 0$, $x \in V_0$, which is deduced from the formula

$$\partial^* x = -\sum_{e \in E_{0x}} e.$$

Indeed

$$\langle \partial^* x, \omega \rangle = -\left(\sum_{e \in E_{0x}} e, \omega \right) = -\sum_{e \in E_{0x}} \omega(e) = 0.$$

Here comes the question: Why did we give the name "harmonic 1-form" to a function ω satisfying (10.3) and (10.4)? The reason will become apparent if we use the *coboundary operator*

$$d : C^0(X_0, \mathbb{R}) \longrightarrow C^1(X_0, \mathbb{R}),$$

introduced in Notes (I) in Chap. 4, where

$$C^0(X_0, \mathbb{R}) = \{ f : V_0 \longrightarrow \mathbb{R} \},$$
$$C^1(X_0, \mathbb{R}) = \{ \omega : E_0 \longrightarrow \mathbb{R} |\ \omega(\bar{e}) = -\omega(e)\ (e \in E_0) \}.$$

What we need to keep in mind is that the coboundary operator is considered a discrete analogue of the *exterior differentiation* on a manifold.

Let d^* be the adjoint of d with respect to the inner products:

$$\langle f_1, f_2 \rangle_0 = \sum_{x \in V} f_1(x) f_2(x) \quad (f_i \in C^0(X_0, \mathbb{R})),$$

$$\langle \omega_1, \omega_2 \rangle_1 = \frac{1}{2} \sum_{e \in E} \omega_1(e) \omega_2(e) \quad (\omega_i \in C^1(X_0, \mathbb{R})),$$

that is, $\langle df, \omega \rangle_1 = \langle f, d^* \omega \rangle_0$. Then

$$(d^* \omega)(x) = -\sum_{e \in E_{0x}} \omega(e) \quad (\omega \in C^1(X_0, \mathbb{R})),$$

so that ω is a harmonic 1-form if and only if $d^* \omega = 0$. The equation $d^* \omega = 0$ is an analogue of the equation for harmonic 1-forms on a Riemannian manifold (exactly speaking, we must add the equation $d\omega = 0$ in the case of manifolds, which is understood to be "trivially" satisfied in the case of graphs since graphs are 1-dimensional). Thus the name "harmonic 1-form" is justified.

We have more about relationships between the standard realization and harmonic 1-forms. If we express $\mathbf{v}_{ab}(e)$ by a column vector ${}^t\big(\omega_1(e), \ldots, \omega_b(e)\big)$ ($b = b_1(X_0)$) using an orthonormal basis of $H_1(X_0, \mathbb{R})$, then ω_i is a harmonic 1-form in view of the equation $\sum_{e \in E_{0x}} \mathbf{v}_{ab}(e) = \mathbf{0}$. Moreover $\{\omega_1, \ldots, \omega_b\}$ is a basis of the space of harmonic

1-forms because of the fact that $\{\mathbf{v}_{ab}(e)\}_{e\in E_0}$ spans $H_1(X_0,\mathbb{R})$. The property $\sum_{e\in E_0}\langle \mathbf{x},\mathbf{v}_{ab}(e)\rangle^2 = 2\|\mathbf{x}\|^2$ is equivalent to $\{\omega_1,\ldots,\omega_b\}$ being an orthonormal basis [see (7.10)].

A fact worth knowing is that the volume of $J(X_0)$ is expressed in terms of the *tree number* $\kappa(X_0)$, a significant graph-theoretic invariant, which is defined to be the number of spanning trees in X_0. In fact,

$$\mathrm{vol}\bigl(J(X_0)\bigr) = \kappa(X_0)^{1/2} \qquad (10.6)$$

as will be proved in Sect. 10.4.[7]

Remark. Using (10.6), we get the formula for the energy of Φ^{ab}:

$$\mathcal{E}(\Phi^{ab}) = 2b_1(X_0)\kappa(X_0)^{-1/b_1(X_0)}.$$

For example, in the case of $X_0 = D_{d+1}$, the base graph of the d-dimensional topological diamond, we have $b_1(X_0) = d$, $\kappa(X_0) = d+1$ so that $\mathcal{E}(\Phi^{ab}) = 2d(d+1)^{-1/d}$. □

10.3 Discrete Abel's Theorem

We go further by introducing the notion of *discrete Jacobian*, a direct discrete analogue of Jacobian in algebraic geometry.

Let us use the notation $H_1(X_0,\mathbb{Z})^{\#}$ for the dual lattice[8] of $H_1(X_0,\mathbb{Z})$ in $H_1(X_0,\mathbb{R})$:

$$H_1(X_0,\mathbb{Z})^{\#} = \{\alpha \in H_1(X_0,\mathbb{R})|\ \langle \alpha,\beta\rangle \in \mathbb{Z} \text{ for every } \beta \in H_1(X_0,\mathbb{Z})\}.$$

The lattice $H_1(X_0,\mathbb{Z})$ is *integral* in the sense that $\langle \alpha,\beta\rangle \in \mathbb{Z}$ for every $\alpha,\beta \in H_1(X_0,\mathbb{Z})$. Thus $H_1(X_0,\mathbb{Z}) \subset H_1(X_0,\mathbb{Z})^{\#}$. The *discrete Jacobian* $\mathcal{J}(X_0)$ is the factor group

$$H_1(X_0,\mathbb{Z})^{\#}/H_1(X_0,\mathbb{Z}).$$

This is a finite subgroup of the Jacobian torus $J(X_0)$.

We now observe that $\Phi_0^{ab}(V_0) \subset \mathcal{J}(X_0)$. Recall P_{ab}, the orthogonal projection of $C_1(X_0,\mathbb{R})$ onto $H_1(X_0,\mathbb{R})$. For any $e \in E_0$ and $\alpha \in H_1(X_0,\mathbb{Z})$, we find that $\langle P_{ab}(e),\alpha\rangle = \langle e,P_{ab}(\alpha)\rangle = \langle e,\alpha\rangle \in \mathbb{Z}$, and so

$$\mathbf{v}_{ab}(e) = P_{ab}(e) \in H_1(X_0,\mathbb{Z})^{\#}.$$

---

[7] The notion of tree numbers originated in the work of A. Cayley (1889) who tried to count all trees with given n vertices (the answer is n^{n-2}, which is nothing but the tree number of the complete graph K_n over n vertices; see Sect. 10.4).
[8] See Sect. 9.2.

10.3 Discrete Abel's Theorem

Therefore $\Phi^{ab}(x) \in H_1(X_0,\mathbb{Z})^{\#}$ for every $x \in V_0^{ab}$, which immediately leads to $\Phi_0^{ab}(V_0) \subset \mathfrak{J}(X_0)$.

We call the restriction $\Phi_0^{ab}|V_0 : V_0 \longrightarrow \mathfrak{J}(X_0)$ the *discrete Albanese map* (cf. [70]).

The discrete Albanese map is a "harmonic function" on V_0 with values in $\mathfrak{J}(X_0)$. Here in general, a function $f : V_0 \longrightarrow A$ with values in an additive group A is said to be *harmonic* if

$$\sum_{e \in E_{0x}} \left[f(t(e)) - f(o(e)) \right] = 0.$$

This is, of course, a direct generalization of a harmonic function with values in $\mathbb{R}$; see Notes (II) in Chap. 7.

We have more about $\mathfrak{J}(X_0)$ and the discrete Albanese map.

Theorem 10.1. *1.* $\Phi_0^{ab}(V_0)$ *generates* $\mathfrak{J}(X_0)$.
2. $|\mathfrak{J}(X_0)| = \kappa(X_0)$.

Proof. To prove (1), take a spanning tree T of X_0, and let

$$e_1, \ldots, e_b, \overline{e_1}, \ldots, \overline{e_b} \quad (b = b_1(X_0))$$

be all edges not in T. The vectors $\mathbf{v}_{ab}(e_1), \ldots, \mathbf{v}_{ab}(e_b)$ constitute a $\mathbb{Z}$-basis of $H_1(X_0,\mathbb{Z})^{\#}$. This is so because we may create a $\mathbb{Z}$-basis of the lattice $H_1(X_0,\mathbb{Z})$ consisting of circuits $c_1, \ldots, c_b$ in X_0 such that c_i contains e_i, and

$$\langle c_i, \mathbf{v}_{ab}(e_j) \rangle = \langle c_i, P_{ab}(e_j) \rangle = \langle P_{ab}(c_i), e_j \rangle = \langle c_i, e_j \rangle = \delta_{ij},$$

namely $\{\mathbf{v}_{ab}(e_1), \ldots, \mathbf{v}_{ab}(e_b)\}$ is the dual basis of $\{c_1, \ldots, c_b\}$. From this, our assertion immediately follows.

We shall show (2). To this end, we take a look at the exact sequence

$$0 \to \mathfrak{J}(X_0) \to J(X_0) \to H_1(X_0,\mathbb{R})/H_1(X_0,\mathbb{Z})^{\#} \to 0.$$

We therefore have the following formula for the order of $\mathfrak{J}(X_0)$.

$$|\mathfrak{J}(X_0)| = \mathrm{vol}(J(X_0))/\mathrm{vol}\big(H_1(X_0,\mathbb{R})/H_1(X_0,\mathbb{Z})^{\#}\big).$$

We also have, by virtue of (9.4),

$$\mathrm{vol}\big(H_1(X_0,\mathbb{R})/H_1(X_0,\mathbb{Z})^{\#}\big) = \mathrm{vol}(J(X_0))^{-1}.$$

Consequently $|\mathfrak{J}(X_0)| = \mathrm{vol}(J(X_0))^2$, and hence (10.6) gives us $|\mathfrak{J}(X_0)| = \kappa(X_0)$, as claimed. □

So far we have handled the discrete Jacobian $\mathfrak{J}(X_0)$ as a finite abelian group. We may equip $\mathfrak{J}(X_0)$ with a natural graph structure as follows.

Let E_0^o be an orientation of X_0. The proof of Theorem 10.1 says that $\{\mathbf{v}_{ab}(e)\}_{e \in E_0^o}$ generates $\mathcal{J}(X_0)$. Thus the map $i : E_0^o \longrightarrow \mathcal{J}(X_0)$, defined by

$$i(e) = \mathbf{v}_{ab}(e) \mod H_1(X_0, \mathbb{Z}),$$

yields the Cayley–Serre graph $X(\mathcal{J}(X_0), E_0^o)$ [see Notes (II) in Chap. 5], which we denote by $\mathcal{J}^{gr}(X_0)$. The graph structure does not depend on the choice of an orientation. Assigning the edge $(\Phi_0^{ab}(o(e)), e)$ of $\mathcal{J}^{gr}(X_0)$ to each $e \in E_0^o$, and forgetting the orientation, we get a morphism $\Phi_0^{ab} : X_0 \longrightarrow \mathcal{J}^{gr}(X_0)$.

Remark. $\mathcal{J}(X_0)$ has another appendage. A non-degenerate symmetric bilinear form on $\mathcal{J}(X_0)$ with values in $\mathbb{Q}/\mathbb{Z}$ is induced from the inner product on $H_1(X_0, \mathbb{R})$. Thinking of this form as an analogue of "polarization", one may ask whether the *Torelli type theorem* holds in the discrete realm. More specifically, one asks whether two finite graphs X_1 and X_2, with the same degree, are isomorphic when there exists a group isomorphism between $\mathcal{J}(X_1)$ and $\mathcal{J}(X_2)$ preserving polarizations.[9] See Artamkin [1] and Caporaso and Viviani [18] for other formulations of the Torelli type theorem. □

Now we define the *group of divisors of degree zero* in the discrete category by setting

$$\mathrm{Div}^0(X_0) = \Big\{ \sum_{x \in V_0} a_x x \in C_0(X_0, \mathbb{Z}) \,\Big|\, \sum_{x \in V_0} a_x = 0 \Big\}$$

(if we use the homomorphism $\varepsilon_0 : C_0(X_0, \mathbb{Z}) \longrightarrow \mathbb{Z}$ introduced in Sect. 4.2, this is nothing but $\mathrm{Ker}\,\varepsilon_0$). On the other hand, the *group of principal divisors* is defined by

$$\mathrm{Prin}(X_0) = (\partial \circ \partial^*)(C_0(X_0, \mathbb{Z})).$$

Here the concrete shape of $\partial \circ \partial^*$ is given by

$$(\partial \circ \partial^*)\Big(\sum_{x \in V_0} a_x x \Big) = -\sum_{x \in V_0} a_x \sum_{e \in E_{0x}} (t(e) - o(e)). \tag{10.7}$$

Obviously $\mathrm{Prin}(X_0)$ is a subgroup of $\mathrm{Div}^0(X_0)$.

Again we have the issue of naming. The reader might wonder why $\mathrm{Prin}(X_0)$ defined above is an analogue of $\mathrm{Prin}(S)$; its formulation seems quite a bit far from the original form. To explain the rationale, we identify $C_0(X_0, \mathbb{R})$ with $C(V_0)$, the space of real-valued functions on V_0, via the correspondence

$$\sum_{x \in V_0} a_x x \implies f \in C(V_0); f(x) = a_x.$$

---

[9] This problem was suggested by Kenichi Yoshikawa. If we would remove the conditions on polarizations, there are many examples of X_1, X_2 with $\mathcal{J}(X_1) \cong \mathcal{J}(X_2)$ [3].

10.3 Discrete Abel's Theorem

Then $\partial \circ \partial^* : C_0(X_0, \mathbb{R}) \longrightarrow C_0(X_0, \mathbb{R})$ is identified with

$$-\Delta : C(V_0) \longrightarrow C(V_0),$$

where Δ is the discrete Laplacian introduced in Sect. 7.4. Thus a similarity between $\text{Prin}(X_0)$ and $\text{Prin}(S)$ is somehow explained by the formula

$$\Delta \log |f| = 2\pi \sum \text{ord}_p(f) \delta_p$$

which holds for a meromorphic function f on S. Here δ_p stands for the Dirac delta function with support p, and a divisor $\sum a_p p$ is identified with $\sum a_p \delta_p$.

It is now natural to define the *discrete Picard group* by

$$\text{Pic}(X_0) = \text{Div}^0(X_0)/\text{Prin}(X_0).$$

As will be proved in the next section, the order $|\text{Pic}(X_0)|$ coincides with the tree number $\kappa(X_0)$; therefore $|\text{Pic}(X_0)| = |\mathcal{J}(X_0)|$.

We are ready to introduce a discrete version of Abel–Jacobi map. Imitating the case of algebraic curves, we define the *discrete Abel–Jacobi map* $\Phi_0^{\text{aj}} : V_0 \longrightarrow \text{Pic}(X_0)$ by

$$\Phi_0^{\text{aj}}(x) \equiv x - x_0 \qquad (\text{mod } \text{Prin}(X_0)).$$

An important fact is that the discrete Abel–Jacobi map has the following "universal" property (cf. [5]).

Theorem 10.2. *Given a harmonic function $f : V_0 \longrightarrow A$ with $f(x_0) = 0$, there exists a unique homomorphism $\psi : \text{Pic}(X_0) \longrightarrow A$ such that $\psi \circ \Phi_0^{\text{aj}} = f$.*

Proof. Recalling that $\{x - x_0 | \ x \in V_0, \ x \neq x_0\}$ is a $\mathbb{Z}$-basis of $\text{Div}^0(X_0)$, we find a unique homomorphism $\varphi : \text{Div}^0(X_0) \longrightarrow A$ which is characterized by $\varphi(x - x_0) = f(x)$ (note that $\varphi(x - x_0) = f(x)$ holds for $x = x_0$). Rewriting (10.7) as

$$(\partial \circ \partial^*)\left(\sum_{x \in V_0} a_x x\right) = -\sum_{x \in V_0} a_x \sum_{e \in E_{0x}} \left[(t(e) - x_0) - (o(e) - x_0)\right],$$

we have

$$\varphi\left((\partial \circ \partial^*)\left(\sum_{x \in V_0} a_x x\right)\right) = -\sum_{x \in V_0} a_x \sum_{e \in E_{0x}} [f(t(e)) - f(o(e))] = 0,$$

which implies that $\text{Prin}(X_0) \subset \text{Ker } \varphi$, and hence φ induces a homomorphism $\psi : \text{Pic}(X_0) \longrightarrow A$. From the definition of Φ_0^{aj}, it follows that $\psi \circ \Phi_0^{\text{aj}} = f$ as desired. $\square$

The following theorem tells us that the discrete Abel–Jacobi map is essentially the same as the discrete Albanese map.

Theorem 10.3 (A Discrete Version of Abel's Theorem). *There exists a unique isomorphism φ of $\mathrm{Pic}(X_0)$ onto $\mathcal{J}(X_0)$ such that $\varphi \circ \Phi_0^{aj} = \Phi_0^{ab}$.*

The proof is carried out as follows. Use the universal property of discrete Abel–Jacobi maps to find a unique homomorphism $\varphi : \mathrm{Pic}(X_0) \longrightarrow \mathcal{J}(X_0)$ satisfying $\varphi \circ \Phi_0^{aj} = \Phi_0^{ab}$. The image $\Phi_0^{ab}(V_0)$ generates $\mathcal{J}(X_0)$, and hence φ is surjective. The claim that φ is an isomorphism is a consequence of the fact that $|\mathcal{J}(X_0)| = \kappa(X_0) = |\mathrm{Pic}(X_0)|$.

Applying the discussion above, we may prove non-degeneracy of Φ^{ab} under a mild condition.

Theorem 10.4. 1. *If X_0 has no separating edges, then $\Phi^{ab} : V_0^{ab} \longrightarrow H_1(X_0, \mathbb{R})$, the normalized standard realization restricted to the set of vertices, is injective.*
2. *Suppose that the graph obtained from X_0 by removing any geodesic of length two is connected. Then Φ^{ab} is non-degenerate.*

Proof. 1. It suffices to verify that $\Phi_0^{aj} : V_0 \longrightarrow \mathrm{Pic}(X_0)$ is injective because if this is true, then $\Phi_0^{ab} : V_0 \longrightarrow \mathcal{J}(X_0) \subset J(X_0)$ is injective in view of Abel's theorem so that if $\Phi^{ab}(x) = \Phi^{ab}(y)$, then there exists $\alpha \in H_1(X_0, \mathbb{Z})$ with $y = \alpha x$, and hence $\Phi^{ab}(y) = \Phi^{ab}(x) + \alpha$. Thus $\alpha = 0$ and $x = y$.

We shall show that Φ_0^{aj} is an injection (cf. [5]). Let x, y be two distinct vertices of X_0. Under the assumption on X_0, there exists a circuit $c = (e_1, \ldots, e_n)$ containing x and y; say, $x = o(e_i)$, $y = o(e_j)$ (this is a consequence of Menger's theorem; see [105]). Define the function $f : V_0 \longrightarrow \mathbb{Z}/n\mathbb{Z}$ by setting

$$f(z) = \begin{cases} k & \text{if } z = o(e_k) \text{ for some } k \\ 0 & \text{otherwise,} \end{cases}$$

Clearly $h = f - f(x_0)$ is harmonic, and $h(x) \neq h(y)$. Applying the universal property to h, we find that $\Phi_0^{aj}(x) \neq \Phi_0^{aj}(y)$, as desired.

2. Under the assumption on X_0, X_0 has no separating edges, and hence $\Phi^{ab} : V^{ab} \longrightarrow H_1(X_0, \mathbb{R})$ is injective. Let e_1, e_2 be distinct edges in X_0 with the same origin x. Suppose $\mathbf{v}_{ab}(e_1) = a\mathbf{v}_{ab}(e_2)$ for a positive number a. Then $e_1 - ae_2 \in H_1(X_0, \mathbb{R})^\perp$ because

$$\mathbf{0} = \mathbf{v}_{ab}(e_1) - a\mathbf{v}_{ab}(e_2) = P_{ab}(e_1 - ae_2).$$

If $\overline{e_2} = e_1$, then e_1 is a loop edge (so that $e_1 \in H_1(X_0, \mathbb{Z})$) and

$$e_1 - ae_2 = (1 + a)e_1.$$

Since $0 = \langle e_1 - ae_2, e_1 \rangle = 1 + a$, we have a contradiction. Thus $\overline{e_2} \neq e_1$, which implies $(\overline{e_2}, e_1)$ is a geodesic of length two. From the assumption one can find a circuit c containing e_1 and $\overline{e_2}$. We then have $\langle e_1 - ae_2, c\rangle = 1 + a$, but $\langle e_1 - ae_2, c\rangle = 0$ since $e_1 - ae_2 \in H_1(X_0, \mathbb{R})^\perp$, thereby again a contradiction. Thus the correspondence $e \in E_{0x} \longmapsto \mathbf{v}_{ab}(e)/\|\mathbf{v}_{ab}(e)\|$ is injective, as claimed. □

We finally see that as far as the symmetry of maximal topological crystals is concerned, the state of things is quite satisfactory.

Theorem 10.5. *If X_0 has no separating edges, then* $\mathrm{Aut}(X_0^{ab})$ *is isomorphic to a crystallographic group. The homology group $H_1(X_0,\mathbb{Z})$ as a subgroup of* $\mathrm{Aut}(X_0^{ab})$ *is maximal abelian. The factor group* $\mathrm{Aut}(X_0^{ab})/H_1(X_0,\mathbb{Z})$ *is isomorphic to* $\mathrm{Aut}(X_0)$.

Proof. Remember that X_0^{ab} has neither loop edges nor parallel edges (Theorem 6.6). Therefore combining Theorem 10.4 with Theorem 9.5, we conclude that $\mathrm{Aut}(X_0^{ab})$ is isomorphic to a crystallographic group. In Sect. 6.3, we proved that $H_1(X_0,\mathbb{Z})$ is maximal abelian in $\mathrm{Aut}(X_0^{ab}/X_0)$, and $\mathrm{Aut}(X_0^{ab}) = \mathrm{Aut}(X_0^{ab}/X_0)$. This completes the proof. □

10.4 Intersection Matrix and Tree Number

This section is devoted to the proofs for the claims about tree number stated in the previous sections. The method we rest on is adopted from *algebraic graph theory*.[10]

In general, we denote by $M_0(n,n;\mathbb{Z})$ the set of invertible matrices $S \in M(n,n)$ with integral entries, and by $GL_n(\mathbb{Z})$, the group consisting of integral square matrices A with $\det A = \pm 1$ (thus A^{-1} also has integral entries). Define the equivalence relation $\sim$ on $M_0(n,n;\mathbb{Z})$ by setting

$$S \sim T \iff \text{there exist } P,Q \in GL_n(\mathbb{Z}) \text{ with } T = PSQ.$$

The theory of elementary divisors tells us that for any $S \in M_0(n,n;\mathbb{Z})$, one can find $P,Q \in GL_n(\mathbb{Z})$ and positive integers $k_1,\ldots,k_n$ such that

$$PSQ = \begin{pmatrix} k_1 & 0 & 0 & \cdots & 0 \\ 0 & k_2 & 0 & \cdots & 0 \\ \cdots & \cdots & & & \\ 0 & 0 & 0 & \cdots & k_n \end{pmatrix}, \tag{10.8}$$

and k_i divides k_{i+1} ($i = 1,\ldots,n-1$). The array $\mathbf{k}(S) = (k_1,\ldots,k_n)$ depends only on the equivalence class of S, and $\det S = \pm k_1 \cdots k_n$.

The expression (10.8) is equivalent to the statement that there exist two $\mathbb{Z}$-bases $\{\mathbf{e}_1,\ldots,\mathbf{e}_n\}$ and $\{\mathbf{f}_1,\ldots,\mathbf{f}_n\}$ of $\mathbb{Z}^n$ with $S\mathbf{e}_i = k_i \mathbf{f}_i$ ($i=1,\ldots,n$). Therefore $\mathbb{Z}^n/S(\mathbb{Z}^n)$

---

[10] Algebraic graph theory is a field in which we make use of algebraic techniques to study combinatorial properties of graphs. This field has been developed since the first edition of Biggs' book [11] was published in 1974.

is isomorphic to $\mathbb{Z}_{k_1} \times \cdots \times \mathbb{Z}_{k_n}$. Indeed, the correspondence

$$\sum_{i=1}^{n} x_i \mathbf{f}_i \in \mathbb{Z}^n \longmapsto (x_1,\ldots,x_n) \in \mathbb{Z}_{k_1} \times \cdots \times \mathbb{Z}_{k_n}$$

yields a surjective homomorphism $\varphi : \mathbb{Z}^n \longrightarrow \mathbb{Z}_{k_1} \times \cdots \times \mathbb{Z}_{k_n}$ whose kernel is $S(\mathbb{Z}^n)$. In particular, $|\mathbb{Z}^n/S(\mathbb{Z}^n)| = |\det S|$.

Let $X_0 = (V_0, E_0)$ be a connected finite graph, and let $\alpha_1, \ldots, \alpha_b$ be a $\mathbb{Z}$-basis of $H_1(X_0, \mathbb{Z})$ ($b = \operatorname{rank} H_1(X_0, \mathbb{Z})$). The intersection matrix $A = (a_{ij})$, $a_{ij} = \langle \alpha_i, \alpha_j \rangle$, is symmetric, positive definite, and has integral entries. If $B = (\langle \beta_i, \beta_j \rangle)$ is the intersection matrix for another $\mathbb{Z}$-basis $\{\beta_1, \ldots, \beta_b\}$ of $H_1(X_0, \mathbb{Z})$, then

$$\alpha_i = \sum_{j=1}^{b} r_{ij} \beta_j$$

for some $R = (r_{ij}) \in GL_b(\mathbb{Z})$, and hence $A = RB^tR$. In particular, $A \sim B$.

Theorem 10.6. *1. Let A be the intersection matrix for a $\mathbb{Z}$-basis of the homology group $H_1(X_0, \mathbb{Z})$, and let $\mathbf{k}(A) = (k_1, \ldots, k_b)$ be the array associated with A. Then $\mathbf{k}(A)$ does not depend on the choice of a $\mathbb{Z}$-basis, and hence depends only on the graph structure of X_0 (therefore one may use the symbol $\mathbf{k}(X_0)$ instead of $\mathbf{k}(A)$).*
2. The discrete Jacobian $\mathfrak{J}(X_0)$ is isomorphic to $\mathbb{Z}_{k_1} \times \cdots \times \mathbb{Z}_{k_b}$.

The first assertion is a direct consequence of the discussion above. To prove the second assertion, we take the $\mathbb{Z}$-basis $\{c_1, \ldots, c_b\}$ of $H_1(X_0, \mathbb{Z})$ and the $\mathbb{Z}$-basis $\{\mathbf{v}_{ab}(e_1), \ldots, \mathbf{v}_{ab}(e_b)\}$ of $H_1(X_0, \mathbb{Z})^{\#}$ which we used in the proof of Theorem 10.1. Remember that $e_1, \ldots, e_b, \overline{e_1}, \ldots, \overline{e_b}$ are all edges that are not in a spanning tree T, and that c_i is a circuit containing e_i. Define the matrix $S = (s_{ij}) \in M_0(b, b; \mathbb{Z})$ by expressing c_i as a linear combination of $\mathbf{v}_{ab}(e_j)$ ($j = 1, \ldots, b$):

$$c_i = \sum_{j=1}^{b} s_{ji} \mathbf{v}_{ab}(e_j).$$

Then $s_{ij} = \langle c_i, c_j \rangle$ since $\langle c_i, \mathbf{v}_{ab}(e_j) \rangle = \delta_{ij}$. Thus S is the intersection matrix associated with $\{c_1, \ldots, c_b\}$. Moreover, the isomorphism

$$\psi : \mathbb{Z}^b \longrightarrow H_1(X_0, \mathbb{Z})^{\#}$$

defined by $\psi({}^t(x_1, \ldots, x_b)) = \sum_{i=1}^{b} x_i \mathbf{v}_{ab}(e_i)$ satisfies

$$\psi(S({}^t(y_1, \ldots, y_b))) = \sum_{j=1}^{b} \sum_{k=1}^{b} s_{jk} y_k \mathbf{v}_{ab}(e_j) = \sum_{k=1}^{b} y_k c_k,$$

10.4 Intersection Matrix and Tree Number

which implies that $\psi(S(\mathbb{Z}^b)) = H_1(X_0,\mathbb{Z})$. Thus the discrete Jacobian $\mathcal{J}(X_0) = H_1(X_0,\mathbb{Z})^\#/H_1(X_0,\mathbb{Z})$ is isomorphic to the factor group $\mathbb{Z}^b/S(\mathbb{Z}^b)$, thereby completing the proof.

Theorem 10.6 (2) is useful when we determine the structure of $\mathcal{J}(X_0)$. For instance, one can show, using this idea, that $\mathcal{J}(K_n) = (\mathbb{Z}_n)^{n-2}$ and $\mathcal{J}(D_d) = \mathbb{Z}_{d+1}$.

We shall establish a relationship between the intersection matrix and the tree number.

Theorem 10.7. *Let A be the intersection matrix associated with a $\mathbb{Z}$-basis of $H_1(X_0,\mathbb{Z})$. Then*

$$\det A = \kappa(X_0),$$

or equivalently, $\kappa(X_0) = k_1 \cdots k_b$ *where* $\mathbf{k}(X_0) = (k_1,\ldots,k_b)$.

Thus we get the following corollary which we used in the previous section.

Corollary 10.1. $\mathrm{vol}(J(X_0)) = \kappa(X_0)^{1/2}$.

The proof of Theorem 10.7 is quite long and needs several lemmas. In the discussion, we retain the notations introduced in the previous section.

Lemma 10.4.1. *Let $\{\gamma_1,\ldots,\gamma_s\}$ be a $\mathbb{Z}$-basis of $C_1(X_0,\mathbb{Z})$. Then*

$$\det(\langle \gamma_i, \gamma_j \rangle) = 1.$$

Proof. $\det(\langle \gamma_i, \gamma_j \rangle)^{1/2}$ is the volume of the flat torus $C_1(X_0,\mathbb{R})/C_1(X_0,\mathbb{Z})$, which does not depend on the choice of a $\mathbb{Z}$-basis $\{\gamma_1,\ldots,\gamma_s\}$. We thus conclude that $\det(\langle \gamma_i, \gamma_j \rangle) = 1$ because the matrix associated with the $\mathbb{Z}$-basis $\{e|\ e \in E_0^o\}$, E_0^o being an orientation of X, is the identity matrix. □

We shall show that $\partial \circ \partial^* : \mathrm{Div}^0(X_0) \longrightarrow \mathrm{Div}^0(X_0)$ is injective. Suppose $(\partial \circ \partial^*)\alpha = 0$ for $\alpha \in \mathrm{Div}^0(X_0)$. Then $\partial^*\alpha = 0$. Writing

$$\alpha = \sum_{x \in V_0} a_x x \quad (a_x \in \mathbb{Z} \text{ and } \sum_{x \in V_0} a_x = 0),$$

and using the definition of ∂^*, we have

$$0 = \partial^*\alpha = -\sum_{x \in V_0} a_x \sum_{e \in E_{0x}} e = -\sum_{e \in E_0} a_{o(e)} e,$$

from which it follows that $a_{o(e)} = a_{t(e)}$ for every $e \in E_0$. Since X_0 is connected, a_x turns out to be constant, and so $a_x \equiv 0$.

Choosing a $\mathbb{Z}$-basis of $\mathrm{Div}^0(X_0)$, we express $(\partial \circ \partial^* | \mathrm{Div}^0(X_0))$ by a matrix. Since the determinant of this matrix does not depend on the choice of $\mathbb{Z}$-basis, it is legitimate to write $\det(\partial \circ \partial^* | \mathrm{Div}^0(X_0))$ for this determiant.

Lemma 10.4.2. $\det(\partial \circ \partial^* | \mathrm{Div}^0(X_0)) = |V_0| \cdot \det A$.

Proof. Let $V_0 = \{x_0, \ldots, x_N\}$ ($N+1 = |V_0|$), and put $v_i = x_i - x_0 \in \mathrm{Div}^0(X_0)$. Note that $\{v_1, \ldots, v_N\}$ forms a basis for $\mathrm{Div}^0(X_0)$ and

$$W := (\langle v_i, v_j \rangle) = \begin{pmatrix} 2 & 1 & \cdots & 1 \\ 1 & 2 & \cdots & 1 \\ \cdots & \cdots & & \\ 1 & 1 & \cdots & 2 \end{pmatrix}.$$

Because Image $\partial = \mathrm{Ker}\,\varepsilon = \mathrm{Div}^0(X_0)$ as observed in Sect. 4.2, one may choose $\beta_1, \ldots, \beta_N \in C_1(X_0, \mathbb{Z})$ such that $\partial \beta_i = v_i$. Then

$$\{\alpha_1, \ldots, \alpha_b, \beta_1, \ldots, \beta_N\}$$

forms a $\mathbb{Z}$-basis of $C_1(X_0, \mathbb{Z})$. We put

$$\begin{pmatrix} (\langle \alpha_i, \alpha_j \rangle) & (\langle \alpha_i, \beta_k \rangle) \\ (\langle \beta_h, \alpha_i \rangle) & (\langle \beta_h, \beta_k \rangle) \end{pmatrix} = \begin{pmatrix} A & {}^tC \\ C & B \end{pmatrix}.$$

Lemma 10.4.1 says that the determinant of this matrix is equal to 1. We now write

$$\partial^* v_i = \sum_{h=1}^{b} d_{ih} \alpha_h + \sum_{k=1}^{N} e_{ik} \beta_k$$

to get two matrices $D = (d_{ih}) \in M(N, b)$ and $E = (e_{ik}) \in M(N, N)$. Then

$$(\partial \circ \partial^*) v_i = \sum_{k=1}^{N} e_{ik} v_k,$$

so that the matrix E representing the endomorphism $\partial \circ \partial^* | \mathrm{Div}^0(X_0)$ is invertible. Next, we show that $\det E = (N+1) \det A$, which is nothing but our claim. Since

$$0 = \langle v_i, \partial \alpha_j \rangle = \langle \partial^* v_i, \alpha_j \rangle = \sum_{h=1}^{b} d_{ih} \langle \alpha_h, \alpha_j \rangle + \sum_{k=1}^{N} e_{ik} \langle \beta_k, \alpha_j \rangle,$$

and $C = (\langle \beta_k, \alpha_j \rangle)$, we have $DA + EC = O$, and $C = -E^{-1}DA$. An easy computation leads to

$$\begin{pmatrix} I & O \\ E^{-1}D & I \end{pmatrix} \begin{pmatrix} A & {}^tC \\ C & B \end{pmatrix} = \begin{pmatrix} A & {}^tC \\ O & E^{-1}D{}^tC + B \end{pmatrix}.$$

Hence by taking the determinant of both sides, we obtain

10.4 Intersection Matrix and Tree Number

$$1 = \det \begin{pmatrix} A & {}^tC \\ C & B \end{pmatrix} = \det A \det(E^{-1}D^tC + B)$$

$$= \det A \det E^{-1} \det(D^tC + EB).$$

On the other hand,

$$\sum_{h=1}^{n} d_{ih}\langle \alpha_h, \beta_j \rangle + \sum_{k=1}^{N} e_{ik}\langle \beta_k, \beta_j \rangle = \langle \partial^* v_i, \beta_j \rangle = \langle v_i, \partial \beta_j \rangle = \langle v_i, v_j \rangle,$$

which is neatly expressed as

$$D^tC + EB = W.$$

Using $\det W = N+1$, we get

$$\det E = \det(D^tC + EB) \det A = (N+1) \det A,$$

as claimed. □

To go further, we fix an orientation $E_0^o = \{e_1, \ldots, e_s\}$, and write

$$\partial e_i = \sum_{h=1}^{N} f_{ih} v_h, \qquad \partial^* v_k = \sum_{j=1}^{s} g_{kj} e_j$$

to obtain the matrices $F = (f_{ih}) \in M(s,N)$ and $G = (g_{kj}) \in M(N,s)$. We then have $E = GF$. Using the relation $\langle v_k, \partial e_i \rangle = \langle \partial^* v_k, e_i \rangle$, we get

$$\sum_{h=1}^{N} f_{ih}\langle v_h, v_k \rangle = \sum_{j=1}^{s} g_{kj}\langle e_j, e_i \rangle = g_{ki},$$

or equivalently, $FW = {}^tG$. This implies that $E = GF = W {}^tFF$. In particular, $\det E = (N+1) \det({}^tFF)$. Therefore to complete the proof of Theorem 10.7, it suffices to prove $\det({}^tFF) = \kappa(X_0)$.

Let $T = (V_T, E_T)$ be a spanning tree of X_0. Then the number of edges in $E_T \cap E_0^o$ is N since $1 = \chi(T) = N+1 - |E_T \cap E_0^o|$ (recall that $\chi(\cdot)$ expresses the Euler number). Conversely, if $|E_T \cap E_0^o| = N$ for a subtree $T = (V_T, E_T)$, then $|V_T| = 1 + |E_T \cap E_0^o| = 1 + N = |V_0|$. Thus T is a spanning tree.

Lemma 10.4.3. *Let $I = (i_1, \ldots, i_N)$ be a multi-index with $1 \leq i_1 < \cdots < i_N \leq s$, and put*

$$F_I = \begin{pmatrix} f_{i_1 1} & \cdots & f_{i_1 N} \\ & \cdots & \\ f_{i_N 1} & \cdots & f_{i_N N} \end{pmatrix}.$$

Then $\det F_I$ is either 0 or ± 1. Moreover, $\det F_I = \pm 1$ if and only if the set of edges $\{e_{i_1}, \ldots, e_{i_N}\}$ forms a spanning tree of X_0.

Proof. Let X_1 be the (possibly disconnected) graph formed by the edges $e_{i_1}, \ldots, e_{i_N}$. The restriction of ∂ to $C_1(X_1, \mathbb{Z})$ coincides with the boundary operator $\partial_{X_1} : C_1(X_1, \mathbb{Z}) \longrightarrow \mathrm{Div}^0(X_1)$ ($\subset C_0(X_1, \mathbb{Z})$) for X_1. We find that F_I is the matrix representing ∂_{X_1}.

Suppose that X_1 is a tree (hence a spanning tree of X_0). This being the case, $C_0(X_0, \mathbb{Z}) = C_0(X_1, \mathbb{Z})$ and $\mathrm{Ker}\, \partial_{X_1} = \{0\}$. Therefore ∂_{X_1} is an isomorphism of $C_1(X_1, \mathbb{Z})$ onto $\mathrm{Div}^0(X_1)$, and $\det F_I = \pm 1$.

Next suppose that X_1 is not a tree. We first check that X_1 has a circuit. If not, then X_1 is a *forest*, i.e., a disjoint union of subtrees $T_1, \ldots, T_l$. If T_i has N_i vertices, then it has $N_i - 1$ undirected edges. Thus $N = N_1 + \cdots + N_l - l \leq N + 1 - l$, which implies that $l = 1$, or equivalently, that X_1 is a tree. This contradicts the assumption that X_1 is not a tree. Therefore there is a circuit c in X_1. Since $\partial_{X_1}\langle c \rangle = 0$, we find that F_I is singular, and $\det F_I = 0$. $\square$

Lemma 10.4.4. $\det({}^t FF) = \kappa(X_0)$.

Proof. Recall the Binet–Cauchy formula: for a matrix $A \in M(m,n)$ and a matrix $B \in M(n,m)$ with $m \leq n$,

$$\det(AB) = \sum_{1 \leq i_1 < \cdots < i_m \leq n} \det \begin{pmatrix} a_{1i_1} & \cdots & a_{1i_m} \\ & \cdots & \\ a_{mi_1} & \cdots & a_{mi_m} \end{pmatrix} \det \begin{pmatrix} b_{i_1 1} & \cdots & b_{i_1 m} \\ & \cdots & \\ b_{i_m 1} & \cdots & b_{i_m m} \end{pmatrix}.$$

Apply this formula to the case $A = {}^t F$, $B = F$ and use Lemma 10.4.3 to obtain the conclusion. $\square$

Putting all three lemmas together, we complete the proof of Theorem 10.7.

Finally we prove the following theorem mentioned in the previous section, which is equivalent to the theorem of Kirchhoff–Trent (cf. [17, 74, 75]).

Theorem 10.8. $|\mathrm{Pic}(X_0)| = \kappa(X_0)$.

Remember $\mathrm{Pic}(X_0) = \mathrm{Div}^0(X_0)/(\partial \circ \partial^*)(C_0(X_0, \mathbb{Z}))$. To prove the theorem, we introduce the following subgroup of $C_0(X_0, \mathbb{Z})$:

$$C_0(X_0, \mathbb{Z})_0 = \Big\{ \sum_{x \in V_0} a_x x \in C_0(X_0, \mathbb{Z})\, \Big|\, \sum_{x \in V_0} a_x \equiv 0 \pmod{N+1} \Big\}.$$

Since $C_0(X_0, \mathbb{Z})_0$ is the kernel of the surjective homomorphism given by

$$\sum_{x \in V_0} a_x x \longmapsto \sum_{x \in V_0} a_x \in \mathbb{Z}/(N+1)\mathbb{Z},$$

10.4 Intersection Matrix and Tree Number

we find that

$$\left|C_0(X_0,\mathbb{Z})/C_0(X_0,\mathbb{Z})_0\right| = N+1.$$

On the other hand, if we put $v_0 = \sum_{x \in V_0} x \in C_0(X_0, \mathbb{Z})$, then $C_0(X_0,\mathbb{Z})_0$ is decomposed into the direct sum

$$C_0(X_0,\mathbb{Z})_0 = \mathrm{Div}^0(X_0) \oplus \mathbb{Z}v_0.$$

This is actually deduced from the expression

$$\sum_{x \in V_0} a_x x = \left[\sum_{x \in V_0} a_x x - \frac{1}{N+1}\left(\sum_{x \in V_0} a_x\right)v_0\right] + \frac{1}{N+1}\left(\sum_{x \in V_0} a_x\right)v_0.$$

Therefore $(\partial \circ \partial^*)(\mathrm{Div}^0(X_0)) = (\partial \circ \partial^*)(C_0(X_0,\mathbb{Z})_0)$. Recalling $\mathrm{Ker}\,\partial \circ \partial^* = \mathbb{Z}v_0$, we observe that $(\partial \circ \partial^*)(C_0(X_0,\mathbb{Z}))$ and $(\partial \circ \partial^*)(C_0(X_0,\mathbb{Z})_0)$ are isomorphic to the factor groups $C_0(X_0,\mathbb{Z})/\mathbb{Z}v_0$ and $C_0(X_0,\mathbb{Z})_0/\mathbb{Z}v_0$, respectively, and hence

$$(\partial \circ \partial^*)(C_0(X_0,\mathbb{Z}))/(\partial \circ \partial^*)(C_0(X_0,\mathbb{Z})_0)$$

is isomorphic to $C_0(X_0,\mathbb{Z})/C_0(X_0,\mathbb{Z})_0$. Using this fact and taking a look at the exact sequence

$$0 \to (\partial \circ \partial^*)(C_0(X_0,\mathbb{Z}))/(\partial \circ \partial^*)(C_0(X_0,\mathbb{Z})_0)$$
$$\to \mathrm{Div}^0(X_0)/(\partial \circ \partial^*)(C_0(X_0,\mathbb{Z})_0)$$
$$\to \mathrm{Div}^0(X_0)/(\partial \circ \partial^*)(C_0(X_0,\mathbb{Z})) \to 0,$$

we get

$$\left|\mathrm{Pic}(X_0)\right| = \frac{1}{N+1}\left|\mathrm{Div}^0(X_0)/(\partial \circ \partial^*)(\mathrm{Div}^0(X_0))\right|.$$

On the other hand, using the equality $|\mathbb{R}^n/S(\mathbb{R}^n)| = |\det S|$ mentioned at the beginning of the present section, we have

$$\left|\mathrm{Div}^0(X_0)/(\partial \circ \partial^*)(\mathrm{Div}^0(X_0))\right| = \det\left((\partial \circ \partial^*)|\mathrm{Div}^0(X_0)\right)$$
$$= \det E = |V_0| \cdot \det A = (N+1)\kappa(X_0),$$

from which the claim follows.

10.5 Notes

(I) Another classical result in algebraic geometry of curves is the *Riemann–Roch theorem*. This theorem gives a relation among numerical invariants of a divisor D on a curve S; say,

$$l(D) - l(K - D) = \deg(D) - g + 1, \tag{10.9}$$

where $g = \frac{1}{2} \dim H_1(S, \mathbb{R})$ stands for the genus of S, and the symbol $\deg(D)$ denotes the degree of the divisor D, i.e., the sum of the coefficients occurring in D. The number $l(D)$ is the dimension (over $\mathbb{C}$) of the vector space of meromorphic functions f on S, such that all the coefficients of $(f) + D$ are non-negative. The symbol K denotes the canonical divisor, i.e., the divisor of a global meromorphic 1-form.

We shall give a brief glimpse into an analogue of (10.9) in the graph-theoretical set-up [5]. To this end, first define an equivalence relation $\sim$ on the group of divisors $\text{Div}(X_0)(= C_0(X_0, \mathbb{Z}))$ by declaring that $D_1 \sim D_2$ (called *linearly equivalent*) if and only if $D_1 - D_2 \in \text{Prin}(X_0)$ (recall that this equivalence relation restricted to $\text{Div}^0(X_0)$ yields the Picard group). The *linear system* $|D|$ associated with $D \in \text{Div}(X_0)$ is the set $\{D' | D' \geq 0, D' \sim D\}$, where $D' \geq 0$ means that every coefficient in D' as a 0-chain is non negative (such E is called *effective*). The *degree* $\deg(D)$ is nothing but $\varepsilon_0(D)$ introduced in Sect. 4.2.

The number $l(D)$ for the linear system $|D|$ is defined by setting $l(D)$ equal to 0 if $|D| = \emptyset$, and then declaring that for each integer $s \geq 0$, $l(D) \geq s + 1$ if and only if $|D - D'| \neq \emptyset$ for all effective divisors D' of degree s. The quantity $l(D)$ depends only on the linear equivalence class of D.

The *canonical divisor* K on X_0 is given by

$$K = \sum_{x \in V_0} [(\deg x) - 2] x.$$

Obviously $\deg(K) = 2b_1(X_0) - 2$.

Theorem 10.9 (Riemann–Roch Theorem for Graphs). *For a divisor D on X_0, the following equality holds:*

$$l(D) - l(K - D) = \deg(D) - b_1(X_0) + 1.$$

This theorem has an application to the existence or non-existence of a winning strategy for a certain chip-firing game played on the vertices of a graph.

Remark. The quantity $k(x) = 2 - (\deg x)$ appearing in K is an analogue of the scalar curvature for which we have the "Gauss–Bonnet formula":

$$\sum_{x \in V_0} k(x) = 2\chi(X_0).$$

10.5 Notes

This formula, in view of the algebraic curve's case, somehow justifies the name "canonical divisor". □

(II) The tree number shows up in the "special value" of the Ihara zeta function [48]. Recall the definition of the zeta function [see Notes (I) in Chap. 6].

$$Z(u) = \prod_{\mathfrak{p} \in \mathcal{P}} \left(1 - u^{|\mathfrak{p}|}\right)^{-1}.$$

If X_0 is a non-circuit graph, then $u = 1$ is a pole of $Z(u)$ of order b ($= \dim H_1(X_0, \mathbb{R})$), and

$$\lim_{u \to 1} (1-u)^{-b} Z(u)^{-1} = 2^b \chi(X_0) \kappa(X_0).$$

It is interesting to compare this with the famous conjecture due to B. Birth and P. Swinnerton–Dyer for the L-functions associated with elliptic curves.

An analogue of the *Riemann hypothesis* for $Z(u)$, another interesting item in discrete geometric analysis, has something to do with the efficiency of communication networks.

(III) What we have described in this chapter is generalized to the category of "tropical curves", objects in *tropical algebraic geometry* [7,67]. Roughly speaking, tropical algebraic geometry is based on the algebraic system with operations $x \oplus y$ and $x \otimes y$ in $\mathbb{R}$ defined by

$$x \oplus y = \min\{x, y\}, \quad x \otimes y = x + y.$$

Tropical (algebraic) geometry,[11] a relatively new field, is a skeletonized version of algebraic geometry. In this context, a tropical curve is represented by an edge-weighted graph (metric graph).

(IV) In view of Theorem 10.5, if X_0 has no separating edges, then we have the exact sequence:

$$0 \to H_1(X_0, \mathbb{Z}) \to \mathrm{Aut}(X_0^{\mathrm{ab}}) \to \mathrm{Aut}(X_0) \to 1.$$

---

[11] The adjective "tropical" is given in honor of the Brazilian mathematician Imre Simon who pioneered this area.

Fig. 10.1 2-Cocysle

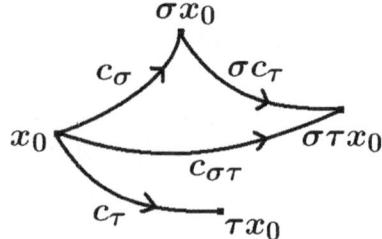

In other words, Aut(X_0^{ab}) is an *extension* of $H_1(X_0, \mathbb{Z})$ by the finite group Aut(X_0), and hence its isomorphism class is determined by a group cohomology class $\Theta \in H^2\big(\text{Aut}(X_0), H_1(X_0, \mathbb{Z})\big)$ (see [16]). More explicitly, Θ is represented by the 2-cocycle

$$\theta \in C^1\big(\text{Aut}(X_0), H_1(X_0, \mathbb{Z})\big)$$

defined by

$$\theta(\sigma, \tau) = \langle c_\sigma \cdot \sigma c_\tau \cdot \overline{c_{\sigma\tau}} \rangle, \quad (\sigma, \tau \in \text{Aut}(X_0))$$

where c_σ is a path in X_0 joining x_0 and σx_0 (Fig. 10.1).

Finale

If we take for granted the belief that beauty is bound up with symmetry, the standard realization may deserve to be called the most "beautiful" among all periodic realizations of a given topological crystal. For instance, the models of the diamond and K_4 crystals, typical standard realizations with large symmetry, appeal personally to my taste in beauty. Especially, the artistic (and geometric) structure of the K_4 crystal has intrigued me for some time after I rediscovered it by chance in the midst of my study of random walks. However, one's aesthetic sense is not that simple to take the belief above as a complete agreement. Some cynics might say that "symmetry is death" since, if something has symmetry, it is static, unchanging, frozen as in death.

Yet "reality favors symmetry", said J.L. Borges (1899–1986), an Argentine writer, essayist, and poet.[1] The famous ancient Roman architect Vitruvius (80 BC–15 BC) pronounced in his monumental book *De Architectura* dedicated to Augustus that nature's designs are based on universal laws of proportion and symmetry,[2] and thus in order to exhibit the *venustas* (beauty) of structures, we need to understand the proportions of the human body,[3] the greatest work of nature. Having a naive understanding of reality leads us to be convinced that there is truth to some extent in what Borges and Vitruvius asserted. Thus just like "golden numberism" in aesthetics claiming that the golden ratio[4] was there before the artifact, we are tempted to claim that standard realizations are seen everywhere, even in unexpected places.

---

[1] It is interesting that in his novel *Library of Babel*, Borges's narrator (a librarian) describes his library as a crystal-net-like architecture.

[2] "Symmetry" in Greek means "to measure together" or "harmony of proportions".

[3] Leonardo da Vinci correctly illustrated the proportions outlined in Vitruvius' idea (*Vitruvian Man*, 1513).

[4] The *golden ratio* (1.618033...), a root of $x^2 = x+1$, is a number often encountered when taking the ratios of distances in simple geometric figures such as the pentagon and dodecahedron. Much is claimed about this number in the history of art, architecture, sculpture and anatomy, but many claims of its significance are exaggerated.

Fig. 1 Other realizations of the hexagonal and quadrangle lattice

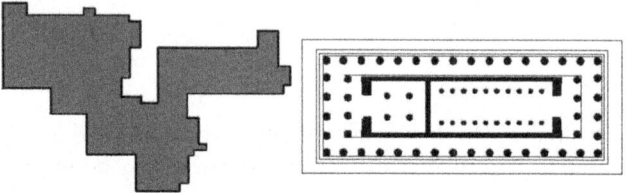

Fig. 2 Plans of Katsura Imperial Villa and Parthenon

However, although it is true that a few beautiful structures in nature and art are explained in terms of standard realizations, it is dangerous to leap from a few to general statements without any definite evidence.

Figure 1 illustrates periodic realizations of the hexagonal lattice and quadrangle lattice, which are different from the honeycomb lattice and the square lattice. Some people may think that these realizations are more interesting than the honeycomb and square lattices even though they are less symmetric.

And what is more, one's aesthetic sense depends heavily on one's cultural background. I remember what Bruno Taut (1880–1938), a German architect, wrote in his essay. He expressed, by revealing the beauty of the Katsura Imperial Villa,[5] his appreciation of Japanese culture, in comparing the *asymmetric* feature of Japanese architecture with the so-called Greco-Roman tradition that strictly obeys the rule of symmetry (Fig. 2). The Japanese taste for asymmetry is also seen in Bonsai and flower arrangements[6]; both are disciplined arts in which nature and humanity are brought together.

Apart from an aesthetic sense, it is rather usual to find asymmetry or the *breaking of symmetry* in physical phenomena, despite what Borges said to the contrary. For instance, the direction of time is asymmetric in the sense that the future is intrinsically distinguished from the past. In quantum physics, symmetry-breaking

---

[5]The villa, also called Katsura Detached Palace, was constructed during the early and mid-seventeenth century by Toshihito Shinno and Toshitada Shinno in the western suburbs of Kyoto, Japan.

[6]"Ikebana" in Japanese; literally "flowers kept alive".

is a serious issue. "Parity non-conservation" occurring in the weak interaction is a typical example. However, even if the really interesting things involve asymmetry or symmetry-breaking,[7] we first have to understand the nature of symmetry itself. In other words, whether or not one rejects symmetric objects as being unrealistic, one cannot talk about asymmetry without a deep knowledge of symmetry. In any case, history vindicates that symmetry gives rise to a never-ending source of interesting problems in mathematics. In this connection, I quote what Weyl wrote in his book [110] as a final remark of this book:

> Symmetry, as wide or narrow as you may define its meaning, is one idea by which man through the ages has tried to comprehend and create order, beauty, and perfection.
>
> Symmetry is a vast subject, significant in art and nature. Mathematics lies at its root, and it would be hard to find a better one on which to demonstrate the working of the mathematical intellect.

---

[7] Currie claimed that *the asymmetry makes the phenomenon* (1894).

Appendix

The purpose of this appendix is to explain some basic concepts and terminology in modern mathematics that are indispensable for reading through Part I and II. We also give a brief review of free groups and crystallographic groups.

1 Sets and Maps

"Set and map" are fundamental terms in the language of modern mathematics without which it is almost impossible to explain what mathematicians are doing nowadays.[1]

A *set* is a collection of distinct objects. Objects are called *elements*. When x is an element of a set A, we say that x belongs to A and write $x \in A$. The set consisting of elements x having a property P is expressed as $\{x|\ P(x)\}$ where $P(x)$ stands for the fact that element x has property P. The braces are also used to contain lists of the elements of a set. If every $x \in A$ belongs to another set B, then A is said to be a *subset* of B, and we write $A \subset B$. If $A \subset B$ and $B \subset A$, i.e., A and B have exactly the same elements, then they are said to be *equal*; symbolically $A = B$. When $A \subset B$ and $A \neq B$, we say that B is a *proper subset* of A. The *empty set* $\emptyset$ is the unique set having no elements. A finite set is a set with a finite number of elements. The notation $|A|$ means the number of elements in A.

The *union* of A and B, symbolically $A \cup B$, is the set of all elements belonging to either A or B. The *intersection* $A \cap B$ is the set of all elements belonging to both A and B simultaneously. If $A \cap B = \emptyset$, i.e., A and B have no elements in common, then A and B are said to be *disjoint*. Union and intersection are defined for a family consisting

---

[1] Cantor (1845–1918) is the founder of set theory. He established the concept of "infinite" in the framework of his theory (1874–1884). Though a loose use of the notion of sets leads us to a serious contradiction like Russell's paradox, we may be content with the "naive" theory of sets as far as we are working in a reasonable mathematical framework.

of more than two sets. For a family of sets $\{A_i\}_{i \in I}$, its union and intersection are denoted by $\bigcup_{i \in I} A_i$ and $\bigcap_{i \in I} A_i$, respectively.

The collection of elements in A that are not in B is expressed as $A \backslash B$ or $A - B$, and is called the *difference set* of A and B.

The *cartesian product* (*direct product* or simply *product*) $A_1 \times \cdots \times A_n$ of a family of sets $A_1, \ldots, A_n$ is a collection of sequences of the form $(a_1, \ldots, a_n)$ with $a_i \in A_i$, where $(a_1, \ldots, a_n) = (b_1, \ldots, b_n)$ if and only if $a_i = b_i$ for $i = 1, \ldots, n$. A cartesian product $A_1 \times \cdots \times A_n$ in which $A_1 = \cdots = A_n = A$ is denoted by A^n.

A *map* (or *mapping*) of a set A into another set B is a rule assigning to each element of A (the domain) an element in B (the range).[2] The expression $f : A \longrightarrow B$ (or $A \xrightarrow{f} B$) means a map assigning $f(a) \in B$ to $a \in A$. We frequently use the notation $a \in A \mapsto f(a) \in B$ to represent the map f. Two maps f and g are *equal* if their domains coincide and $f(a) = g(a)$ for every a in the common domain.

For a map $f : A \longrightarrow B$ and a subset C of A, the subset $\{f(c)| \ c \in C\}$ is called the *image* of C under f, and denoted by $f(C)$. When $C = A$, we occasionally write Image f instead of $f(A)$. For a subset D of B, the *inverse image* of D under f is defined to be $\{a \in A| \ f(a) \in D\}$, and denoted by $f^{-1}(D)$.

The *restriction* of $f : A \longrightarrow B$ to a subset C of A, symbolically written as $f|C : C \longrightarrow B$, is defined by $f|C(c) = f(c)$ $(c \in C)$. For a map $g : C \longrightarrow B$, a map $f : A \longrightarrow B$ with $f|C = g$ is called an *extension* of g to A.

Given two maps $f_1 : A_1 \longrightarrow B_1$ and $f_2 : A_2 \longrightarrow B_2$, we define $f_1 \times f_2 : A_1 \times A_2 \longrightarrow B_1 \times B_2$ by setting

$$(f_1 \times f_2)(a_1, a_2) = (f_1(a_1), f_2(a_2)).$$

A map $f : A \longrightarrow B$ is *injective* (or an *injection*) if $f(a_1) = f(a_2)$ implies $a_1 = a_2$. If $f(A) = B$ (or equivalently, if, for every $b \in B$, there exists $a \in A$ such that $f(a) = b$), then f is said to be *surjective* (or a *surjection*). A *bijection*, or a *one-to-one correspondence* is a map which is both injective and surjective at the same time.

For a subset A of B, we have the *inclusion map* $i : A \longrightarrow B$ defined by $i(a) = a$. In particular, $i : A \longrightarrow A$ is the *identity map* which we denote by I_A or simply by I.

For two maps $f : A \longrightarrow B$, $g : B \longrightarrow C$, the composition $g \circ f$ of f and g is the map assigning $g(f(a))$ to $a \in A$. For another map $h : C \longrightarrow D$

$$[h \circ (g \circ f)](a) = h[(g \circ f)(a)] = h[g(f(a))] = (h \circ g)(f(a)) = [(h \circ g) \circ f](a),$$

and so $h \circ (g \circ f) = (h \circ g) \circ f$ (the associative law for maps). For $f : A \longrightarrow A$, we write f^n for $f \circ \cdots \circ f$ (n-times).

---

[2] This is a rather vague explanation. To be more precise, a map is defined to be a subset R of $A \times B$ complying with a certain condition (we have in mind the notion of graphs of functions).

2 Groups and Homomorphisms

The concept of "group" originating in the work of Evariste Galois (1811–1832) is one of the great unifying ideas of mathematics, and is required whenever we wish to talk about symmetry in a rigorous way. The reader should refer to [72] for details.

A set G having a special element e together with two maps $\varphi : G \times G \longrightarrow G$ and $\psi : G \longrightarrow G$ is called a *group* if the following three conditions are satisfied:

1. $\varphi(\varphi(a,b),c) = \varphi(a,\varphi(b,c))$ for every $a,b,c \in G$ (the associative law),
2. $\varphi(a,e) = \varphi(e,a) = a$ for every $a \in G$,
3. $\varphi(a,\psi(a)) = \varphi(\psi(a),a) = e$ for every $a \in G$.

When we write ab for $\varphi(a,b)$ and a^{-1} for $\psi(a)$, the above conditions are expressed as $(ab)c = a(bc)$, $ae = ea = a$, $aa^{-1} = a^{-1}a = e$. We call ab the *multiplication* or *product* of a and b. We also call a^{-1} the *inverse* of a. The special element e is said to be the *identity element*, for which we often use the symbol 1, in spite of the fact that different groups generally have different identity elements. A *trivial group* is a group consisting only of the identity element. A *finite group* is a group whose underlying set is finite. The number of elements in a finite group G is called the *order* of G.

If $ab = ba$ holds for every $a,b \in G$, we say that G is an *abelian group* (or a commutative group[3]). In this case, we often use the additive form $a+b$ instead of the multiplicative form ab, and also use the symbol 0 (*zero*) for the identity element. We write $-a$ for the inverse a^{-1}. A group G with additive operation is referred to as an *additive group*. The set of integers $\mathbb{Z}$ is a typical additive group.

Given two groups G_1 and G_2, the cartesian product $G_1 \times G_2$ has a natural group structure with the identity element $(1,1)$, the group operations being defined by

$$(g_1,g_2)(h_1,h_2) = (g_1 h_1, g_2 h_2), \quad (g_1,g_2)^{-1} = (g_1^{-1}, g_2^{-1}).$$

In the case of additive groups, $G_1 \times G_2$ is called the *direct sum*, and is written as $G_1 \oplus G_2$. The product of groups is defined for a family consisting of more than two groups.

A non-empty subset H of a group G is said to be a *subgroup* if H is "closed" under the group operations, i.e., (i) $a,b \in H \Longrightarrow ab \in H$, (ii) $a \in H \Longrightarrow a^{-1} \in H$. If, in addition, $ghg^{-1} \in H$ for every $h \in H$ and $g \in G$, the subgroup H is said to be *normal*.

For a subgroup H of G, the *normalizer* $N_G(H)$ of H is defined to be $\{g \in G \mid gHg^{-1} = H\}$. Evidently $N_G(H)$ is a subgroup of G, and H is a normal subgroup of $N_G(H)$.

---

[3]The term "abelian" is used in honor of Abel whose work in 1827 and 1828 on a certain special type of algebraic equations was later interpreted in terms of commutative groups.

As for the definition of *coset spaces* and *factor groups*, see Example 2.3 in Chap. 2.

Theory of matrices is a rich repository of examples of groups.

Example 1 *The general linear group $GL_n(\mathbb{R})$ is the group consisting of invertible matrices $A \in M(n,n)$; that is,*

$$GL_n(\mathbb{R}) = \{A \in M(n,n)|\ \det A \neq 0\}.$$

The group multiplication is given by matrix multiplication.

The set $GL_n(\mathbb{Z})$ of matrices $A \in M(n,n)$ with integral entries such that $\det A = \pm 1$ is a subgroup of $GL_n(\mathbb{R})$. The set $O(n)$ of $A \in M(n,n)$ satisfying ${}^t\!AA = I_n$ is also a subgroup of $GL_n(\mathbb{R})$, which is called the orthogonal group. The matrices $A \in O(n)$ with determinant 1 form a normal subgroup of $O(n)$ known as the special orthogonal group (or rotation group), denoted $SO(n)$. □

For a subgroup H of G, if the coset space G/H is a finite set, then H is said to be *of finite index* in G. We call $|G/H|$ the index of H.

For a family of subgroups $\{H_i\}_{i \in I}$ of G, the intersection $\bigcap_{i \in I} H_i$ is also a subgroup. For a non-empty subset A of G, the subgroup *generated by A* is the intersection H of all subgroups containing A. It is easy to see that H coincides with the totality of elements of the form $a_1^{\pm 1} \cdots a_k^{\pm 1}$ ($a_i \in A$, $k = 1,2,\ldots$); that is, H consists of all elements of G that can be obtained by multiplying finitely many elements of A and their inverses together. We call A a *set of generators* for H. A group with a finite set of generators is said to be *finitely generated*. A *cyclic group* is a group with a single generating element.

For two groups G and G_1, a map $\rho : G \longrightarrow G_1$ is called a *homomorphism* if $\rho(gh^{-1}) = \rho(g)\rho(h)^{-1}$ ($g,h \in G$), or equivalently, $\rho(1) = 1$ and $\rho(gh) = \rho(g)\rho(h)$.

The image of a homomorphism ρ is a subgroup of G_1, and the inverse image of the identity element of G_1 is a normal subgroup of G, which we call the *kernel* of ρ and write $\mathrm{Ker}\,\rho$. A useful fact is that $\mathrm{Ker}\,\rho = \{1\}$ if and only if ρ is injective. When ρ is a bijection, ρ is called an *isomorphism*, and we say that G and G_1 are *isomorphic*. When $G = G_1$, an isomorphism ρ is called an *automorphism* of G. The set of automorphisms of G, symbolically $\mathrm{Aut}(G)$, becomes a group in a natural manner. A cyclic group is isomorphic to $\mathbb{Z}$ or $\mathbb{Z}_n = \mathbb{Z}/n\mathbb{Z}$ (n being a positive integer).

The following theorem is fundamental in group theory. Above all, the first statement has been used several times in the main text of this book.

Theorem 1 1. *First isomorphism theorem: Let $f : G \longrightarrow H$ be a homomorphism. Then the image of f is isomorphic to the factor group $G/\mathrm{Ker}\,f$. In particular, if f is surjective, then H is isomorphic to $G/\mathrm{Ker}\,f$.*
2. *Second isomorphism theorem: Let G be a group, and let H be a subgroup of G. If N is a normal subgroup of G, then*

$$HN(= \{hn|\ h \in H,\ n \in N\}$$

is a subgroup of G. Moreover the intersection $H \cap N$ is a normal subgroup of H, and the quotient groups $(HN)/N$ and $H/(H \cap N)$ are isomorphic.

3. **Third isomorphism theorem.** Let N and K be normal subgroups of G, with $K \subset N \subset G$; then the factor group N/K is a normal subgroup of the factor group G/K, and the factor group $(G/K)/(N/K)$ is isomorphic to G/N.

Given two elements a, b of a group G, we write $[a,b] = aba^{-1}b^{-1}$ and call this the *commutator* of a and b. The subgroup generated by all commutators is called the *commutator subgroup* and denoted by $[G,G]$.

We shall show[4]

Theorem 2 *1. $[G,G]$ is a normal subgroup of G,*
2. The factor group G/N modulo a normal subgroup N is abelian if and only if $[G,G] \subset N$. In particular, $G/[G,G]$ is an abelian group.
3. Let $p : G \longrightarrow G/[G,G]$ be the canonical homomorphism. Any homomorphism f of G into an abelian group A descends to a homomorphism $f' : G/[G,G] \longrightarrow A$ in the sense that $f' \circ p = f$.

Proof. Evidently $[a,b]^{-1} = [b,a]$ and $c[a,b]c^{-1} = [cac^{-1}, cbc^{-1}]$. Therefore

$$c[a_1, b_1]^{\pm 1} \cdots [a_k, b_k]^{\pm 1} c^{-1} = (c[a_1, b_1]c^{-1})^{\pm 1} \cdots (c[a_k, b_k]c^{-1})^{\pm 1},$$

which says that $c[a_1, b_1]^{\pm 1} \cdots [a_k, b_k]^{\pm 1} c^{-1}$ belongs to $[G,G]$. This proves (1).

Since $p(a)p(b)p(a)^{-1}p(b)^{-1} = p([a,b]) = 1$, we have $p(a)p(b) = p(b)p(a)$, which says that $G/[G,G]$ is abelian.

If $[G,G] \subset N$, we have a surjective homomorphism of $G/[G,G]$ onto G/N, and hence G/N is abelian. Conversely, suppose that G/N is abelian. For the canonical homomorphism $\varphi : G \longrightarrow G/N$, we have $\varphi([a,b]) = [\varphi(a), \varphi(b)] = 1$ so that $[G,G] \subset \operatorname{Ker} \varphi = N$. Thus (2) is proved.

To prove (3), it suffices to note that $[G,G] \subset \operatorname{Ker} f$ in view of (2). □

For a homomorphism $\rho : G \longrightarrow G_1$, we find $\rho([G,G]) \subset [G_1, G_1]$ so that we obtain a homomorphism $\rho' : G/[G,G] \longrightarrow G_1/[G_1, G_1]$ such that the following diagram is commutative:

$$\begin{array}{ccc} G & \xrightarrow{\rho} & G_1 \\ \downarrow & & \downarrow \\ G/[G,G] & \xrightarrow[\rho']{} & G_1/[G_1, G_1] \end{array}$$

where the vertical arrows stand for the canonical homomorphisms.

We now look at abelian groups in detail. Suppose that an abelian group A is generated by a subset Λ. Thus, in the additive form, every $a \in A$ is written as

---

[4] We freely use the facts stated in Example 2.3.

$$\sum_{\lambda \in \Lambda} n_\lambda \lambda \qquad (2.1)$$

with integers n_λ such that $n_\lambda = 0$ for all but finitely many λ. If the expression (2.1) is unique, then A is said to be a *free abelian group* with $\mathbb{Z}$-*basis* Λ.

Given a set Λ, we may construct a free abelian group $A(\Lambda)$ with $\mathbb{Z}$-basis Λ. In fact, as (2.1) suggests, $A(\Lambda)$ is defined to be the totality of formal finite sums $\sum_{\lambda \in \Lambda} n_\lambda \lambda$ obeying the condition:

$$\sum_{\lambda \in \Lambda} n_\lambda \lambda = 0 \text{ if and only if } n_\lambda = 0 \text{ for all } \lambda.$$

Addition and subtraction are defined by

$$\sum_{\lambda \in \Lambda} m_\lambda \lambda \pm \sum_{\lambda \in \Lambda} n_\lambda \lambda = \sum_{\lambda \in \Lambda} (m_\lambda \pm n_\lambda) \lambda.$$

A subgroup B of an abelian (additive) group A is called a *direct summand* of A if there exists another subgroup B' such that the correspondence

$$(b, b') \in B \oplus B' \longmapsto b + b' \in A$$

is an isomorphism (we write $A = B \oplus B'$). The subgroup B' is called a *complement* of B. It is easy to check that $A = B \oplus B'$ if and only if $A = B + B'$ and $B \cap B' = \{0\}$.

If B is a direct summand of A, then any homomorphism $f : B \longrightarrow C$ into an abelian group C extends to a homomorphism $F : A \longrightarrow C$. In fact, writing $a \in A$ as $a = b + b'$ with $b \in B$, $b' \in B'$, define F by $F(a) = f(b)$.

We shall list several fundamental facts about finitely generated abelian groups:

1. If a finitely generated abelian group A has no torsion element, then A is free abelian, where $a \in A$ is said to be a *torsion element* if $na = 0$ for some integer $n > 0$.
2. If B is a subgroup of a finitely generated abelian group, then B is also finitely generated,
3. If A is finitely generated, then A is a direct sum of a free abelian group (called a *free part* of A) and a finite abelian group. More precisely, the *torsion part* $\mathrm{Tor}(A) = \{a \in A | \ na = 0 \text{ for an integer } n > 0\}$ is a finite subgroup and a direct summand of A such that $A/\mathrm{Tor}(A)$ is free abelian.

A non-trivial finitely generated free abelian group A is isomorphic to $\mathbb{Z}^n$. The number n, which coincides with the number of elements in a $\mathbb{Z}$-basis, is called the *rank* of A, and expressed as $\mathrm{rank}\, A$. For $A = \{0\}$, we put $\mathrm{rank}\, A = 0$. The rank of a finitely generated abelian group is defined to be the rank of a free part of A.

The following theorem is used in the main text.

Theorem 3 *1. Let B be a subgroup of an abelian group A such that A/B is free abelian. Then B is a direct summand.*

2. Let A be a finitely generated free abelian group, and let $f : A \longrightarrow C$ be a surjective homomorphism onto a free abelian group C with $\mathbb{Z}$-basis $\{f(a_1), \ldots, f(a_s)\}$. If we take a $\mathbb{Z}$-basis $\{a_{s+1}, \ldots, a_t\}$ of $\mathrm{Ker}\, f$, then

$$\{a_1, \ldots, a_s, a_{s+1}, \ldots, a_t\}$$

is a $\mathbb{Z}$-basis of A.

Proof. 1. Let $f : A \longrightarrow A/B$ be the canonical homomorphism, and $\{c_s\}_{s \in S}$ be a $\mathbb{Z}$-basis of A/B. Take $b'_s \in A$ such that $f(b_s) = c_s$, and let B' be the subgroup of A generated by $\{b_s\}$. Then for any $a \in A$, writing $f(a) = \sum_{s \in S} n_s c_s$, we have

$$f\left(a - \sum_{s \in S} n_s b'_s\right) = f(a) - \sum_{s \in S} n_s c_s = 0,$$

so that $a \in B + B'$. On the other hand, if $a \in B \cap B'$, then one can write $a = \sum_{s \in S} n_s b_s$, and $0 = f(a) = \sum_{s \in S} n_s c_s$. Thus $n_s = 0$ for every $s \in S$, and $a = 0$.

2. If we put $B = \mathrm{Ker}\, f$, then C is isomorphic to A/B (the first isomorphism theorem). Thus one may apply the idea in the proof of (1). But for completeness, we give a proof.

Take an arbitrary $a \in A$. Then there exist integers $n_1, \ldots, n_s$ such that $f(a) = \sum_{i=1}^{s} n_i f(a_i)$. Since $a - \sum_{i=1}^{s} a_i \in \mathrm{Ker}\, f$, we find integers $m_{s+1}, \ldots, m_t$ such that

$$a - \sum_{i=1}^{s} n_i a_i = \sum_{j=s+1}^{t} m_j a_j.$$

This implies that $\{a_1, \ldots, a_s, a_{s+1}, \ldots, a_t\}$ generates A.

Next suppose $\sum_{i=1}^{t} n_i a_i = 0$. Then

$$\sum_{i=1}^{s} n_i f(a_i) = f\left(\sum_{i=1}^{t} n_i a_i\right) = 0.$$

Since $\{f(a_1), \ldots, f(a_s)\}$ is a $\mathbb{Z}$-basis, we conclude that $a_1 = \cdots = a_s = 0$, and $\sum_{j=s+1}^{t} n_j a_j = 0$. Therefore $a_{s+1} = \cdots = a_t = 0$, as desired. □

3 Vector Spaces and Linear Operators

Taking it that the reader will have studied matrices and determinants already, we shall explain some rudiments of *linear algebra*, the discipline handling vector spaces and linear operators abstractly. The book [65] by S. Lang is a basic reference.

A *vector space* (or *linear space*) W over $\mathbb{R}$ is a set with two operations called *vector sum* $\mathbf{x}+\mathbf{y}$ and *scalar product* $a\mathbf{x}$ ($\mathbf{x},\mathbf{y}\in W, a\in\mathbb{R}$) satisfying:

(a) The vector sum makes W an additive group; thus W has a special element $\mathbf{0}$ (*zero vector*) such that $\mathbf{x}+\mathbf{0}=\mathbf{x}$ ($\mathbf{x}\in W$).
(b) $1\mathbf{x}=\mathbf{x}$;
(c) $a(b\mathbf{x})=(ab)\mathbf{x}$
(d) $(a+b)\mathbf{x}=a\mathbf{x}+b\mathbf{x}$;
(e) $a(\mathbf{x}+\mathbf{y})=a\mathbf{x}+a\mathbf{y}$.

Elements of W are called *vectors*. A *linear combination* of $\mathbf{x}_1,\ldots,\mathbf{x}_n\in W$ is a vector of the form $a_1\mathbf{x}_1+\cdots+a_n\mathbf{x}_n$. If

$$a_1\mathbf{x}_1+\cdots+a_n\mathbf{x}_n=\mathbf{0}\implies a_1=\cdots=a_n=0,$$

then $\mathbf{x}_1,\ldots,\mathbf{x}_n$ are called *linearly independent* (otherwise *linearly dependent*). If, in addition to linear independence, every vector of W is expressed as a linear combination of $\mathbf{x}_1,\ldots,\mathbf{x}_n$, then $\{\mathbf{x}_1,\ldots,\mathbf{x}_n\}$ are said to be a *basis* of W. Here the number n, which does not depend on the choice of a basis, is the *dimension* of W, and symbolically written as $\dim W$.

Example 2 *For a finite set V with n elements, the set $C(V)$ of real-valued functions f on V is endowed with the structure of an n-dimensional vector space with the operations defined by*

$$(f+g)(x)=f(x)+g(x),$$
$$(af)(x)=af(x).$$

If we define $\delta_x\in C(V)$ by setting

$$\delta_x(y)=\begin{cases}1 & (x=y)\\ 0 & (x\neq y),\end{cases}$$

then $\{\delta_x\}_{x\in V}$ gives a basis of $C(V)$. □

From now on vector spaces are supposed to be finite dimensional.

A *subspace* W_1 of a vector space W is a non-empty subset which is closed under the vector sum and scalar product. For two subspaces W_1, W_2 of W, the intersection $W_1\cap W_2$ and the sum

3 Vector Spaces and Linear Operators

$$W_1 + W_2 := \{\mathbf{x}_1 + \mathbf{x}_2 |\ \mathbf{x}_1 \in W_1,\ \mathbf{x}_2 \in W_2\}$$

are also subspaces of W. For a subset A of W, the subspace *spanned by A* is the intersection of all subspaces containing A. When $A = \{\mathbf{x}_1, \ldots, \mathbf{x}_n\}$, this coincides with the set consisting of all linear combinations of $\mathbf{x}_1, \ldots, \mathbf{x}_n$.

For vector spaces $W_1, \ldots, W_s$, one can define the *direct sum* $W_1 \oplus \cdots \oplus W_s$ in a similar way as the direct sum of additive groups.

$\mathbb{R}^n$ is an n-dimensional vector space with vector sum

$$^t(x_1, \ldots, x_n) + {}^t(y_1, \ldots, y_n) = {}^t(x_1 + y_1, \ldots, x_n + y_n)$$

and scalar product

$$a^t(x_1, \ldots, x_n) = {}^t(ax_1, \ldots, ax_n).$$

As a basis of $\mathbb{R}^n$, one can take the *standard basis* $\{\mathbf{e}_1, \ldots, \mathbf{e}_n\}$ defined by

$$\mathbf{e}_1 = {}^t(1,0,0,\ldots,0),\ \mathbf{e}_2 = {}^t(0,1,0,\cdots,0), \ldots,\ \mathbf{e}_n = {}^t(0,0,0,\ldots,1).$$

A *linear operator* (or *linear map*) $T: W_1 \longrightarrow W_2$ between vector spaces is a map with the property

$$T(a\mathbf{x} + b\mathbf{y}) = aT(\mathbf{x}) + bT(\mathbf{y}) \quad (a, b \in \mathbb{R},\ \mathbf{x}, \mathbf{y} \in W_1).$$

Abbreviating parenthese in $T(\mathbf{x})$, we often write $T\mathbf{x}$. The *kernel*.

$$\mathrm{Ker}\ T = \{\mathbf{x} \in W_1|\ T\mathbf{x} = \mathbf{0}\}$$

and the *image*

$$\mathrm{Image}\ T = \{T\mathbf{x}|\ \mathbf{x} \in W_1\}$$

are subspaces of W_1 and W_2, respectively. For a subspace W_1 of W, the inclusion map $i: W_1 \longrightarrow W$ is linear.

If $\{\mathbf{x}_1, \ldots, \mathbf{x}_n\}$ and $\{\mathbf{y}_1, \ldots, \mathbf{y}_m\}$ are bases of W_1 and W_2, respectively, then there exists a (unique) matrix $A = (a_{ij})$ of size (m, n) such that

$$\mathbf{y}_i = \sum_{j=1}^{n} a_{ji} \mathbf{x}_j \quad (i = 1, \ldots, m).$$

The matrix A is called the *matrix associated with T* (when bases of W_1 and W_2 are fixed).

For $W_1 = \mathbb{R}^n$, $W_2 = \mathbb{R}^m$, a linear map T is expressed as

$$T({}^t(x_1, \ldots, x_n)) = {}^t\left(\sum_{j=1}^{n} a_{1j} x_j, \ldots, \sum_{j=1}^{n} a_{mj} x_j\right),$$

where $A = (a_{ij})$ is the matrix associated with the fundamental bases of $\mathbb{R}^n$ and $\mathbb{R}^m$. Thus $T\mathbf{x} = A\mathbf{x}$.

A linear operator $T: W_1 \longrightarrow W_2$ is said to be a *linear isomorphism* of W_1 onto W_2 if T is bijective and the inverse T^{-1} is linear. When $\dim W_1 = \dim W_2$, a surjective or injective linear operator $T: W_1 \longrightarrow W_2$ is a linear isomorphism. If $\{\mathbf{x}_1, \ldots, \mathbf{x}_n\}$ is a basis of W, then the correspondence

$$a_1 \mathbf{x}_1 + \cdots + a_n \mathbf{x}_n \mapsto {}^t(a_1, \ldots, a_n)$$

gives rise to a linear isomorphism of W onto $\mathbb{R}^n$.

The *dual space* W^* of W is the space of linear operators $\mathbf{f}: W \longrightarrow \mathbb{R}$ which carries a structure of vector space in a natural manner. Occasionally we write $(\mathbf{f}, \mathbf{x})$ for $\mathbf{f}(\mathbf{x})$. If we put $T_\mathbf{x} \mathbf{f} = (\mathbf{f}, \mathbf{x})$, then $T_\mathbf{x} \in (W^*)^*$, and $\mathbf{x} \mapsto T_\mathbf{x}$ is a linear isomorphism of W onto $(W^*)^*$. Given a basis $\{\mathbf{x}_1, \ldots, \mathbf{x}_n\}$ of W, one can find its *dual basis* $\{\mathbf{f}_1, \ldots, \mathbf{f}_n\}$ of W^* characterized by

$$(\mathbf{f}_i, \mathbf{x}_j) = \mathbf{f}_i(\mathbf{x}_j) = \begin{cases} 1 & (i = j) \\ 0 & (i \neq j). \end{cases}$$

For a subspace W_1 of W, a *complement* of W_1 is a subspace W_2 such that $W = W_1 + W_2$ and $W_1 \cap W_2 = \{\mathbf{0}\}$. The correspondence $(\mathbf{x}_1, \mathbf{x}_2) \mapsto \mathbf{x}_1 + \mathbf{x}_2$ yields a linear isomorphism of $W_1 \oplus W_2$ onto W. We write $W = W_1 \oplus W_2$.

An *inner product* on a vector space W is an operation which assigns a real number $\langle \mathbf{x}, \mathbf{y} \rangle$ to two vectors $\mathbf{x}, \mathbf{y}$ such that

(a1) $\langle \mathbf{x}, \mathbf{y} \rangle = \langle \mathbf{y}, \mathbf{x} \rangle$.
(a2) $\langle a\mathbf{x} + b\mathbf{y}, \mathbf{z} \rangle = a \langle \mathbf{x}, \mathbf{z} \rangle + b \langle \mathbf{x}, \mathbf{z} \rangle$.
(a3) $\langle \mathbf{x}, \mathbf{x} \rangle \geq 0$, and $\langle \mathbf{x}, \mathbf{x} \rangle = 0$ if and only if $\mathbf{x} = \mathbf{0}$.

The *norm* of $\mathbf{x}$ is $\|\mathbf{x}\| = \langle \mathbf{x}, \mathbf{x} \rangle^{1/2}$. The inequalities $|\langle \mathbf{x}, \mathbf{y} \rangle| \leq \|\mathbf{x}\| \|\mathbf{y}\|$ (*Cauchy–Schwarz inequality*) and $\|\mathbf{x} + \mathbf{y}\| \leq \|\mathbf{x}\| + \|\mathbf{y}\|$ (*triangle inequality*) are checked.

A basis $\{\mathbf{x}_1, \ldots, \mathbf{x}_n\}$ of W with an inner product is called an *orthonormal basis* if

$$\langle \mathbf{x}_i, \mathbf{x}_j \rangle = \begin{cases} 1 & (i = j) \\ 0 & (i \neq j). \end{cases}$$

Every vector space with an inner product has an orthonormal basis (use the *Gram–Schmidt process*).

The vector space $\mathbb{R}^n$ has the standard inner product defined by

$$\langle {}^t(x_1, \ldots, x_n), {}^t(x_1, \ldots, x_n) \rangle = x_1 y_1 + \cdots + x_n y_n,$$

for which the fundamental basis of $\mathbb{R}^n$ is an orthonormal basis.

For a subspace W_1 of W, define the *orthogonal complement* $W_1^\perp$ by setting $W_1^\perp = \{\mathbf{x} \in W | \langle \mathbf{x}, \mathbf{y} \rangle = 0 \text{ for every } \mathbf{y} \in W_1\}$. Obviously $W_1^\perp$ is a subspace of W and $(W^\perp)^\perp = W$. Also we have $W = W_1 \oplus W_1^\perp$.

3 Vector Spaces and Linear Operators

It is a fundament fact that for any $\mathbf{f} \in W^*$, there exists a unique vector $\mathbf{y} \in W$ such that $\mathbf{f}(\mathbf{x}) = \langle \mathbf{x}, \mathbf{y} \rangle$ for every $\mathbf{x} \in W$. For a linear operator $T : W_1 \longrightarrow W_2$ between vector spaces with inner products, applying this fact to $\mathbf{f} \in W_1^*$ defined by $\mathbf{f}(\mathbf{x}) = \langle T\mathbf{x}, \mathbf{y} \rangle$, we find a unique $\mathbf{z} \in W_1$ such that $\langle T\mathbf{x}, \mathbf{y} \rangle = \langle \mathbf{x}, \mathbf{z} \rangle$. Then define $T^*\mathbf{y}$ to be $\mathbf{z}$. The map $T^* : W_2 \longrightarrow W_1$, called the *adjoint operator* of T, is a linear operator, which is, by definition, characterized by

$$\langle T\mathbf{x}, \mathbf{y} \rangle = \langle \mathbf{x}, T^*\mathbf{y} \rangle \qquad (\mathbf{x} \in W_1, \ \mathbf{y} \in W_2).$$

It is shown that $(T^*)^* = T$. For the matrix A associated with T, the matrix associated with T^* is the transpose ${}^t A$. A linear operator $T : W \longrightarrow W$ is said to be *symmetric* if $T^* = T$. The matrix A associated with a symmetric operator is a symmetric matrix, i.e., ${}^t A = A$. It is checked that

$$(\operatorname{Ker} T)^\perp = \operatorname{Image} T^*, \quad (\operatorname{Ker} T^*)^\perp = \operatorname{Image} T.$$

In particular, if T is symmetric, then $(\operatorname{Ker} T)^\perp = \operatorname{Image} T$ [this fact is applied to the discrete Poisson equation in Sect. 7.4 and Notes (II) of Chap. 7].

An *orthogonal transformation* of W is an operator $U : W \longrightarrow W$ satisfying $\langle U\mathbf{x}, U\mathbf{y} \rangle = \langle \mathbf{x}, \mathbf{y} \rangle$ for every $\mathbf{x}, \mathbf{y} \in W$, or equivalently $U^*U = U^*U = I$.

An *orthogonal projection* of W is a symmetric operator $P : W \longrightarrow W$ satisfying $P^2 = P$. We find $W = \operatorname{Ker} P \oplus \operatorname{Image} P$.

A vector space over $\mathbb{C}$ (together with basis, subspaces, linear operators, dual spaces, inner products, adjoint operator, and orthogonal projections) is defined in a similar way. In this case, instead of (a1), we take up the following condition for an inner product.

(a1') $\overline{\langle \mathbf{x}, \mathbf{y} \rangle} = \langle \mathbf{y}, \mathbf{x} \rangle$.

A typical vector space over $\mathbb{C}$ is $\mathbb{C}^d$ with the standard inner product

$$\langle {}^t(z_1, \ldots, z_n), {}^t(w_1, \ldots, w_n) \rangle = z_1 \overline{w_1} + \cdots + z_n \overline{w_n}.$$

Another example is the space of complex-valued functions on a finite set V. This has the inner product defined by

$$\langle f_1, f_2 \rangle = \sum_{x \in V} f_1(x) \overline{f_2(x)}.$$

A linear operator $T : W \longrightarrow W$ with $T^* = T$ is called a *hermitian operator* (or *self-adjoint operator*). An *isometry* (or a *unitary map*) $U : W_1 \longrightarrow W_2$ is a surjective linear operator satisfying $\langle U\mathbf{x}, U\mathbf{y} \rangle = \langle \mathbf{x}, \mathbf{y} \rangle$. When $W_1 = W_2$, an isometry U is called a *unitary transformation*.

An *eigenvalue* of $T : W \longrightarrow W$ is a complex number λ such that there exists a non-zero vector $\mathbf{x}$ such that $T\mathbf{x} = \lambda \mathbf{x}$. The vector $\mathbf{x}$ is called an *eigenvector*. The *eigenspace* for an eigenvalue λ is $E_\lambda(T) = \{\mathbf{x} \in W | \ T\mathbf{x} = \lambda \mathbf{x}\}$.

If T is hermitian, every eigenvalue is a real number, and

$$W = E_{\lambda_1}(T) \oplus \cdots \oplus E_{\lambda_s}(T),$$

where $\lambda_1, \ldots, \lambda_s$ are distinct eigenvalues of T. Furthermore $E_{\lambda_i}(T)$ and $E_{\lambda_j}(T)$ are perpendicular for $i \neq j$ in the sense that $\langle \mathbf{x}, \mathbf{y} \rangle = 0$ for $\mathbf{x} \in E_{\lambda_i}(T)$ and $\mathbf{y} \in E_{\lambda_j}(T)$. Therefore one can find an orthonormal basis consisting of eigenvectors. In terms of matrices, this implies that for a hermitian matrix A, one can find a unitary matrix U such that ${}^t U A U$ is a diagonal matrix with real entries.

4 Free Groups

The notion of free groups occupies a significant position in combinatorial group theory. In this appendix we only give the definition together with a concrete construction of a free group.

A *free group* F with basis Λ ($\subset F$) is a group having the property: if $f : \Lambda \longrightarrow G$ is any map into a group G, then there is a unique homomorphism $\varphi : F \longrightarrow G$ such that $\varphi|\Lambda = f$.

Assuming the existence of F, we show that the set Λ generates F. Let H be the subgroup of F generated by Λ, and let $\varphi : F \longrightarrow H$ be the unique extension of the inclusion map $f : \Lambda \longrightarrow H$. For $h \in H$, write $h = \lambda_1^{\pm 1} \cdots \lambda_n^{\pm 1}$ ($\lambda_i \in \Lambda$). Then

$$\varphi(h) = \varphi(\lambda_1^{\pm 1} \cdots \lambda_n^{\pm 1}) = \varphi(\lambda_1)^{\pm 1} \cdots \varphi(\lambda_n)^{\pm 1} = \lambda_1^{\pm 1} \cdots \lambda_n^{\pm 1} = h$$

so that $\varphi|H$ is the identity map of H, and hence φ is surjective. Let $i : H \longrightarrow F$ be the inclusion map, and consider the composition $i \circ \varphi$ which is a homomorphic extension of the inclusion map $\iota : \Lambda \longrightarrow F$. On the other hand, the identity map of F is also a homomorphic extension of ι. Thus by uniqueness, $i \circ \varphi$ must be the identity map of F. This implies that $H = F$.

If F' is another free group with basis Λ, then there is a unique homomorphism $\phi : F \longrightarrow F'$ such that $\phi|\Lambda = I_\Lambda$. Changing the role of F and F', we also have a unique homomorphism $\psi : F \longrightarrow F'$ such that $\psi|\Lambda = I_\Lambda$. The compositions $\psi \circ \phi : F \longrightarrow F$ and $\phi \circ \psi : F' \longrightarrow F'$ satisfy $(\psi \circ \phi)|\Lambda = I_\Lambda = I_F|\Lambda$ and $(\phi \circ \psi)|\Lambda = I_\Lambda = I_{F'}|\Lambda$. Therefore by uniqueness, we conclude that $\psi \circ \phi = I_F$ and $\phi \circ \psi = I_{F'}$, and so a free group with basis Λ is uniquely determined up to isomorphisms.

Now we shall construct a free group with basis Λ. The reader may see that our construction has a similarity to the definition of the fundamental group of a graph; actually what we are going to do is the same as the description of the fundamental group of the bouquet graph.

Let S be a set. A *word* with letters in S is either void (denoted by $\emptyset$) or a finite sequence $(s_1, s_2, \ldots, s_n)$ of elements of S, and we denote the set of all words by W_S. The *length* of a word w, symbolically $|w|$, is the number of letters in w ($|\emptyset| = 0$). The product ww' of two words w and w' is defined by concatenating w with w'.

Let $\overline{\Lambda}$ be the disjoint copy of Λ. As the set S, we take $S = \Lambda \cup \overline{\Lambda}$. Denote by $\overline{\lambda} \in \overline{\Lambda}$ the twin of $\lambda \in \Lambda$. For a word $w = (s_1, s_2, \ldots, s_n) \in W_S$, define the *inverse* $\overline{w}$ by $(\overline{s}_n, \ldots, \overline{s}_2, \overline{s}_1)$, where $(\overline{\overline{\lambda}}) = \lambda$ by convention. For two words $w, w' \in W_S$, we write $w \succ w'$ when w has successive letters $\lambda, \overline{\lambda}$ or $\overline{\lambda}, \lambda$ ($\lambda \in \Lambda$), and the word w' is obtained from w by deleting such letters. We write $w \equiv w'$ if there exists a finite sequence of words $w = w_0, w_1, \ldots, w_n = w'$ such that for each i ($1 \leq i \leq n$), either $w_{i-1} \succ w_i$ or $w_i \succ w_{i-1}$. This relation is an equivalence relation and is compatible with the multiplication and the inverse operation. Therefore, one may define a multiplication and an inverse operation for the quotient set $F(\Lambda)$ of W_S by this equivalence relation. The set $F(\Lambda)$ forms a group whose identity element is the equivalence class of the void word. In the usual convention, we use the symbols Λ^{-1} and λ^{-1} for $\overline{\Lambda}$ and $\overline{\lambda}$, respectively. Regarding $\lambda \in \Lambda$ as a word, we have an injective map from Λ into $F(\Lambda)$ by assigning the corresponding class to each element of Λ. We identify Λ with the image of this injection map.

We shall show that $F(\Lambda)$ is a free group with basis Λ. Let G be any group, and $f : \Lambda \longrightarrow G$ be any map. Define a map of W_S into G, which we denote again by f, by setting

$$f(s) = \begin{cases} f(s) & \text{if } s \in \Lambda \\ f(\overline{s})^{-1} & \text{if } s \in \overline{\Lambda} \end{cases}$$

and

$$f\big((s_1, \ldots, s_n)\big) = f(s_1) \cdots f(s_n), \qquad (s_i \in S).$$

We easily find that $f(ww') = f(w)f(w')$. If $w \equiv w'$, then $f(w) = f(w')$ so that f induces a homomorphism $\varphi : F(\Lambda) \longrightarrow G$. This φ is a (unique) homomorphic extension of f.

A word $w \neq \emptyset$ is said to be *reduced* if it has neither successive letters λ, $\overline{\lambda}$ nor $\overline{\lambda}$, λ. Each equivalence class contains a unique reduced word (hence the set of reduced words together with $\emptyset$ is identified with $F(\Lambda)$).

$F(\Lambda)/[F(\Lambda), F(\Lambda)]$ is isomorphic to the free abelian group $A(\Lambda)$ with $\mathbb{Z}$-basis Λ. In fact the natural map $\phi : \Lambda \longrightarrow A(\Lambda)$ induces a homomorphism of $F(\Lambda)$ onto $A(\Lambda)$, which yields a surjective homomorphism $\Phi : F(\Lambda)/[F(\Lambda), F(\Lambda)] \longrightarrow A(\Lambda)$. To show that Φ is an isomorphism, define $\Psi : A(\Lambda) \longrightarrow F(\Lambda)/[F(\Lambda), F(\Lambda)]$ by setting $\Psi(\lambda) = \lambda \, [F(\Lambda), F(\Lambda)]$ ($\lambda \in \Lambda$). It is easy to check that $\Psi \circ \Phi$ is the identity map of $F(\Lambda)/[F(\Lambda), F(\Lambda)]$. Clearly Φ is injective, and hence an isomorphism.

5 Crystallographic Groups

This section gives a brief look into crystallographic groups, a protagonist in classical crystallography, although they play a subordinate role in this book.

We use the notation $\mathbf{M}(d)$ for the motion group (group of congruent transformations) of $\mathbb{R}^d$; that is, $T \in \mathbf{M}(d)$ is expressed as

$$T(\mathbf{x}) = A\mathbf{x} + \mathbf{a}$$

with an orthogonal matrix $A \in O(d)$ and a vector $\mathbf{a} \in \mathbb{R}^d$. If we write $T = (A, \mathbf{a})$, then for $T_1 = (A_1, \mathbf{a}_1), T_2 = (A_2, \mathbf{a}_2)$, we have

$$T_1 T_2(\mathbf{x}) = T_1(A_2 \mathbf{x} + \mathbf{a}_2) = A_1(A_2 \mathbf{x} + \mathbf{a}_2) + \mathbf{a}_1 = A_1 A_2 \mathbf{x} + A_1 \mathbf{a}_2 + \mathbf{a}_1.$$

Therefore $(A_1, \mathbf{a}_1)(A_2, \mathbf{a}_2) = (A_1 A_2, A_1 \mathbf{a}_2 + \mathbf{a}_1)$. We also have

$$(A, \mathbf{a})^{-1} = (A^{-1}, -A^{-1}\mathbf{a}),$$
$$(A, \mathbf{a})(B, \mathbf{b})(A, \mathbf{a})^{-1} = (ABA^{-1}, -ABA^{-1}\mathbf{a} + A\mathbf{b} + \mathbf{a}).$$

In particular,

$$(A, \mathbf{a})(I, \mathbf{b})(A, \mathbf{a})^{-1} = (I, A\mathbf{b}), \tag{5.1}$$

$$(I, \mathbf{a})(B, \mathbf{b})(I, \mathbf{a})^{-1} = (B, -B\mathbf{a} + \mathbf{b} + \mathbf{a}), \tag{5.2}$$

where $I = I_d$ is the identity matrix. Equation (5.1) implies that the subgroup of $\mathbf{M}(d)$ consisting of parallel translations $(I, \mathbf{a})$ is a normal subgroup. We identify this subgroup with $\mathbb{R}^d$. The map $p : \mathbf{M}(d) \longrightarrow O(d)$ defined by setting $p(A, \mathbf{a}) = A$ is a homomorphism, and $\operatorname{Ker} p = p^{-1}(I) = \mathbb{R}^d$ ($\mathbf{M}(d)$ is the semi-direct product of $\mathbb{R}^d$ and $O(d)$).

A subgroup $\mathfrak{G}$ of $\mathbf{M}(d)$ is said to be a d-dimensional *crystallographic group* (or *space group*) if it is discrete and has a compact quotient $\mathbb{R}^d / \mathfrak{G}$.

The following theorem is due to Bieberbach.

Theorem 4 *Let $\mathfrak{G}$ be a d-dimensional crystallographic group.*

1. $\mathfrak{H} = \mathfrak{G} \cap \mathbb{R}^d$ *is a lattice group in $\mathbb{R}^d$.*
2. $\mathfrak{H}$ *is a normal subgroup of $\mathfrak{G}$ such that the factor group $\mathfrak{G}/\mathfrak{H}$ is finite (in particular, $\mathfrak{G}$ is virtually abelian). Furthermore $\mathfrak{H}$ is maximal as an abelian subgroup in the sense that if L is an abelian subgroup of $\mathfrak{G}$ with $\mathfrak{H} \subset L$, then $L = \mathfrak{H}$.*
3. *If two d-dimensional crystallographic groups $\mathfrak{G}_1, \mathfrak{G}_2$ are isomorphic, then there exists an affine transformation T of $\mathbb{R}^d$ such that $\mathfrak{G}_2 = T \mathfrak{G}_2 T^{-1}$.*
4. *There are only finitely many isomorphism classes of d-dimensional crystallographic groups.*

We shall prove (2). See Wolf [111] and Charlap [19] for the proof of (1), (3), (4).

First, $\mathfrak{H}$ is a normal subgroup of $\mathfrak{G}$. Indeed, for $(A, \mathbf{a}) \in \mathfrak{G}$ and $(I, \mathbf{b}) \in \mathfrak{H}$, we find that

$$(A, \mathbf{a})(I, \mathbf{b})(A, \mathbf{a})^{-1} = (I, A\mathbf{b}) \in G \cap \mathbb{R}^d = \mathfrak{H}.$$

Assuming (1), we show that $\mathfrak{H}$ is maximal as an abelian subgroup. Suppose that $(A, \mathbf{a}) \in \mathfrak{G}$ commutes with every element of $\mathfrak{H}$. Then for any $(I, \mathbf{b}) \in \mathfrak{H}$, we have $(I, A\mathbf{b}) = (I, \mathbf{b})$ by using again $(A, \mathbf{a})(I, \mathbf{b})(A, \mathbf{a})^{-1} = (I, A\mathbf{b})$, which implies $A\mathbf{b} = \mathbf{b}$ for every $\mathbf{b} \in \mathfrak{H}$. Since the lattice group $\mathfrak{H}$ contains a basis of $\mathbb{R}^d$, we conclude that $A = I$ and $(A, \mathbf{a}) = (I, \mathbf{a}) \in \mathfrak{H}$. Thus the claim is proved.

Next we prove finiteness of $\mathfrak{G}/\mathfrak{H}$. If we put $p(\mathfrak{G}) = \mathfrak{K}$, then the kernel of the surjective homomorphism $p : \mathfrak{G} \longrightarrow \mathfrak{K}$ is $\mathfrak{H}$, and hence $\mathfrak{G}/\mathfrak{H}$ is isomorphic to $\mathfrak{K}$. Therefore it suffices to prove that $\mathfrak{K}$ is finite. Given $A \in \mathfrak{K}$, take $(A, \mathbf{a}) \in \mathfrak{G}$ with $p(A, \mathbf{a}) = A$. Then we find $A(\mathfrak{H}) = \mathfrak{H}$ using $(A, \mathbf{a})(I, \mathbf{b})(A, \mathbf{a})^{-1} = (I, A\mathbf{b})$. Taking a $\mathbb{Z}$-basis of $\mathfrak{H}$, we think of A as an element of $GL_d(\mathbb{Z})$. Thus $\mathfrak{K}$ is identified with a subgroup of $O(d) \cap GL_d(\mathbb{Z})$. By virtue of compactness of the orthogonal group $O(d)$ and discreteness of $GL_d(\mathbb{Z})$, we conclude that $\mathfrak{K}$ is finite.

The finite subgroup $\mathfrak{K}$ is what is termed a *point group* in crystallography.

We shall say more about the subgroup $\mathfrak{H}$.

For a general group G, the *conjugacy class* $[g]$ of $g \in G$ is the set $\{hgh^{-1}|\, h \in G\}$. We put
$$S(G) = \{g \in G|\, |[g]| < \infty\}.$$
The *centralizer* of g is defined to be $C_g = \{h \in G|\, gh = hg\}$. Evidently C_g is a subgroup of G. Then
$$g \in S(G) \iff C_g \text{ is of finite index in } G.$$
In fact, noticing $hgh^{-1} = h'gh'^{-1} \iff h \in h'C_g$, we find that the correspondence $hgh^{-1} \in [g] \mapsto hC_g \in G/C_g$ gives a bijection of $[g]$ onto G/C_g.

$S(G)$ is a normal subgroup of G because
$$hg_1g_2h^{-1} = (hg_1h^{-1})(hg_2h^{-1}), \quad hg^{-1}h^{-1} = (hgh^{-1})^{-1}, \quad [hgh^{-1}] = [g].$$

We shall show that $S(\mathfrak{G}) = \mathfrak{H}$ for a crystallographic group $\mathfrak{G}$. For $h = (I, \mathbf{b}) \in \mathfrak{H}$, the conjugacy class $[h]$ coincides with $\{(I, A\mathbf{b});\, A \in \mathfrak{K}\}$, which is obviously a finite set, and so $h \in S(\mathfrak{G})$. Conversely, if $g = (A, \mathbf{a}) \in S(\mathfrak{G})$, then by using $(I, \mathbf{b})g(I, \mathbf{b})^{-1} = \big(A, (I-A)\mathbf{b}+\mathbf{a}\big)$ [see (5.2)], we see that $\big\{\big(A, (I-A)\mathbf{b}+\mathbf{a}\big)|\, (I, \mathbf{b}) \in \mathfrak{H}\big\}$ is contained in the conjugacy class $[g]$. If $A \neq I$, then $\big|\big\{\big(A, (I-A)\mathbf{b}+\mathbf{a}\big)|\, (I, \mathbf{b}) \in \mathfrak{H}\big\}\big|$ is infinite. Thus $A = I$ and $g \in \mathfrak{H}$.

From what we have proved it follows that $\mathfrak{H}$ is largest among all abelian subgroups of finite index in $\mathfrak{G}$. For, if L is an abelian subgroup of finite index in $\mathfrak{G}$, then for any $g \in L$, we have $L \subset C_g$ so that C_g is of finite index in $\mathfrak{G}$. Hence $g \in S(\mathfrak{G}) = \mathfrak{H}$ and $L \subset S(\mathfrak{G})$.

The theorem due to Auslander and Kuranishi [2] (cf. [19, 111]) tells us that an abstract group $\mathfrak{G}$ is isomorphic to a d-dimensional crystallographic group if and only if $\mathfrak{G}$ has a normal free abelian subgroup $\mathfrak{H}$ with the properties:

1. $\mathfrak{H}$ is of finite index in $\mathfrak{G}$, and rank$\mathfrak{H} = d$.
2. $\mathfrak{H}$ is a maximal abelian subgroup of $\mathfrak{G}$.

We conclude this appendix with a brief history of crystallographic groups.

As mentioned at the beginning of Chap. 6, it is in the nineteenth century that mathematics began to play an important role in crystallography, more specifically in the morphology of crystals.

As a prehistory of the morphology, there have been observations about the angles between the faces of a crystal. That is, as Kepler's study on snow flakes suggests (Notes (III) in Chap. 8), the faces could be of different sizes, depending on the conditions under which the crystal have grown, but the relations between them remain fixed. As a reasoning of this fact, the Dutch mathematician C. Huygens thought of crystals, say the mineral calcite, as being built from small ellipsoidal units (1690). Using this picture, he explained that the resulting faces always preserve the same relationships.

The science of crystallography in the true sense started when R. J. Haüy published his *Essai d'une Théorie sur la Structure des Cristaux* (1784). In this essay he developed Huygen's concept to explain the law of angles between faces. Since then, scientists have studied many physical properties (for instance, optical activity discovered by F. J. D. Arago in 1811) in connection with the morphology of crystals.

The study of morphology led the early crystallographers to a simple classification of all crystals in terms of symmetry. It was in 1830 that the German crystallographer J.F.C. Hessel for the first time investigated the possible types of symmetry. He found that there are 32 types of symmetry. His book *Krystallonomie und Krystallographie* stating this conclusion was published in Leipzig in 1831, and was not read seriously enough by other scientists at that time. A. Bravais in 1849 and A. Gadolin in 1867 rediscovered the same 32 types of symmetry by repeating the derivation. Their work was well known before L. Sohncke, a crystal physicist, found Hessel's earlier book in 1891. Correctly speaking, the notion of group was seldom used until 1860s. C. Jordan is the first to recognize that the classification of crystal symmetry is described in terms of subgroups of $O(3)$ (1869). As for three-dimensional crystallographic groups, after the fundamental work of C. Jordan and L. Sohncke, all 230 isomorphism classes were determined in 1891 by E.S. Fedorov and A. Schoenflies independently.[5]

From a mathematical viewpoint it is natural to study crystallographic groups in general dimension. In 1900, D. Hilbert posed a question on the finiteness of isomorphism classes of crystallographic groups in his famous address at the second International Congress of Mathematics held in Paris. L. Bieberbach (1910–1912) and G. Frobenius (1911) solved affirmatively the problem. Efficient enumeration of crystallographic groups has been carried out for lower dimensional cases by applying the method due to J.J. Burckhardt and H. Zassenhaus. For instance, there are 17 isomorphism classes of two-dimensional crystallographic groups (E. Fedorov in 1891, G. Polya and P. Niggli in the middle 1920s). Interestingly 15 classes out of those 17 classes are found in the famous tile decorations in the Alhambra[6] at Granada which was laid out in fourteenth century. For the four-dimensional case, there are 4,783 isomorphism classes.

---

[5] See Wood [113] more about the history of crystallography.

[6] A palace and fortress complex constructed by the Moorish rulers of the Emirate of Granada in Al-Andalus.

References

1. Artamkin IV (2006) Discrete Torelli theorem. Sbornik: Math 197:1109–1120
2. Auslander L, Kuranishi M (1957) On the holonomy group of locally Euclidean spaces. Ann Math 65:411–415
3. Bacher R, De La Harpe P, Nagnibeda T (1997) The lattice of integral flows and the lattice of integral cuts on a finite graph. Bull Soc Math Fr 125:167–198
4. Bader M, Klee WE, Thimm G (1997) The 3-regular nets with four and six vertices per unit cell. Z Kristallogr 212:553–558
5. Baker M, Norine S (2007) Riemann-Roch and Abel-Jacobi theory on a finite graph. Adv Math 215:766–788
6. Baker M, Norine S (2009) Harmonic morphisms and hyperelliptic graphs. Int Math Res Notices (15):2914–2955
7. Baker M, Faber X (2011) Metric properties of the tropical Abel-Jacobi map. J Algebraic Combin 33:349–381
8. Bass H (1992) The Ihara-Selberg zeta function of a tree lattice. Int J Math 3:717–797
9. Beukemann A, Klee WE (1992) Minimal nets. Z Kristallogr 201:37–51
10. Biggs NL, Lloyd EK, Wilson RJ (1999) Graph theory 1736–1936. Oxford University Press, Oxford
11. Biggs NL (1993) Algebraic graph theory. Cambridge University Press, Cambridge
12. Biggs NL (1997) Algebraic potential theory on graphs. Bull Lond Math Soc 29:641–682
13. Blatov V (2000) Search for isotypism in crystal structures by means of the graph theory. Acta Crystallogr A 56:178–188
14. Bollmann W (1972) The basic concepts of the O-lattice theory. Surf Sci 31:1–11
15. Bollobas B (1998) Modern graph theory. Springer, New York
16. Brown KS (1972) Cohomology of groups. Springer, New York
17. Bryant PR (1967) Graph theory applied to electrical networks. In: Harary F (ed) Graph theory and theoretical physics. Academic, New York, pp 111–137
18. Caporaso L, Viviani F (2010) Torelli theorem for graphs and tropical curves. Duke Math J 153:129–171
19. Charlap LS (1986) Bieberbach groups and flat manifolds. Springer, New York
20. Chung SJ, Hahn T, Klee WE (1984) Nomenclature and generation of three-periodic nets: the vector method. Acta Crystallogr A 40:42–50
21. Conway JH, Burgiel H, Goodman-Strauss C (2008) The symmetries of things. A K Peters Ltd, Wellesley
22. Coxeter HSM (1955) On Laves' graph of girth ten. Can J Math 7:18–23
23. Coxeter HSM (1973) Regular polytopes. Dover, New York
24. Coxeter HSM (1974) Regular complex polytopes. Cambridge University Press, Cambridge
25. Cromwell P (1999) Polyhedra. Cambridge University Press, Cambridge

26. Curtarolo S, Morgan D, Persson K, Rodgers J, Ceder G (2003) Predicting crystal structures with data mining of quantum calculations. Phys Rev Lett 91:135503
27. Dai J, Li Z, Yang J (2010) Boron K_4 crystal: a stable chiral three-dimensional sp^2 network. Phys Chem Chem Phys 12:12420–12422
28. Danzer L. Something about $[10,3]_a$. Unpublished
29. Delgado-Friedrichs O, Dress A, Huson D, Klinowski J, Mackay A (1999) Systematic enumeration of crystalline networks. Nature 400:644–647
30. Delgado-Friedrichs O, O'Keeffe M (2003) Identification of and symmetry computation for crystal nets. Acta Crystallogr A 59:351–360
31. Delgado-Friedrichs O, O'Keeffe M, Yaghin OM (2003) Three-periodic nets and tilings: regular and quasiregular nets. Acta Crystallogr A 59:22–27
32. Delgado-Friedrichs O (2004) Barycentric drawings of periodic graphs. Lect Notes Comput Sci 2912:178–189
33. Delgado-Friedrichs O, Foster MD, O'Keeffe M, Proserpio DM, Treacy MMJ, Yaghi OM (2005) What do we know about three-periodic nets? J Solid State Chem 178:2533–2554
34. Delgado-Friedrichs O, O'Keeffe M (2007) Three-periodic tilings and nets: face-transitive tilings and edge-transitive nets revisited. Acta Crystallogr A 63:344–347
35. Delgado-Friedrichs O, O'Keeffe M (2009) Edge-transitive lattice nets. Acta Crystallogr A 65:360–363
36. Diestel R (2006) Graph theory. Springer, New York
37. Dixmier J (1981) Von Neumann algebras. North-Holland, Amsterdam
38. Ebeling W (1994) Lattices and codes. Vieweg, Wiesbaden
39. Eells J, Sampson JH (1964) Harmonic mappings of Riemannian manifolds. Am J Math 86:109–160
40. Eells J, Fuglede B (2001) Harmonic maps between Riemannian polyhedra. Cambridge University Press, Cambridge
41. Eon J-G (1998) Geometrical relationships between nets mapped on isomorphic quotient graphs: examples. J Solid State Chem 138:55–65
42. Eon J-G (1999) Archetypes and other embeddings of periodic nets generated by orthogonal projection. J Solid State Chem 147:429–437
43. Eon J-G (2011) Euclidean embeddings of periodic nets: definition of a topologically induced complete set of geometric descriptors for crystal structures. Acta Crystallogr A 67:68–86
44. Eon J-G, Klee WE, Souvignier B, Rutherford JS (2012) Graph-theory in crystallography and crystal chemistry. Oxford University Press with IUCr (to be published)
45. Greenberg M (1971) Lectures on algebraic topology. Benjamin, Menlo Park
46. Gromov M (1999) Metric structures for Riemannian and non-Riemannian spaces. Birkhäuser, Basel
47. Harper PG (1955) Single band motion of conduction electrons in a uniform magnetic field. Proc Phys Soc Lond A 68:874–878
48. Hashimoto K (1990) On zeta and L-functions of finite graphs. Int J Math 1:381–396
49. Hörmander L (1983) The analysis of linear partial differential operators *I*. Springer, New York
50. Hyde ST, O'Keeffe M, Proserpio DM (2008) A short history of an elusive yet ubiquitous structure in chemistry, materials, and mathematics. Angew Chem Int Ed 47:7996–8000
51. Ihara Y (1966) On discrete subgroups of the two by two projective linear group over p-adic fields. J Math Soc Jpn 18:219–235
52. Itoh M, Kotani M, Naito H, Kawazoe Y, Adschiri T (2009) New metallic carbon crystal. Phys Rev Lett 102:055703
53. Jost J (1996) Generalized harmonic maps between metric spaces. In: Geometric analysis and calculus of variations. International Press, Cambridge, pp 143–174
54. Katsuda A, Sunada T (1990) Closed orbits in homology classes. Publ Math IHES 71:5–32
55. Klein H-J (1996) Systematic generation of models for crystal structures. Math Model Sci Comput 6:325–330
56. Koch E, Fischer W (1995) Sphere packings with three contacts per sphere and the problem of the least dense sphere packing. Z Kristallogr 210:407–414

57. Kotani M, Sunada T (2000) Zeta functions of finite graphs. J Math Sci Univ Tokyo 7:7–25
58. Kotani M, Sunada T (2000) Standard realizations of crystal lattices via harmonic maps. Trans Am Math Soc 353:1–20
59. Kotani M, Sunada T (2000) Jacobian tori associated with a finite graph and its abelian covering graphs. Adv Appl Math 24:89–110
60. Kotani M, Sunada T (2000) Albanese maps and off diagonal long time asymptotics for the heat kernel. Commun Math Phys 209:633–670
61. Kotani M, Sunada T (2003) Spectral geometry of crystal lattices. Contemporary Math 338:271–305
62. Kotani M, Sunada T (2006) Large deviation and the tangent cone at infinity of a crystal lattice. Math Z 254:837–870
63. Krámli A, Szász D (1983) Random walks with internal degree of freedom, I. Local limit theorem. Z. Wahrscheinlichkeittheorie 63:85–95
64. Kuchment P (1993) Floquet theory for partial differential operators. Birkhäuser, Basel
65. Lang S (1987) Linear algebra. Springer, Berlin
66. Magnus W, Karrass A, Solitar D (1976) Combinatorial group theory. Dover, New York
67. Mikhalkin G, Zharkov I (2008) Tropical curves, their Jacobians and theta functions. In: Alexeev V et al (eds) Curves and abelian varieties. International conference, 2007. Contemporary Math 465:203–230
68. Milnor J (1969) Morse theory. Princeton University Press, Princeton
69. Nagano T, Smith B (1975) Minimal varieties and harmonic maps in tori. Commun Math Helv 50:249–265
70. Nagnibeda T (1997) The Jacobian of a finite graph. Contemporary Math 206:149–151
71. Neukirch J (1999) Algebraic number theory. Springer, Berlin
72. Newman P, Stoy G, Thompson E (1994) Groups and geometry. Oxford University Press, Oxford
73. Oganov A (ed) (2010) Modern methods of crystal structure prediction. Wiley-VCH, Berlin
74. Oda T, Seshadri CS (1979) Compactifications of the generalized Jacobian variety. Trans Am Math Soc 253:1–90
75. Oda T (2011) Voronoi tilings hidden in crystals—the case of maximal abelian coverings arXiv:1204.6555 [math.CO]
76. O'Keeffe M (1991) N-dimensional diamond, sodalite and rare sphere packings. Acta Crystallogr A 47:748–753
77. O'Keeffe M, Peskov MA, Ramsden SJ, Yaghi OM (2008) The reticular chemistry structure resource (RCSR) database of, and symbols for, crystal nets. Acc Chem Res 41:1782–1789
78. Peresypkina E, Blatov V (2000) Molecular coordination numbers in crystal structures of organic compounds. Acta Crystallogr B 56:501–511
79. Radin C (1987) Low temperature and the origin of crystalline symmetry. Int J Mod Phys B 1:1157–1191
80. Radin C (1991) Global order from local sources. Bull AMS 25:335–364
81. Rangnathan S (1966) On the geometry of coincidence-site lattices. Acta Crystallogr 21:197–199
82. Resnikoff H, Wells Jr R (1998) Wavelet analysis. Springer, Heidelberg
83. Scott L (2012) A primer on ice (in preparation)
84. Serre JP (1980) Trees. Springer, Berlin
85. Shubin M, Sunada T (2006) Mathematical theory of lattice vibrations and specific heat. Pure Appl Math Q 2:745–777
86. Strong R, Packard CJ (2004) Systematic prediction of crystal structures: an application to sp^3-hybridized carbon polymorphs. Phys Rev B 70:045101
87. Sunada T (1984) Geodesic flows and geodesic random walks. In: Geometry of geodesics and related topics (Tokyo, 1982). Advanced Studies in Pure Mathematics, vol 3. North-Holland, Amsterdam, pp 47–85
88. Sunada T (1985) Riemannian coverings and isospectral manifolds. Ann Math 121:169–186

89. Sunada T (1986) L-functions in geometry and some applications. In: K. Shiohama, T Sakai, T. Sunada (ed) Proceedings of the 17th International Taniguchi symposium, 1985. Curvature and topology of Riemannian manifolds. Lecturer notes in mathematics, vol 1201. Springer, Berlin, pp 266–284
90. Sunada T (1988) Fundamental groups and Laplacians. In: T. Sunada (ed) Proceedings of the Taniguchi symposium, 1987. Geometry and analysis on manifolds. Lecture notes in mathematics, vol 1339. Springer, Berlin, pp 248–277
91. Sunada T (1989) Unitary representations of fundamental groups and the spectrum of twisted Laplacians. Topology 28:125–132
92. Sunada T (1994) A discrete analogue of periodic magnetic Schrödinger operators. Contemporary Math 173:283–299
93. Sunada T (2006) Why do diamonds look so beautiful? Springer, Tokyo (in Japanese)
94. Sunada T (2008) Crystals that nature might miss creating. Notices Am Math Soc 55:208–215
95. Sunada T (2008) Discrete geometric analysis. In: Exner P, Keating JP, Kuchment P, Sunada T, Teplyaev A (eds) Geometry on Graphs and Its Applications, Proceedings of symposia in pure mathematics, vol 77, pp 51–86
96. Sunada T (2012) Lecture on topological crystallography. Jpn J Math 7:1–39
97. Sunada T (2012) Commensurable Euclidean lattices (in preparation)
98. Tanaka R (2011) Large deviation on a covering graph with group of polynomial growth. Math Z 267:803–833
99. Tanaka R (2011) Hydrodynamic limit for weakly asymmetric exclusion processes in crystal lattices. arXiv:1105.6220v1 [math.PR]
100. Tate T, Sunada T (2012) Asymptotic behavior of quantum walks on the line. J Funct Anal 262:2608–2645
101. Terras A (2010) Zeta functions of graphs: a stroll through the garden. Cambridge Studies in Advanced Mathematics, Cambridge
102. Tutte WT (1960) Convex representations of graphs. Proc Lond Math Soc 10:304–320
103. Tutte WT (1963) How to draw a graph. Proc Lond Math Soc 13:743–767
104. Uralawa H (2000) A discrete analogue of the harmonic morphism and Green kernel comparison theorems. Glasgow Math J 42:319–334
105. van Lint JH, Wilson RM (1992) A course in combinatorics. Cambridge University Press, Cambridge
106. van der Schoot A (2001) Kepler's search for forms and proportion. Renaissance Stud 15: 59–78
107. Vick JW (1994) Homology theory, 2nd edn. Springer, New York
108. Wells AF (1954) The geometrical basis of crystal chemistry. Acta Crystallogr 7:535
109. Wells AF (1977) Three dimensional nets and polyhedra. Wiley, New York
110. Weyl H (1983) Symmetry. Princeton University Press, Princeton
111. Wolf JA (1967) Spaces of constant curvature. McGraw-Hill, New York
112. Woess W (2000) Random walks on infinite graphs and groups. Cambridge University Press, Cambridge
113. Wood EA (1977) Crystals and light, an introduction to optical crystallography, 2nd revised edn. Dover, New York

Index

Symbols
0-chain, 38, 198
1-chain, 39, 47, 81, 95, 125
1-cochain, 95
A_3-crystal, 151
A_d-crystal, 136
B_n, 25, 44, 51
D_d, 132, 193
G-set, 14
K_4 crystal, 2, 140
K_n, 25, 44, 150, 186, 193
K_{33}, 144, 150
L-crystal, 147
L-function, 90
X_0^{ab}, 77, 126, 183
X_0^{uni}, 63
$\mathbb{Z}$-basis, 16, 38, 43, 210
$\mathcal{S}_n$, 14
d-dimensional diamond, 132, 136
d-dimensional kagome lattice of type II, 138
d-dimensional standard lattice, 83
d-dimensional topological diamond, 186
n-step transition probability, 156, 171
ThSi$_2$, 144
bcu, 135
fcu, 136
srs, 150
3D kagome lattice of type I, 137, 151
3D kagome lattice of type II, 137, 151
3D maximal topological crystal, 140

A
Abel's theorem, 189
abelian covering graph, 76
abelian group, 207
abstract period lattice, 81, 86, 95, 97, 125, 158
action without inversion, 27, 34, 56
additive group, 207
adjacency of vertices, 22
adjoint action, 15
adjoint operator, 215
affine transformation, 108
al-Khowârizmi, 37
Albanese map, 182–184
algebraic graph theory, 191
archetype embedding, 5
archetypical representation, 6
Archimedes, 149
associative law, 207
automorphism (of groups), 208
automorphism group, 24, 49, 171
automorphism of a graph, 24

B
back-tracking part, 31
band structure, 176
barycentric drawing, 5
base graph, 54, 81, 86, 95, 99
base point, 33, 60
base point change, 60, 63
basis (of a vector space), 212
Bieberbach, 218
bijection, 206
Binet–Cauchy formula, 196
bipartite, 47, 80, 85, 157, 163, 170, 176
bipartite graph, 33, 34
bipartition, 33, 34, 170
body-centered cubic lattice, 151
bond, 1, 75
boundary operator, 39, 184, 196
bouquet graph, 25, 44, 69, 135, 147
Bragg, 74

branched covering map, 69
Bravais, 220
Bravais lattice, 2
building block, 95, 97, 125

C
cage compound, 85
canonical homomorphism, 15
canonical morphism, 27, 57
canonical placement, 5, 109
canonical projection, 12
carbon nanotube, 84, 100
cartesian product, 206
Cayley–Serre graph, 69, 179, 188
cell complex, 22
centralizer, 219
circuit, 31, 32, 43, 44, 128, 187
circuit graph, 30, 49, 89
circular cylinder, 17, 100
closed geodesic, 31
closed path, 30, 40
coboundary operator, 50, 185
cohomology group, 50
coincidence symmetry group, 147
coincidence-site lattice, 146
color, 33
combinatorial graph, 22
commensurable, 83, 114, 146
commutator, 209
commutator subgroup, 76, 209
complement, 210
complement (of a subspace), 214
complete graph, 25, 44, 140
composition of morphisms, 24
conjugacy class, 15, 219
conjugate, 15
connected, 30
copper, 136
coset, 15
coset space, 15, 208
counterclockwise parametrization, 44
covalent bonding, 75
covering graph, 54
covering homotopy theorem, 61, 64
covering map, 54
covering transformation group, 56, 57, 64, 68, 75
crystal, 73
crystal net, 95
crystallographic group, 88, 171, 218
cubic lattice, 83, 151
cyclic group, 208
cyclic permutation, 49

D
degree, 116
densest sphere packing, 148
diamond crystal, 2, 100, 134, 151
diamond twin, 150
dice lattice, 29, 121, 132
difference set, 206
dimension (of a topological crystal), 81
dimension (of vector spaces), 212
direct integral, 176
direct sum of abelian groups, 207
direct sum of vector spaces, 213
direct summand, 81, 210
directed edge, 22
directed graph, 23
discrete Abel–Jacobi map, 189
discrete Albanese map, 187
discrete Green operator, 164
discrete Jacobian, 186, 192
discrete Laplacian, 107, 117, 158, 164
discrete Picard group, 189
discrete Schrödinger operator, 179
disjoint, 205
divisor, 198
dual basis, 160, 214
dual lattice, 159, 186
dual space, 214

E
eigenvalue, 215
eigenvector, 215
electric circuit, 39, 119
element, 205
Elements, 21, 73
empty set, 205
energy, 104, 105, 121, 122, 186
equilateral tetrahedron, 103, 134, 145
equilibrium placement, 5, 105
equivalence class, 12
equivalence relation, 12, 22, 33, 60, 217
equivariant, 158
equivariant map, 14, 18, 27, 94, 100
Euclid, 21, 73
Euclidean space, 16
Euler, 21
Euler number, 47, 55, 195
exact sequence, 87
extension, 206

F
face-centered cubic lattice, 136
factor group, 15

Index 227

fiber, 54
finite index, 208
finite-fold covering, 54
finitely generated group, 69, 208
first isomorphism theorem, 15, 42, 59, 208, 211
flat metric, 17
flat torus, 17, 99, 124, 183
forest, 32, 128
free abelian group, 38, 210
free action, 18
free action on graphs, 27
free group, 60, 67–69, 76, 216, 217
Fullerene, 137
fundamental group, 60, 61, 63, 64
fundamental set, 13, 159, 160, 173

G
Galois, 207
general linear group, 208
generalized flat cylinder, 17
geodesic, 31
girth, 31, 116
graph, 22, 38
graph distance, 177
graphene, 28, 84
graphite, 2
graphite-like, 29
graphite-like realization, 93, 143
Gromov–Hausdorff limit, 178
group, 207
group action, 13, 18
group of divisors, 188
group of principal divisors, 188

H
harmonic 1-form, 184
harmonic function, 117, 187
harmonic maps, 121
harmonic oscillator, 104
harmonic realization, 5, 105, 127
height, 82
hermitian operator, 215
Hessel, 220
hexagonal arrangement, 148
hexagonal lattice, 28
Hilbert, 21, 220
homogeniety, 149
homology class, 47
homology group, 2, 39, 43, 44, 51, 63, 76, 82, 86, 90, 126, 128, 129, 146, 151
homology group with real coefficients, 42
homomorphism, 208

homotopic, 33
homotopy class, 33, 60
honeycomb, 7, 99, 100
honeycomb lattice, 28, 34, 99, 100, 131, 134, 143, 150, 151
Hurewicz homomorphism, 63, 76
hypercubic lattice, 83, 132

I
ice, 112
identity element, 207
identity map, 206
Ihara zeta function, 90, 199
image, 206
incidence map, 22, 27
incidence map for a directed graph, 23
inclusion map, 206
index, 15
induced homomorphism, 61
induced map, 13
inearly independent, 212
infinite-fold covering, 54
injection, 206
injective, 206
inner product, 214
integral lattice, 186
intersection, 205
intersection matrix, 129, 192
invariant set, 14
inverse, 207
inverse image, 206
inverse morphism, 24
inversion map, 22
inversion of a path, 30
inversion of an edge, 22
isomorphism (of groups), 208
isomorphism of graphs, 24
isotropic, 150

J
Jacobian torus, 183

K
kagome lattice, 34, 84, 139
Kepler, 148
kernel, 208
Kirchhoff, 119
Kirchhoff's current law, 39, 119
Kirchhoff's voltage law, 97, 119
kissing number, 116
Kronecker's Dream, 152

L

Laplace method, 161
Laplace–de Moivre theorem, 156
large deviations, 177
lattice group, 16, 42, 75, 97, 218
Laue, 3, 74
left action, 15
length, 30
length of a path, 30
length of words, 216
lifting criterion, 61, 65, 77
lifting of a morphism, 55–57
lifting of a path, 54, 61, 62, 65, 77
linear combination, 212
linear isomorphism, 214
linear operator, 213
local central limit theorem, 176
local limit formula, 177
locally finite graph, 23
Lonsdaleite, 2, 29, 100, 112, 144
loop, 33, 60
loop edge, 22, 83, 129

M

magnetic transition operator, 179
map, 206
maximal abelian covering graph, 2, 77, 81
maximal symmetry, 150, 174
maximal topological crystal, 81, 126, 128, 129, 181, 183
minimizer, 5
morphism of graphs, 24
morphology, 74, 219
motion group, 113, 171, 217

N

non-commutative crystal, 178
non-degenerate, 99, 190
norm, 214
normal subgroup, 15, 58, 76, 171, 207
normalized standard realization, 115, 127–129, 146, 164, 171, 181, 183, 190
normalizer, 87, 207

O

octet truss, 137
orbit, 14
orbit space, 14, 27
orientation, 23, 39, 43, 44, 67, 76, 115, 126, 127, 132, 135, 188, 195
origin of an edge, 22

orthogonal complement, 214
orthogonal group, 208
orthogonal projection, 215
orthogonal transformation, 215
orthonormal basis, 214

P

Pappus, 149
parallel edge, 22, 82, 99, 132, 138, 174
parity of a 1-chain, 47
path, 30
pathological realization, 98, 116
period homomorphism, 95, 97, 100, 105, 108, 127, 128, 159
period lattice, 2, 17, 75, 94, 99, 128, 146, 159
period lattice for a periodic realization, 95
periodic graph, 3
periodic realization, 95, 100, 159
permutation, 14, 34, 41
Petersen graph, 79
piecewise linear, 95
plane graph, 44
Plato, 73
Poincaré, 3, 37
point group, 219
point path, 30, 40, 60, 61, 64
Poisson equation, 107, 164
power of a closed path, 90
prime cycle, 89, 91, 152
primitive cubic lattice, 83, 132, 150
product of two paths, 31
proper subset, 205
Pyrite, 73
Pythagoras, 73

Q

quadrangle lattice, 83
quantum walk, 180
quotient graph, 11, 27, 56, 67, 75, 83
quotient set, 12, 16, 217

R

random walk, 155
rank, 210
rational, 146
rational projection, 100, 114, 128
realization of a graph, 22
reduced product, 67, 68
reduced word, 217
reduction, 102, 114, 129
reflection, 145, 147

Index 229

regular convex polyhedra, 73
regular covering map, 56–58, 60, 76
regular graph, 23
regular kagome lattice, 131, 151
regular triangular lattice, 131, 135, 151
relation, 12
representative, 13
restriction of a map, 206
rhombille tiling, 29
Riemann, 37, 181
Riemann–Hurwitz formula, 70
Riemann–Roch theorem, 198
ring, 31
root lattice, 136, 147

S
scalar product, 212
second variation formula, 123
segment graph, 30
semi-direct product, 26, 218
semi-regular polyhedron, 149
separating edge, 31, 32, 48, 49, 128, 190
set, 205
similar transformation, 105, 109
similarity invariant, 105
simple random walk, 156, 157
simply transitive, 18, 56
space group, 218
spanned by, 213
spanning tree, 31, 43, 44, 48, 62, 67, 76, 186, 187, 195
spectrum, 176
square lattice, 34, 85, 131, 151
standard realization, 5, 7, 109, 124, 125, 157, 171
standard unit cell, 16, 102, 104, 160
strong isotropic property, 149
subcovering map, 58
subgraph, 23, 31
subgroup, 207
subset, 205
subspace, 212
subtree, 31
surjection, 206
surjective, 206
symmetric operator, 215
symmetric random walk, 180
symmetry, 25, 131, 173, 201
symmetry group, 14, 34, 41
SYSTRE, 5, 130

T
terminus of an edge, 22
tight frame, 119
topological crystal, 4, 81, 86, 95, 146
topological crystallography, 3
topological cubic lattice, 82
topological diamond, 82, 85, 93, 103
Torelli type theorem, 188
torsion element, 210
torus, 16
transitive, 56
transitive action, 18
tree, 31, 61, 67
tree number, 186, 193, 199
triangular lattice, 34, 83
triangular lattice of general dimension, 135
tropical curve, 199

U
undirected edge, 22
union, 205
unique lifting theorem, 55–58, 64, 65
unique path-lifting property, 54, 65
unit cell, 16
unitary character, 159
unitary map, 215
unitary transformation, 215
universal covering graph, 63, 64
universal covering map, 63, 66

V
valency, 75
vanishing subgroup, 81, 85, 86, 128, 129
vector space, 212
vector sum, 212
vertex, 22
virtually abelian group, 88, 218

W
weighted graph, 117
well-definedness, 13
Weyl, 86, 203
word, 216

Z
zeolite, 85

The manufacturer's authorised representative in the EU is Springer Nature Customer Service Centre GmbH, Europaplatz 3, 69115 Heidelberg, Germany. If you have any concerns regarding our products, please contact ProductSafety@springernature.com

Printed and bound by CPI Group (UK) Ltd, Croydon, CR0 4YY
25/03/2026
02078222-0002